Biodiversity of the Sundarbans

Ethics of Conservation Ecology (1770-2022)

By

Jayanta Kumar Mallick

Biodiversity of the Sundarbans: Ethics of Conservation Ecology (1770-2022)

By Jayanta Kumar Mallick

This book first published 2023

Ethics International Press Ltd, UK

British Library Cataloguing in Publication Data

A catalogue record for this book is available from the British Library

Print Book ISBN: 978-1-80441-101-8

eBook ISBN: 978-1-80441-102-5

This book is dedicated to my late wife Krishna

Table Of Contents

PREFACE

The Sundarban mangrove wetlands, lying a little south to the Tropic of Cancer, occupy the western part of the lower Ganga-Brahmaputra delta. The area was evolved during 5,000 to 1,800 years. This changing landscape consists of 251 (disjunct) islands over 11, 455 km² in 2015-2016 as against 253 (11,903 km²) in 1904-1924, 250 (11,663 km²) in 1967 and 244 in 2001 (11,506 km²), where the trend of reduction was 2.55-times higher in the western (Indian) segment of the region compared to that of the eastern (Bangladesh) section (Bandyopadhya *et al.*, 2022). The region is separated by about 400 interconnected tidal rivers, creeks and canals. Thus, the Sundarban ecosystem has become an isolated and dynamic wildlife refuge, boxed in by humans, inland waterbodies and outer sea in the districts of Satkhira, Khulna and Bagerhat (99%) and partly in Patuakhali and Barguna (1%), in Bangladesh (6,017 km²) and India (4,220 km²) in South and North 24-Parganas districts.

Earlier, a geoinformatics-based study has explored changes in the Indian and Bangladesh Sundarbans over two and half century (1773-2016) (Hussain *et al.*, 2017), where the area coverage has been decreased in different time-scales. The total area was 22,354 km² in 1773, including 12,290 km² and 10,064 km² in Bangladesh and India respectively. In 1873, the area was reduced to 16,431 km² (8,537 km² and 7,894 km² respectively in Bangladesh and India) whereas in 1973, the reduced figure was 10,802 km² (6,525 km² and 4,277 km² in Bangladesh and India respectively) and in 2016 was estimated as 9728 km² (6,152 km² and 3,576 km² in Bangladesh and India respectively). Thus, the total mangrove forest has decreased by 12,626 km² (56.48%) including 10,695 km² of forest land and 1,931 km² of river bodies. The spatial change took place on all sides. The southern part was lost due to erosion and the other three sides for land reclamation. The decreasing trend is more in the Indian Sundarban compared to the Bangladesh Sundarbans. The average rate of decreasing in India 2.68% more than Bangladesh per decade. During 46 years, yearly average loss in Bangladeshi part is 0.35% as against 1.02% in Indian part. This decreasing trend has disrupted the mangrove forest and its ecosystem considerably.

The Sundarban delta has undergone rapid changes caused by neotectonic activity over the past millennium, and geomorphic processes of sediment accretion and erosion have influenced its extent.

Here, the landforms are classified into two broad morphotypes (including sub-types)-

I.   Coastal area (5%)

   (i)   Beach in the outer (seaward) strand of the southern islands,
   (ii) Dune (Aeolian) behind the beaches,
   (iii) Beach-bank transitional (tidal/fluviotidal)
   (iv) Estuary bank (tidal/fluviotidal) along the sides of the estuaries and tidal channels; and

II. Interior island area (95%)

   (v)   Inter-distributary estuarine swamp (biotidal), where mangroves or marshes are present;
   (vi) Inter-creek reclaimed deltaic plains (bio-tidal/anthropogenic modifications) including almost whole of the supratidal interior areas.

Although erosion of the estuary margins and sea face is continuing for many decades, interior channels, especially in the west, are getting silted up due to sediment reworking in a flood-tide dominated environment that is greatly intervened by reclamation efforts initiated in 1770. In 1920-21, when the island was under forests, there were hardly any water bodies in the island. During its subsequent reclamation, a number of tidal creeks were blocked at their entrances and a marginal embankment was constructed all around the island and along the major channels. This transformed the free-flowing water courses to stagnant water bodies. With time, these water bodies were subjected to eutrophication and/or land-filling for expansion of farmlands or for habitation use. In contrast, none of the creeks in the adjacent non-reclaimed areas lost their original density (Nishat et al., 2019).

Coastal retreat in the Sundarbans is highest between the Saptamukhi and the Hariabhanga estuaries and gradually reduces eastward, almost to reach zero on the west bank of the Baleswar. Here, the Bulchery Island has shown relentless erosion through different years. Like other sea-facing islands in the region, its area was reduced by 50% within last 100 years and is estimated to get obliterated within the next 100 years or so. A number of interior creeks and estuaries are also getting silted, mostly (partly) in the western (northern) section.

Analysis of the noninvasively-collected tiger samples from the Sundarbans for mitochondrial and microsatellite markers in comparison to the mainland (northern and peninsular) Bengal tiger populations in India, the Sundarban tigers were found to be genetically distinct and had lower genetic variation in comparison to other mainland tiger populations (Singh, 2017). Demographic analysis indicated recent historical isolation (600–2000 years ago) of the Sundarbans tiger from the mainland. Both historical and genetic evidences supported that the Sundarbans tiger was genetically connected to other mainland tigers until recently. Conclusively, the genetic isolation from the mainland tiger population and unique ability and skill (not inherent but learnt with the passage of time), adapting to the changing factors, conditions or environments of the mangrove ecosystem, might have jointly shaped the genetic architecture of the Sundarbans tiger. Since the Sundarban is totally cut off from the northern and south-western tiger populations and gene pool, the Sundarban tiger may be managed efficiently as an evolutionary significant unit (ESU) under the adaptive evolutionary conservation (AEC) criteria and the largest single transboundary keystone population along with its associates in the changing natural habitat through trans-boundary (India-Bangladesh) collaboration.

The ecological history of Sundarbans dates back to late 17th century, when Kolkata was its northern boundary. The reclamation process started during 1770s to establish settlements in this fertile area considered 'wasteland'. During the early colonial period, the estimated area of Sunderbans was around 43,252 km² (Sharma, 2013), but reduced to 25,000 km² (14,600 km² freshwater swamp forest and 20,400 km² Mangroves) due to rapid reclamation. The present study has revealed that 10,119 km² or 49.60 percent of this ecosystem is impact area or locality covering about 20,400 km² and the remaining 10,281 km² or 50.40 percent is forest land.

During the initial two hundred years of imperial and post-colonial forest management, the primary focus was large-scale clearing of forests and promotion of ruthless hunting for huge revenue and production forestry, violating the traditional ethical issues of living by the 'rules of nature'. Mukherjee (1975: 14) has described some of the consequences of such deforestation:

"The reclamation of the land which rose from the mud and clay by deforestation and human settlement has upset the ecology, resulting in the disappearance of a major part of the wildlife. What exists today in these cultivated tracts are some common forms of birds and aquatic fauna of the tidal creeks, common to both the reclaimed and the forest areas. From the northern part of the district some animals have immigrated and have established themselves in the reclaimed area, for example the jackal, fox, civet cat, mongoose and rats. Freshwater fishes have been introduced in the freshwater (sweet-water) tanks, and various insect pests have appeared on cultivated crops which were not known when these areas were covered with virgin forest."

With the conversion of the swamp grassland to cultivation and human settlements, loss and degradation of the habitats during the intervening period has caused historical species extinction of the herbivorous lesser one-horned rhinoceros, wild buffalo, swamp deer, sambar and hog deer and population decline of the other species. The barking deer has also become extinct in the western Sundarbans and critically endangered in the eastern Sundarbans. Extirpation of multiple prey species has affected the tiger population. In addition to the absence of large herbivores, the tiger population continued to decline because of surreptitious poaching for sports, safety and illegal trade in tiger bones, meat, pelt, eyeballs, teeth, whiskers, tongue, brain, fat, genital organ, paw or nails. Five groups were identified who are involved in tiger killing: villagers, poachers, local hunters, trappers and pirates. The villagers kill straying tigers in the village predominantly for safety, while other groups kill inside the forest professionally or opportunistically. The poachers kill tigers just for money, but other groups hunt for excitement, profit or social status. The pirates, on the other hand, not only kill tigers for profit and but also for personal safety in the forests. There are many streams and rivers, which are used by the offenders as their entry and exit ways, particularly during the high tide because of easy navigability.

In the Bangladesh Sundarbans, the tiger killers locally tan the skin using local ingredients and bury the rest of the body to collect the bones later. The price range of a skin varies between BDT 40,000-90,000 (USD520-1,169); for bones BDT 1,500-3,000/kg (USD20-39) and for a canine BDT 1,000-7,000 (USD13-91) and the outsiders buy the bones from the tiger killers (Saif, 2016).

According to Seidensticker and Hai (1983), the problems for conservation and management arise not only from the physical alteration of the vegetation type but also invasion by alien species and these animals, through competition for food or competitive exclusion in a number of forms, threaten the survival of the forest animals. In the past, the animals living in the forest were buffered by a broad transitional belt of habitat that man destroyed in the process of reclamation. In the process of transforming this belt, a considerable segment of the fauna of the basin was lost, including the more spectacular forms and the animals left with can survive in the southern portion of the forest if conserved well, a number of threats must be contained.

After independence in 1947, the need for wildlife conservation made a shift in the government policy. Here, 'wildlife' refers to the mammals, birds, fishes, frogs, lizards and snakes, flies and beetles, crabs and lobsters, sponges and jellyfish, and all the little bacteria and microbes. All of the wildlife on our planet needs plants and other wildlife to survive. They depend on each other for food and habitats. These inter-dependent wildlife and plants may be termed 'biodiversity', which is the spice of all life-forms and has three levels- genetic diversity, species diversity and ecosystem diversity. The Sundarban is an ideal cradle of such diversity.

The 1970's bought with it two landmark events of utmost ethical importance that were to influence the wildlife conservation in India. The first was introduction of the stringent Wildlife (Protection) Act in 1972 banning hunting and establishment of PAs. The other event was launching of the Project Tiger, the largest wildlife conservation initiative in 1973, including the Sundarbans. Biodiversity conservation within and beyond the PAs is the 1990s manifestation based on the ethical issues. In all, 15 PAs have been set up in the Sundarbans- five in India and 10 in Bangladesh. But in spite of creation of a PA Network, the Indian Sundarbans were considered endangered in the 2020 assessment under the IUCN Red List of Ecosystems framework.

There is no substitute for the Sundarbans to protect the environment and biodiversity as well as human life in coastal areas. Therefore, SNP (1,330.10 km^2) in India was declared as a 'World Heritage Site' on the basis of two criteria (ix)(x) as follows:

Criterion (ix): The Sundarbans is the largest area of mangrove forest in the world and the only one that is inhabited by the tiger. The land area in the Sundarbans is constantly being changed, moulded and shaped by the action of the tides, with erosion processes more prominent along estuaries and deposition processes along the banks of inner estuarine waterways influenced by the accelerated discharge of silt from sea water. Its role as a wetland nursery for marine organisms and as a climatic buffer against cyclones is a unique natural process.

Criterion (x): The mangrove ecosystem of the Sundarbans is considered to be unique because of its immensely rich mangrove flora and mangrove-associated fauna. Some 78 species of mangroves have been recorded in the area making it the richest mangrove forest in the world. It is also unique as the mangroves are not only dominant as fringing mangroves along the creeks and backwaters, but also grow along the sides of rivers in muddy as well as in flat, sandy areas.

1,395 km² in Bangladesh Sundarbans was so declared in 1997 on the basis of same criteria mentioned above.

The Sundarban is also one of the most important wetlands under the World Wetlands (Ramsar) certificate in the UNESCO list of the United Nations Educational and Cultural Organization. 10,247 km² (6,017 km² in Bangladesh and 4,230 km² in India was declared in 1992 and 2019 respectively). The Sundarban met four of the nine criteria required for the status of 'Wetland of International Importance'- presence of rare species and threatened ecological communities, biological diversity, significant and representative fish and fish spawning ground and migration path.

According to the Ramsar authorities,

"The mangrove forests protect the hinterland from storms, cyclones, tidal surges, and the seepage and intrusion of saltwater inland and into waterways. They serve as nurseries to shellfish and finfish and sustain the fisheries of the entire eastern coast. The STR is situated within the Site and part of it has been declared a 'critical tiger habitat' under national law and also a 'Tiger Conservation Landscape' (39) of global importance. The Sundarban is the only mangrove habitat which supports a significant population of tigers and they have unique aquatic hunting skills. The Site is also home to a large number of rare and globally threatened species such as the critically endangered northern river terrapin *(Batagur baska)*, the endangered Irrawaddy dolphin *(Orcaella brevirostris)*, and the vulnerable fishing cat *(Prionailurus viverrinus)*. Two of the world's four horseshoe crab species and eight of India's 12 species of kingfisher are also found here. The uniqueness of the habitat and its biodiversity and the many tangible and intangible, local, regional and global services they provide, makes the Site's protection and management a conservation priority".

The supra-tidal area of Bangladesh Sundarbans was bigger than the Indian counterpart [1914: 3,693.5 (+30.36%), i.e. 92.1 km^2 (2.55%); 2016: 3,601.4 km^2 (+28.54%)]. Now, India holds 34.82% and Bangladesh 65.18% of the total area. During the initial two hundred years of imperial and post-colonial forest management, the primary focus was revenue and production forestry, whereas biodiversity conservation within and beyond the protected areas is the 1990s manifestation based on the ethical issues- "biodiversity, aesthetic values and integrity" and management of ecological balances challenged by development works and anthropogenic activities.

Involvement of local people/community in the forest and wildlife management in the form of co-management is now obvious for the protection, conservation and management of the Sundarbans.

A published book highlighting the ethical, ecological and historical (250 years) perspectives on the Great Sundarbans is not readily available. In this context, this book is a novel one. This book contains well-researched (literature/field surveys) and 47 years' experience-oriented information on the species (micro-macro) diversity, first-ever reported from the world's largest mangrove ecosystem and quarter-millennial reclaimed hinterland, struggling against the natural and anthropogenic threats.

This monograph is ideal for the researchers, postgraduate and graduate students in zoology, botany, ecology and conservation. This comprehensive treatise will serve the professionals, such as foresters, environmentalists, conservationists, resource managers, planners, government agencies, academic institutions, NGOs and naturalists.

ACKNOWLEDGEMENTS

Firstly, I am grateful to two of my 'gurus' (mentors) in the Wildlife Wing, Forest Directorate, West Bengal, who opened my insights on the Sundarbans- (late) Dr. Ramakrishna Lahiri (author of first Management Plan of STR) and (late) Amal Bhushan Chaudhuri, one of the authors of *Mangroves of the Sundarbans, Volume one: India*, IUCN (1994). I also had many learning sessions with (late) Dr. Kalyan Chakraborti, who headed STR after creation. I am also indebted to Mr. Ganesh Behari Thapliyal, who involved me in in-house projects of biodiversity assessment of PAs and entrusted me to develop a wildlife databank, which helped me a lot in writing this book. I also thank all the frontliners and unsung heroes of Sundarbans, who, in spite of their busy schedule of protecting biodiversity resources, gave me valuable field information. I am also thankful to the anonymous reviewers of my book. I also acknowledge the support of my family (deceased wife Krishna, daughter Runa, son Sumanta and daughter-in-law Sagarika).

CHAPTER 1
INTRODUCTION

Study area

During the British rule, the entire Sundarban was administered as a single unit in India. After partition in 1947, the eastern Sundarban eco-region is in Bangladesh, located from the Harinbhanga (West Bengal, India) to the western bank of the Meghna, and is composed of SRF and ECA as well as SIZ surrounding SRF. In India, SBR, i.e. the western Sundarban eco-region, extends from the river Hooghly on the west to the river Matla (South 24-Parganas Forest Division) and from the Matla to the Raimangal-Harinbhanga on the east (STR) as well as the inhabited areas south of the Dampier-Hodges (imaginary) line, passing through South and North 24-Parganas districts. Geographically, the undivided Sundarban tract was extended approximately 260 km west-east along BoB from the Hooghly estuary to the western segment of the Meghna estuary. This region has a coastal stretch of 289.682 km and reaches inland (upstream) for about 112.654 km at its broadest point, comprising the major portions of the 24-Parganas, Khulna and Barisal districts (Curtis, 1933).

Indian Sundarbans

A large part of the South 24-Parganas district falls under this western region (proposed Sundarbans District). 2,788.71 km^2 (27.99 per cent of the geographical area) as per 2019 estimates, including very dense forest (983.10 km^2), medium dense forest (745.03 km^2) and open forest (1,060.58 km^2). This forest land is spread over five community blocks:

(1) Basanti: Herobhanga RF: J.L.No. (Jurisdiction List Number)- 114-116 (between Matla and Bidya rivers);
(2) Gosaba: Sundarbans RF (between Matla and Raimangal-Harinbhanga rivers);
(3) Kakdwip: Sundarbans RF: Small portion south of J.L. Nos. 231 and 232;
(4) Namkhana: Sundarbans RF: Two river (Saptamukhi) banks east of J.L. No. 286 and 290; J.L. Nos. 318 (Lothian Islands); 319 (Henry Island); 320 (Frederick Island); 321 (Susni Island) and 322 (Jambu Island); and

(5) Patharpratima: Sundarbans RF: J.L. Nos. 361, 372, 411 and 414.

A small part of North 24-Parganas district is included in the Sundarbans (proposed Basirhat and Ichamati districts):

(1) Arbesi RF (1-5)= 142.84 (Land 98.35+ Water 44.49) km² and (2) Hingalganj Community Block (238.80 km²) between the two branches of the River Ichamati and SRF.

SBR (1989) - Total area 9,630 km² (Deforested Settlement Area 5,366 km² + SRF and PAs: 4,264 km²) (Govt. of India's Notification No. 16/6/84-CSC dated 19.03.1989).

I. STR (21°51'-22°31'N, 88°10'-89°51'E) (Proclamation 23.12.1973)- Field Management System:

Four Forest Ranges:

1. Basirhat: 5 Beats – Bagna, Burirdabri, Jhingakhali, Khatuajhuri and Harinbhanga;
2. Sajnekhali WLS: 3 Beats – Sajnekhali, Dobanki and Dattar;
3. Sundarban NP (East): 3 Beats – Chamta, Bagmari and Chandkhali;
4. Sundarban NP (West) 3 Beats – Haldibari, Netidhopani and Kendo.

Official records: Area changed over 30 years [2,584.89 (forest 1,680.13+ water 904.76) or 2,585 km² as per 1973's estimate, but has increased to 2,626.36 km² as per 2003 satellite data (forest 1,624.18+water 1,002.18)] with 9 Blocks (9 x 5 = 45 Compartments):

PA 1. Sajnekhali WLS- 376.34 km² (as per Government Notification No. 5396-Forest, dated 24.6.1976)

Boundary: North: Bara Herobhanga Khal, Gomdi Khal and Pitch Khal; East: Duttar Gang; South: Part of Gosaba river and Netidhopani Khal; West : Bidya river

Forest Block: (i) Panchmukhani (1-5)- 176.90 (142.82+34.08) km²; (ii) Pirkhali (1-5)- 199.44 (148.88+50.56) km²;

PA 2. Sundarbans NP (1,330.12 km² as per Government Notification No. 2867-Forest, dated 4.5.1984; increased to 1,520.78 km² by including additional area in 2003)

Boundary: North : Reserve forest blocks and compartments, namely, Netidhopani1, 2 and 3; Chamta-3 and 2; Chandkhali-2 and 4; and Baghmara-1; East: Harinbhanga river adjoining the international boundary with Bangladesh; South : Bay of Bengal; West: Matla river.

(i) Matla (1-5) 189.74 (123.50+66.24) km²;
(ii) Chamta (1-5) 228.36 (169.43+58.93) km²;
(iii) Chotohardi (1-5) 166.11 (78.50+87.61) km²;
(iv) Gosaba (1-5) 173.11 (103.79+69.32) km²;
(v) Gona (1-5) 149.02 (79.20+69.82) km²;
(vi) Bagmara (1-5) 307.23 (155.00+152.23) km²;
(vii) Mayadip (1-5) 307.21 (82.25+224.96) km²;

PA 3. Buffer Zone [Reserved Forest] (6 Blocks 885.27 km²)(6 x 5= 30 Compartments) (as per Government Notification No. 615-Forest, dated 17.2.2009)

(i) Arbesi (1-5) 142.84 (98.35+44.49) km²;
(ii) Jheela (1-5) 120.13 (89.30+30.83) km²;
(iii) Khatuajhuri (1-5) 119.43 (100.00+19.43) km²;
(iv) Harinbhanga (1-5) 116.76 (83.50+33.26) km²;
(v) Netidhopani (1-5) 77.04 (54.74+22.30) km²;
(vi) Chandkhali (1-5) 153.04 (114.92+38.12) km².

As per the 21st century's satellite imagery, the forest cover has decreased by 55.95 km² in STR during last 30 years due to extensive land erosion, while the water area has increased by 97.42 km². Therefore, the total area has increased to 41.47 km².

PA 4. Core or Critical Tiger Habitat- 1,699.62 km² (re-distributed or newly constituted core area as per Government Notification No. 6028-Forest, dated 18.12.2007)

(i) Matla (1-4) [NP] 176.30 km²;
(ii) Chamta (1-3) [RF] 96.32 km²;
(iii) Chamta (4-8) [NP] 124.37 km²;

(iv) Chhotahardi (1-3) [NP] 175.67 km²;
(v) Gosaba (1-4) [NP] 171.73 km²;
(vi) Gona (1-3) [NP] 139.03 km²;
(vii) Bagmara (1) [RF] 24.30 km²;
(viii) Bagmara (2-8) [NP] 269.63 km²;
(ix) Mayadip (1-5) [NP] 273.36 km²;
(x) Netidhopani (1-3) [RF] 93.00 km²;
(xi) Chandkhali (1-4) [RF] 155.91km².

II(A). South 24-Parganas Forest Division: Reserved Forest Area 1,679 km² (4 WLS: 600.40 km² + Other Forests: 1,078.60 sq. km):

1. Halliday Island WLS: Dulivasani- 7 (Part) - 5.95 or about 6 km² (Notification No.5388-Forest dated 24.06.1976)

2. Lothian Island WLS: Saptamukhi 1 (Part) - 38 km² (Notification No.5392-Forest dated 28.06.1976)

3. West Sundarbans WLS: 556.45 km² (Notification no. 1828-Forest dated 11.09.2013) composed of 2 forest blocks

(i) Dulivasani 307.49 km² [Dulivasani 1= 48.85, Dulivasani 2= 48.39, Dulivasani 3= 19.02, Dulivasani 4= 34.24, Dulivasani 5= 35.80, Dulivasani 6= 20.88, Dulivasani 7 (partial)= 42.20, Dulivasani 8= 58.11] and (ii) Chulkati 248.96 km² [Chulkati 1= 23.46, Chulkati 2= 25.14, Chulkati 3= 50.66, Chulkati 4 = 14.96, Chulkati 5 = 60.11, Chulkati 6 = 36.10, Chulkati 7 = 19.96 and Chulkati 8 = 18.57]
Boundaries: In the north, the Ajmalmari river joins the Thakuran river and the Suiya Ganges; BoB to the south; on the east the Matla River; Thakuran river in the west.

4. Chintamani Kar (Formerly Narendrapur) WLS (0.06956 km²) (Notification no.4300-For dt. 21.10.2005)

Boundaries: Adiganga canal flows to the west, roads to the south and east and villages to the north.

II(B). North 24-Parganas Forest Division:

5. Bibhuti Bhushan WLS (formerly Parmadan) 0.64 km² (Block Parmadan-1)(Notification no. 2776-For dt. 19.8.98)

Boundaries: North, South and West by Ichamati river and village to the east.

III. East Calcutta (Kolkata) Ramser Wetlands (No. 1208) Calcutta, North 24 Parganas and South Twenty Four Parganas District: Total area 125 km² (2003) - Fish farming 5,852.14 ha, Agriculture 4,718.56 ha, Waste area 602.78 ha and human habitations 1,326 ha.

IV. CRZ (1991 amended in 2011)- 545.191 km²

The region has ecologically sensitive conservation areas like Sand Dune-Fraserganj, Gangasagar Island, etc., located between HTL and LTL and affected by the tidal action, which includes most of the localities with tidally influenced rivers/creeks and roadside vegetation. Out of the total 2,227.19 km² coastal controlled zone area (mostly in the Sundarbans), mud flats occupy 223.94 km², wetland vegetation occupies 22.59 km² and islands occupy 1,303.60 km². The deforested areas are mainly converted to settlement and agriculture. Of this, about 347.46 km² of brackish water is used for fish farming in the *bheri* (shallow pond).

However, the natural landscape has changed considerably since 1991. Some islands have completely sunk and some new islands have formed. This L-shaped coastal zone of the Sundarbans controls two distinct geographies-

(1) along the Hooghly River (Central Zone) and
(2) east of the Hooghly River (Eastern Zone).

Bangladesh Sundarbans

The protected areas consist of 10 sanctuaries –

1. Sundarbans East WLS (Notified on 29.6.2017)- District Bagerhat (1,22,920.90 ha), Compartment Nos. 1,2,3,4,5,6,7, 8 (Part), 9, 11, 45 (Part) (excluding 170 ha Dudhmukhi Dolphin WLS)

Boundary north from Baga Canal, Miter Bharani, Bara Shiala Canal, Shela Gang, Harintana Camp downstream of Latimara Canal via Betmore Gang to Dudhmukhi Camp and Daburi Bharani Canal to junction of Bhola River; Baleshwar River on the east; west from the Passur River to the Namur River; BoB (except Dublar Char) to the south.

2. Sundarbans South WLS (Notified on 29.6.2017)- District Khulna (75,310.30 ha), Compartment Nos. 18, 19, 43, 44

Boundary on the north via Pataksta Canal upstream of Hansraj River to Pathakatha Bharani, on the east via Morjat River to Numud River, on the west BoB from Arpangasiya River and Bara Panga River to Malancha River, south with Putni Island.

3. Sundarbans West WLS (Notified on 29.6.2017)- District Satkhira (1,19,718.88 ha), Compartment Nos. 49, 51A, 51B, 52, 53, 54, 55

Boundary: On the north by Jamuna River or Mahmuda River to Firingi Gang, on the northwest by Kachikata Canal, from Arpangasiya River and Bara Panga River to Malancha River in the east, Harinbhanga River to Rayamangal River in the west, BoB including Talpatti in the south.

4. Dudhmukhi WLS (Notified on 29.1.2012) Compartment no. 2 and 3, District Bagerhat (170 ha) for dolphins

It is 5 km long, opposite to the Dudhmukhi patrol post, from the junction of Boro Sheola canal to the Bhola River and the junction of the Boro Sheola canal.

Boundary: North: Land area of 2 no. compartment and West bank of Betmore River from 22⁰06′082″ Latitude and 89⁰46′325″ Longitude to East bank of Betmore River, 2 no. compartment 22⁰06′082″ Latitude and 89⁰46′512″ Longitude, also from the adjoining East bank of Bhola River 22⁰04′631″ Latitude and 89⁰48′464″ Longitude to 1 no. compartment adjoining east bank of Betmore River 22⁰02′631″ Latitude and 89⁰48′510″ Longitude; South: Land area of 3 no. Compartment and from the estuary of Betmore River and Boro Shawla khal to the estuary of Bhola River and Boro Shawlakhal on the opposite of Dudhmukhi Patrol Camp; East: Across the Bhola River and Boro Shawla khal and the land area of 1 and 2 no. compartment; West: Betmore River adjoining land area of East part of 12 no. compartment.

5. Chandpai WLS (Notified on 29.1.2012)- 560 ha, Compartment no: 27 and 28. District Bagerhat for dolphins

It is 15 km long, covering Jongra patrol outpost, Nandabala, Chandpai, Mrigmari and Andharmanik patrol outpost.

Boundary: North: Chandpai Range adjoining Joymonirgul and Chandpai check station; south of Mrigmarikhal adjoining land and Passur River adjoining south reserve forest of 31 compartments; South: Adjoining land of Mrigmari Khal and in front of Mrigmari Patrol Post the estuary of Mrigmari and Andharmanik channel of 26 and 28 Compartment; From Jongra Patrol Post west part of Pashur River to Nandabala Patrol Post; East: Mrigmari Khal adjoining 27 Compartment, on the west of Pashur River Joymonirgul adjoining forest area and Mrigmari Khal adjoining west land of 27 compartments; West: Passur River adjoining land of 30 no. compartment.

6. Dhangmari WLS (Notified on 29.1.2012)- 340 ha, Compartment no: 31, District Bagerhat for dolphins

It is 12 km long, stretching from Ghagramari station to Ghagramari patrol camp and covering the entire Dhangmari canal, extending up to the west bank of river Passur in the Karamjal wildlife breeding centre.

Boundary: North: Dhangmari village, latitude 22°26'807" to 89°36'00" between Dhangmari and Mongla; South: From the Jongra patrol post to the middle of Passur River on the East; East: Across the middle of Passur River; West: The whole channel from Dhangmari Station to Ghagramari Patrol Camp and covering Dhangmari and Bhedonkhali village, Karamjal breeding centre adjoining 31 no Compartment and west bank of Passur River.

7. Marine protected area: Ganga Khat or Swatch of No Ground (Swatch of No Ground or SoNG) WLS (Notified on 27.10. 2014) to conserve marine biodiversity (cetacean and other aquatic resources)= 1,738 km^2; depth 900+ m.

Boundary : North: up to Sundarbans (BoB); Latitudes 21°37'35"N and longitudes 89°30'22"E; North-west: Latitudes 21°37'35"N and longitudes 89°21'13"E; North-east: Latitudes 21°37'35"N and longitudes 89°40'30"E; South-west: BoB up to Indian jurisdiction; Latitudes 21°19'57"N and longitudes 89°21'13"E; South-east: BoB; Latitudes 21°20'28"N and longitudes 89°40'30"E; South: Latitudes 21°06'25"N and longitudes 89°31'14"E; East: BoB; West: BoB up to Indian jurisdiction.

In 2020, three more dolphin sanctuaries were declared by MoEFCC in Pankhali, Sibsha and Bhodra, inside the Sundarbans.

8. Pankhali WLS (Dolphin)- 140 ha.

Boundary: North: The boundary of the border of Pankhali Ferry Ghat, from the edge of the Gas cylinder plant to the middle of the Passur River; South: The margin of the border of Pankhali Ferry Ghat, the adjacent waterway of the Pankhali Bazar border and the middle of the River of Pashur from the adjacent land area of the village; East: The boundary of the village of Pankhali and Kamarkhola along the border between Pashur River; West: The boundary between the banks of the Pankhali River, the Pankhali and Kamarkhola village.

9. Shibsa WLS (Dolphin)- 1,650 ha, Compartment no: 33 and 35

Boundary: North: Kalabogi village and adjoining floating settlement; Kalabogi khal adjoining the land of south-west of 32 no. compartment; Reserve forest and Shibsa River adjoining south part of Nalian Bazar; South: Northern part of 34 and 40 no. compartment, Shibsa River adjacent land and reserve forest; East: Floating populations of Kalabogi village, water bodies and reserve forests near Kalabogi canal, Kalabogi Forest Station, waterbody and reserved forest areas adjacent to 33 no. compartments, water body adjacent to Shibsa Forest Camp and reserved forest areas; West: Waterbody and reserve forest adjacent to 35 no. compartment, Sasanangla canal and Arua Shibsa canal adjacent water body and reserve forest.

10. Vodra WLS (Dolphin)- 410 ha. Compartment no: 39

Boundary: North: From Bhadra Forest Camp to Harbaria Forest Camp, the waterbodies of the Passur River and adjacent waterbodies and reserve forest areas, 22nd and 29th compartment of the waterbody and reserve forest; South: 39 and 21 no. compartments adjacent waterbody and reserve forest, waterbody of Passur River and Charaputia Forest Camp adjacent reserve forest; East: 21 and 22 no. compartments, adjacent waterbodies and reserve forest; West: Bhadra forest camp adjacent water body and reserve forest, 39 no. compartment, adjacent reserve forest.

Total Dolphin Sanctuary= 3,270 ha (Present: 1,070 and new: 2,200); Core area 2,200 ha and extended Buffer area 1,227 ha - 3,427 ha.

Range-wise jurisdiction- 6,087 km^2
1. Satkhira range = 1,826 km^2
2. Khulna range = 1,613 km^2
3. Sharankhola range = 1,332 km^2
4. Chandpai range = 1,316 km^2

Buffer areas

1. ECA

Sundarbans in Bangladesh refers to forest area only, but under the Environmental Conservation Act of 1995 the government of Bangladesh had declared a 10-km (ECA over about 17.5 km^2) to the north and east of the forest boundary on 30 August, 1999 (ref no. pa ba ma/4/7/87/99/263) covering five districts (Bagerhat, Khulna, Satkhira, Pirojpur and Barguna), ten upazilas (Bagerhat Sadar, Mongla, Morrelganj, Sarankhola, Dacope, Koyra, Paikgacha, Shymnagar, Mathbaria and Patharghata), 47 unions (27 full and 20 partial) and 1,302 villages with the main objective of providing protection to the SRF and conservation of its biodiversity. This 10 km band is designated as the interface landscape zone in the context of climate change adaptation through value chain and livelihood enterprises and support for environmental and biodiversity conservation. There has been a great deal of change in the land use patterns of the ECA and agricultural lands have been transferred to *ghērī* that are developed for fish and shrimp culture. This interface landscape or sensitive area continues to suffer over-exploitation. Illegal urban development continues, including the use of bulldozers to extract sand at the confluence of Kalagachi and Chuna River at Burigoalini Range, within the Sundarban ECA.

2. SIZ

Another impact area within a 20 km wide radius surrounding the periphery of the Sundarban (SIZ) is often termed as an Ecologically Critical Area. The populations residing within this area are directly dependent on the Sundarban. Five districts of Bangladesh named Bagerhat, Satkhira, Khulna, Barguna and Pirojpur including 20 upa-zillas fall within the SIZ. The southern part of Satkhira, Khulna and Bagerhat and parts of Barguna and Patuakhali form the Bangladesh Sundarban.

ECOLOGICAL TRANSFORMATION AND MANAGEMENT

Ecological history

Although the geological history of 'Sundarban(s)' began probably during the early Pleistocene and the delta formation started from the tertiary period, extensive mangrove forests developed along the NE-SW zone of thickened Holocene. This estuarine land was gradually filled up by sediments brought down by the Himalayan drainage in phases over around 10,000 (BP) years.

Geo-morphologically, this landscape is an outcome of progressive tidal accretion vis-à-vis ecological adaptation starting from growth of the pioneer mangrove grasses like *Porteresia coarctata*, herbs and shrubs, thereby stabilising the intertidal shoals splashed by regular tides and successive changes in mangrove vegetation evolving the climax vegetation above the spring high tide line, but not submerged by bay-water except occasionally by tides augmented by episodic cyclones. It was observed that such accretion and development of an island (e.g. Nayachar formed in the Hooghly estuary; 22.03°N, 88.06°E) took approximately 60 years (1948-2008). The shoreline in between the rivers Saptamukhi and Gosaba has faced maximum erosion at the rate of 40 m per year (base 1914). Such erosion is rather slow beyond this stretch in Bangladesh. Along with accretion, there is evidence of simultaneous land subsidence in the geological sequence of the Sundarban delta, creating major ecological problems of instability. The estuary is still undergoing the dynamic process of formation and disappearance through the combined impact of-

(i) Deposition of silt-load by the freshwater rivers flowing into the estuary, including new island formation,

(ii) Erosion and submergence of a few sea-faced islands, and

(iii) Rise of sea-level and flooding due to global warming.

The history of Sundarbans can be traced back to 200-300 AD. During the reign of the Bengal sultanate (1206-1526), the Sundarban forest tracts were converted to wet-rice cultivation. By the mid-fifteenth century, the reclamation process had brought the southern extent of cultivation to the edges of South Jessore and northern Khulna. The process of bringing virgin forest under cultivation continued unabated in the Mughal era. The Mughal Emperors leased the forests of the Sundarbans to nearby residents. Historical records reveal that the northern boundary of the Sundarbans during the Mughal period (1203-1538) extended from Hariagarh, south of Dimond Harbour on the Hooghly, to Bagerhat in the southern part of Jessore and Haringhata, along the southern portions of Sirkars Satgaon and Khalifatabad. During this time the Ganges changed the course from the original Hooghly channel to combine upstream with the Brahmaputra. As a result, most parts of the 24-Parganas Sundarban faced increased salinity and this gradually affected the biodiversity of the area. The era also witnessed devastating cyclones.

The early management of the Sundarbans forests was confined to the realisation of revenue on the export of forest produce. The first recognition of this as a source of revenue was made by Sultan Shuja in 1658, when revising Todar Mal's original settlement of 1582. During the late 18th century up to early 20th century, there were small salt-farms in many parts of the southern and central Sundarbans. There was no comprehensive Mughal policy on problems of forestry including its preservation, propagation, protection or improvement.

The area was mapped first by the Surveyor General as early as 1769 when the uncleared Sundarban forest was a 'no man's land stretched uninterrupted over 19,200 km². First lease was granted by the East India Company to individuals for land reclamation for cultivation and timber supply in 1770-1773. The Sundarbans was taken to be the property of the state in 1817.

Over the next century the British had relentlessly pursued a policy of deforestation and extension of cultivation in the Sundarban. The forests of

Khulna district were reserved (completely government controlled and protected) in 1875 and 1876, and the remaining forests in the 24-Parganas district were declared protected (partly government controlled) in 1879.

The first working plan that of Mr. R.L. Heinig came into force during 1893-1894 followed by periodical revisions. Hunting and shooting were controlled by the rules in the notification no.839-For. dated the 23rd January 1915, and its various amendments. The close season for deer (1st May to 30th September) was not suitable for Cheetal in the Sundarbans. The stags are mostly in velvet from the middle of October to the end of March, and the rutting season is from about the middle of April to the end of June. Since a large number of tiger (604) had been killed during 1912-13 to 1929-30, the number of deer and wild pigs was increased, whereas the number of human killing by the tiger was 23.72 per year during the same period, but reduced during 1923-24 to 1929-30 varying from 1 to 7. As per local government's letter no.318-T.-R., dated the 27th May 1927, issuance of professional permits was stopped and the tiger population started increasing. During last eight years, 60 woodcutters also died due to crocodile attacks (Curtis, 1933). Fishing or collection of natural resources was allowed subject of permits and payment of revenue to the Forest Department. Entry without permit was prohibited from that time.

Post-independence status

The Sundarbans mangrove forest covers an area of about 12,164.1 km², of which forests in Bangladesh's Khulna Division, extend over Satkhira 1632.00 km², Khulna 1,668.14 km² and Bagerhat 1912.82 km² districts. The belt contains, in addition to the vast mangrove forest, the reclaimed and cultivated lands to the north of it. In West Bengal, the forests extend over 4,235.3 km², mostly across the South 24-Parganas and partly (eastern) North 24-Parganas districts. During 1948-2003, the Tiger Reserve area has been increased by 41.47 km² due to increase in water area by 97.42 km² and reduction of land area by 55.95 km² and the area under South 24-Parganas Forest Division increased by 64.74 km² (land: 169.82 km², water: 234.56 km²). During the first decade of 21st century, forest area was reduced from 2,168.9 km² to 2,132 km² due to erosion and submergence and conversion of forest to saline blanks or salt pans were increased from 38.74 km² to 74.796 km² (Hazra *et al.* 2010). Water area is increased engulfing the forest, although some patches of recolonisation by salt-tolerant mangrove species is evident in some islands as a result of continuous plantation.

SBR has lost 284 km^2 of land in the past 50 years, whereas accretion has been only 84 km^2. During the first decade of present century, the rate of land loss has been increased from 2.85 km^2 per year to 5.5 km^2. In case of Jambudwip Island, for example, the total area lost has been over 50%, while Sagar Island has shrunk by 15% and three other islands, Lohachahara, Suparibhanga and Bedford, completely disappeared, whereas Ghoramara Island has been eroded significantly.

In terms of area 272.9 km^2 (6.05%) was lost in India as against 180.8 km^2 (2.22%) in Bangladesh during the same period. In fact, during 1914 the supratidal area of Bangladesh Sundarbans was bigger than the Indian counterpart and the difference was 3,601.4 km^2 (+28.54%), whereas after about hundred years this difference comes to 3,693.5 (+30.36%), a difference of 92.1 km^2 (2.55%). Therefore, out of the total supratidal area shared by India and Bangladesh (2016), India holds 34.82% and the remaining 65.18% is managed by Bangladesh (contrary to the popular guess estimate of 40% and 60% respectively).

CHAPTER 2
MATERIALS AND METHODS

Secondary data were gathered by using an extensive literature (non-conventional, commercial or academic publishing, distribution channels and grey literature) review, including the official policies, practices and procedures in vogue on the biodiversity resources of the Sundarbans before field verification. Data were also collected from the FD, publications and unpublished documents and collated for preparing a database and facilitating field verification.

Both quantitative and qualitative methods were used for collection of data from time to time. Generally, data were gathered from multiple sources to strengthen reliability and consistency in results. The quantitative tools include surveys, questionnaires and statistical data. The qualitative method involves observation, one-on-one interviews, focus groups (stakeholders) either individually or in a community setting and recorded manually or electronically. All quantitative data is based on and interpreted by qualitative judgement. Intensive case studies also enabled in-depth exploration of intricate phenomena within some specific context at foundation, pre-field, field and finalisation phases.

The island habitat of the Sundarbans is totally different from the terrain found in the mainland. In practice, surveying inside the hostile mangrove habitat, which is approachable and traversed only by the watercrafts, is very difficult due to inaccessibility, absence of any forest road, natural obstacles and lack of logistic infrastructures. The stilt root/pneumatophore-infested and muddy forest floor prevents any transect survey. Hence, the field study was conducted with help of the stationed and mobile FD staff, who are engaged in regular patrolling and monitoring. Direct (sighting) observations with binoculars were made to identify the habitats with surviving species. In addition, the indirect methods (identifying the animal signs, roars, calls, nests, pellets and others) were also followed in the pre-selected accessible niches.

Both up- and down-stream surveys were carried out by using a mechanized boat at a speed of 1.5-3 km/h along the rivers and creeks from 5:30 h to 19:00 h with a break during mid-day. Observations were made from both the front and back of this vessel with a binocular. The boat-survey was, however, suspended at night due to poor visibility and security-risk. At night, observations were made from the wire-caged watch towers by using a search light.

CHAPTER 3
ECOLOGICAL RESOURCES AND BIODIVERSITY CONSERVATION

Ecosystems

There are two types of ecosystems in the Sundarbans- terrestrial and aquatic. The coastal and estuarine forest ecosystem is different from the vegetation of mainland forests. Here, the soil is saline. As salinity increases coastward in the tidal and subtidal areas, there is a transition of dense mangrove vegetation. The soil of Sundarbans is also very muddy. So, it is not suitable for air circulation. So, the branch roots of the plants of this region grow erect and spread along the upper layer of the soil instead of inward penetration. Root tips of these plants bear numerous spores through which atmospheric oxygen enters into the plant body for respiration. They are the producers in this ecosystem. Insects, birds, deer, etc. are primary consumers. Jackals, turtles, cranes etc. are secondary consumers. Tiger, hogs, etc. are among the tertiary consumers. Among them, hogs are omnivorous and now the tigers are opportunistic omnivores (Mallick, 2015).

The aquatic (flowing or stagnant) ecosystem is an ecologically independent and self-regulating unit. The abiotic components like water, dissolved oxygen, carbon dioxide are used directly. Here, the producers are minute floating or suspended small plants, e.g. phytoplanktons. Water weeds, *Eichhornia* etc. are among floating macrophytes. Like minute floating plants, there are also some microscopic zooplanktons like aquatic insects, small fish, mussel snail, etc. which primarily consume the producers. Medium sized fishes, those that survive on eating the primary consumers, are the secondary consumers, whereas the big fish, stork etc. who eat secondary consumers are called tertiary consumers. Bacteria, fungi, etc. decompose the dead organisms. These decomposed substances are again used by the producers of this aquatic ecosystem.

Habitat and niche

In mangroves, the habitat has many niches, encompassing various physical and environmental conditions. The niche-preferences are broadly determined by diverse gradients of salinity, lengths of inundation and severity of drought, among others. The mobility of animal species enables them to adapt to varying environmental conditions, to move along with their specific ecological niche as the niche itself moves. Tidal and other water movements also permit occupation of the same site-specific niche at successive moments in time. Herbivores often interact with carnivores here. When two species with similar niches are placed together in the same ecosystem, one species is more successful than the other species. The better adapted species would be able to use more of the niche. But, in the course of evolution, adaptations that decrease competition will also be advantageous for species whose niches overlap.

Ecological relationships

Symbiosis, Mutualism, Commensalism, Amensalism, Parasitism, Kleptoparasitism, Predation, Herbivory, Familiars, Gregariousness, Competition, Territoriality, Colony, Community and Society.

Terrain

The deltaic Sundarbans may be broadly divided into three distinct morphological units-

1. **Mature deltaic plains** (3-6 m amsl): After being cut off from the main tidal river, these areas have been transformed into sewage-cum-freshwater/brackish water fisheries.

2. **Active deltaic lowland** (2-2.5 m amsl): The islands are covered by estuarine forests within a network of tidal rivers, creeks and channels. Most of the rivers, being completely cut-off from the Ganges, are heavily silted under tidal influence, raising their beds and ultimately blocking the waterflow. The swampy forests developing below the high tide level are submerged twice a day, whereas the higher ones are submerged only during spring or exceptional tides/cyclones.

3. **Coastal plains** (c.2 m): This unstable low-lying tidal zone is a network of estuaries with mudflats, mangrove swamps, marshes, sand banks and dunes, affected by salinity intrusion, cyclones and erosion, sea rise and subsidence.

Ecological resources

Climate

Moist sub-humid climate is characterised by hot summer and mild winter; dominated by mainly south-west monsoon that causes regular or seasonal changes in atmospheric light, temperature, wind speed, direction, rainfall, humidity, air pressure, etc.

Seasons

Extended and extreme summers and short winters are the most striking features of Sundarbans.
Pre-monsoon (March-May): Hot Dry (high temperature and humidity), tidal inundation with increased salinity and evaporation;
Hot Wet (Monsoon: June-September); and
Cold (Post-Monsoon: October-February) with the onset of post-monsoon relatively cool and calm conditions with lesser humidity.

Daylight

Mangrove plants are long-day plants and require high intensity and full sunlight. They flower only when the daylight lasts longer than their critical threshold (around 12 hours), typically in spring or early summer, before the equinox. Flowering for most long-days requires long periods of light. Disruption of the dark period with brief flash- light or darkness has no effect. Short night length allows plants to grow.

Solar radiation

At Sundarban latitude, the surface value is approximately 1,000 W/m^2 (clear day, noon, summer); maximum in June; low in November; very low on cloudy days.

Ectotherms (cold-blooded) regulate their body temperature (TB) externally, primarily through behavioural mechanisms that alter heat exchange between their bodies and the environment. Endotherms (warm-blooded) generate heat internally. Insects and other invertebrates living within the intertidal zone just change position that can affect their TB.

Air temperature

The highest air temperature at 10 m height is obtained during May, whereas the lowest values are obtained during December. On the average, the mean annual day-temperature was recorded >10% than that of night. The post-monsoon showed lowest temperature and humidity compared to that of the monsoon and pre-monsoon. A maximum difference in diurnal temperature was observed in December against a minimum difference in June. Temporal variations of air temperature are recorded from 11.96°C to 38.0°C. The data of the previous centuries revealed 0.6°C-0.1°C increase. Air temperature over BoB rose to 0.019°C/year.

Surface water temperature

It rose in both pre-monsoon and monsoon to 0.5°C/decade (higher in western than eastern sector). The temperature in the sector was a little higher than that of the eastern sector. During the early 21st century, the composite sea surface temperature has increased to about 1°C and the average annual rate of increase was assessed to be +0.0450°C. Rising sea surface temperature is directly related to the increased frequency, severity of cyclonic storms and depression. Rise in temperature affects biology of species at molecular, physiological and biochemical levels, thereby altering distribution patterns as well as community interactions.

Rainfall

Sundarban receives rain mainly from the South-West monsoon (mid-June to September); average 1,625 mm/year (2,000-1, 300 mm). 80% precipitation (major freshwater source) takes place between June and September (highest in August). Rainfall is erratic. Mean annual rainfall varies from 1,920 mm at Jhingakhali (north) to 2,002 mm at Sagar (south). Wet days increase with increasing number of heavy rainfall days (>10 mm). During 1933, highest annual rainfall was calculated to 142% of the normal, while lowest rainfall (62%) took place in 1935. On the average, there were 80 rainy days (2.5 mm/day+). Norwesters are common in summer. An increasing trend of frequency of wet days in North 24-Parganas (0.002 days/year) and South 24-Parganas (0.001 days/year) was recorded during 1990-2015 due to an increase in rainy days. The rainfall decreases from the south-east towards the north-west. The erratic rainfall features delayed monsoon, heavy rains in the beginning, late recession and often heavy precipitation. Retreating monsoon takes place in October and November (c.10% of yearly average during cyclonic storms). The post-monsoon season is relatively dry during the rest of the year.

Relative humidity

Seasonal intensity and availability of solar radiation controls the humidity, varying from 70% to 90% (over 85% during Summer-Monsoon and over 70% during the rest of the year). Capacity of mangrove forests in controlling air humidity depends on the width of mangrove forests and their canopy density. Daytime air humidity is more stable and higher than that of sea surface. It remains low in the morning, but increases after midday. Daytime air humidity in the mangrove forest is higher than air humidity above the vegetated land. At a height of 15 cm above the water surface it is slightly lower than the air humidity at the sea surface. Air humidity influences the quantum of mangrove litter in the form of leaves, twigs and other biomass, which are food sources for the aquatic biota and the nutrients released determine productivity of waters. Logging and opening of mangrove forests lead to decreasing air humidity, increasing evaporation and encouraging occurrence of soil moisture deficit and drought.

Evapo-transpiration

Evaporation is controlled by available energy and stomatal conduction. Transpiration is insignificant at night when stomata remain closed. Evapotranspiration in and around the mangroves is markedly less during winter when cool and dry wind (along with relatively lower wind velocity, temperature and humidity compared to those during rest of the year) blows over the forest canopy and the ground surface is washed with warm tidal water. The rate of evaporation at night is generally more than double than that of the day.

Vapour Pressure Deficit (VPD)

VPD, i.e. difference between amount of moisture in air and potentially during saturation, is crucial in inducing specific plant structure and physiological behaviour as well as plant growth and productivity. In the mangroves, mean annual distribution of VPD depends on water availability and diurnal solar radiation year round. Daytime VPD is regulated by the atmospheric temperature. The annual VPD values, being highest during the post-monsoon season [0.093 and 2.17 kPa (kilopascal)] (Ganguly *et al.*, 2014).

Canopy resistance

Resistance of all stomata of leaves primarily depends on solar radiation, VPD and soil moisture. Ganguly *et al.* (2014) have recorded that annual daytime and nighttime mean of canopy resistance to be 1.78 and 0.68 s·cm^{-1} respectively in mangrove forest with highest value during daytime (3.4 s·cm^{-1}) in pre-monsoon and lowest value of 0.045 s·cm^{-1} at night in monsoon period. During the monsoon, its values were recorded lower both in the day and night compared to those of pre-monsoon and post-monsoon.

Fog

During winter (December-February), dense ground mists occur over land and water surfaces in the early morning reducing visibility to less than 1 km, often persisting until mid-morning and disappearing during midday. Scarcity of rainfall, along with low temperatures, high Relative Humidity (RH) and shallow boundary layer constitute such favourable conditions, having significant impacts on environmental components, such as climate, thermal and radiative budget, air quality, waters, flora and fauna, air-surface interactions, etc. Presence of fog contributes to a reduction of the energy available in the environment and poor photosynthesis. Due to reduced visibility, fogs disturb the normal way of life, disrupt movement of transport, etc. Extension of fog often causes diversion of landing areas of the swimming animals, particularly tigers, from one patch of forest to other or human habitations beyond forests. Similarly, thick fog often confuses the birds' sense of direction.

Topography

Topography is more or less flat (0.9 m-2.11 m amsl.) with general slope towards south and west to east. The upper 100 m. layer is composed of thick clay with occasional clay balls. There occurs unconsolidated sediment at 137-152 m depth composed of sand, silt and clay and gravels of varying colours. This serves as the boundary of the upper aquifer. At about 350 m level, there lies a second aquifer of potable water.

Formation of mangrove ecosystems differs among riverine forest, fringe forest and basin forest. Growth level is comparatively highest in the riverine type mangrove colonies.

Soil

Soil comprises mainly coastal saline alluvium of clay (size <0.002 mm), silt (0.5-0.002 mm), fine sand and coarse sand particles (1-0.5 mm), down to a depth of 1.1 to 1.4 m and thereafter stiff black clay and sand. Soils of the upper deltaic plain are fine loamy in texture and neutral to slightly alkaline. In the lower delta, it is fine textured and acidic to alkaline, while soils of marshes are also fine in texture and show acidic to neutral reaction. Frequency of diurnal saline tidal inundation affects soil salinity. All these soil types show gradual increase in pH with increasing soil depth. The soil of proper Sundarbans is almost devoid of humus (<1%) and supports tangled mass of mangrove. It transpires that the percentage of sand is higher towards the outer estuaries and the percentage of silt and clay is more towards the middle and inner estuaries (creeks and canals), whereas the percentage of fine and coarse sand is comparatively higher in the hinterland.

Twelve types of soil have been recorded from the Sundarbans-

1. Fine Aeric Haplaquepts; 2. Fine loamy, Typic Haplaquepts; 3. Fine Typic Haplaquepts; 4. Fine Arctic Haplaquepts, Fine loamy typic ustorthents; 5. Fine loamy Arctic Haplaquepts, 6. Fine Arctic Haplaquepts, Fine loamy Arctic Haplaquepts; 7. Fine Arctic Haplaquepts, Fine typic Haplaquepts; 8. Fine Arctic Haplaquepts, Fine typic Ustorthents; 9. Fine Arctic Haplaquepts, Fine typic Haplaquepts; 10. Fine Arctic Haplaquepts, Fine, typic Haplaquepts; 11. Fine Arctic Haplaquepts; 12. Fine loamy, typic Haplaquepts, Fine Loamy typic Fluvaquents.

Soil temperature: 13-23°C

Soil moisture or water content: Wet (mean value in 0-50 cm depth)[soil organic matter having strong positive correlations with both soil moisture ($r= 0.88$, $p <0.01$) and hydroperiod, i.e., the length of time portion of the year during which the wetland area is waterlogged ($r= 0.53$, $p < 0.05$)].

Soil pH (sediment): Range 5.5-8 (scale of 0 to 14 to measure degree of acidity or alkalinity). pH7 means sediment is neutral; pH<7 acidic and pH>7 alkaline.

Salinity in saturation extract

Non-saline=<3 gram salt/litre; low saline=3-6 g/l, medium saline=6-12 g/l and highly saline=>12 g/l.

Soil fertility

In general, the soil fertility (including suitable water balance) decreases from east to west and from north to south (Curtis, 1933). In the former region, the annual silting process provides fertile nutrients to make the delta forests more highly productive, whereas in the western Sundarbans due to restricted drainage capacity which help build up more soil salinity and hard and compacted soil. In addition, the compacted and dewatered clays tend to selectively concentrate chlorides, which, at high levels, retard the forest growth, even in case of the salt-tolerant species. Conditions are more saline during February-April due to depletion of soil moisture coupled with reduced freshwater flow from upstream.

In practice, the ecological group classification shows that *Avicennia marina* and *A. officinalis* can tolerate wide range of soil salinity, while *Aegiceras corniculatum, Ceriops decandra, Dalbergia spinosa, Derris trifoliata* and *Excoecaria agallocha* are confined to low salinity areas. Most species have an optimum pH range whereas *Avicennia marina* occurs in varied pH conditions. But *Acanthus ilicifolius* is relatively insensitive to pH and salinity gradient due to its wide ecological amplitudes.

Tides

The Sundarban is a tide-dominated delta. The inner estuaries between the Hooghly and the Raimangal-Harinbhanga are tidal and funnel-shaped, having very wide mouths at the river-confluences with BoB. The Dampier-Hodges Line roughly indicates the northernmost limits of the estuarine zone affected by tidal fluctuations. Northwards these estuaries follow meandering courses with sharp bends and swings, more so in case of the longest estuary like Matla, where the width has reduced to about 96% at a distance of about 100 km north of the sea-face and becoming complex at its upstream head. Hence, the degrees and rates of amplification of the tide over various estuarine stretches are not uniform and follow a complex pattern. The tidal duration asymmetry is modified by the stand. Flow-dominant asymmetry was observed almost universally, with ebb-dominant asymmetry partially over some tidal cycles. The tidal asymmetry and stand have implications not only for the human activities, but also for the life-style of wildlife. The longer persistence of high water levels implies a more likely storm surge with the high tide, leading to a greater chance of destruction. The floods (2-3 hrs) with higher velocity and ebb tides (8-9 hrs) of lesser velocity have a semi-diurnal nature (c.12.5 hrs interval) with two high and two low tides. The downstream flows along the right bank and upstream tide flows along the left bank of the channels. Tidal stages are defined as high ebb, mid ebb, low ebb, low flood, mid flood and high flood. Macro-tide ranges from 3 to 6 m amplitude, unequal, varying in time and the range depending on location and spring to neap tide conditions. Maximum rise and fall depends on spring, which occurs at the vernal equinox (March-April), as there is very little current in the rivers during that period and relatively lower and weaker in winter. During the rains, there is practically no flow-tide above and the difference in tide level is not much. The rise and fall varies only with particular phases of the moon. Unless there is excess river energy from upstream flush, the decantation of traction load sediments takes place. Thus, these backwater channels are getting silted up day-by-day.

Role and functions

Tidal fluctuations play a major role in mangrove habitats, as most of the mangroves grow well in between the Mean High Water Spring Tide and Mean Sea Level.

- Help growth and formation of mangroves, its canopy and formation of donation in the mangrove ecosystem and associated diversity.
- Help maintain mudflats that provide flora and fauna suitable habitats.
- Those mangrove plants growing in the lower part of the tidal range have their root systems covered by water twice a day and remain waterlogged for a longer period.
- The terrestrial plants obtain oxygen through the root systems.

- The ebb and flow of the tide controls formation of the islands, and by scouring action of the drainage during the ebb, keeps the rivers and creeks open.
- Tides transport sediment, euryhaline marine organisms, replenish nutrients, flush out wastes and mix fresh and salt waters.
- Affect reproductive activities of fish and aquatic plants; floating plants and animals ride the tidal currents between the breeding areas and deeper waters.
- Dolphins, porpoises and pilot whales often ascend the rivers along with the tides. Their movement pattern and foraging behaviour are correlated with the tidal state, i.e. between the flood and ebb tides of high water because many of their prey species move concurrently to the tidal changes, enticing the cetaceans to have a strategy of feeding while swimming.
- Moderate temperature by stirring currents and producing more habitable climate conditions in the estuaries.
- Tides create an important intertidal zone, which has silt flats often mixed with mud to varying degrees.
- Tides influence the salinity in the estuaries.
- Tides correspond to the rising msl at the river mouths in February, reaching peak in September and falling down in the winter.
- During May to October, tidal inundation regulates the hydrology of Sundarban.
- Help remove pollutants and circulate nutrients required for survival of aquatic plants and animals.
- Owing to the combination of tidal wave and current, it is possible, when travelling in a launch from the sea to inland, moving with the flow-tide takes several hours; when going in the direction of ebb, a launch has to traverse the tidal wave and the current will assist it only for 2-3 hours.

Tidal bores

The tides are occasionally so strong that it gives rise to a bore ('*Ban Daka*' meaning 'calling of the flood', very bold with very high water and soft with slow speed, depending upon various other factors like wind speed, direction, etc). The higher range during the monsoon periods and sometimes during the pre-monsoon and post-monsoon accompanied by the cyclonic storm originated by the intense low-pressure over BoB accelerates damages caused by storms or bank tidal floods. The southern villages adjacent to the coast were found more vulnerable to storm surge, while those located at higher elevation in the central part showed low vulnerability. Some northern villages were under high and very high vulnerability due to presence of low lying and waterlogged wetlands.

Current speed

The current changes its direction about every six hours (140–180 cm/sec). The tidal waves travel from sea-face to the interior at a speed >48 km/hr to half in the upper reaches. The tidal current in the large rivers varies from >3 km/hr near the sea-face to about 6.5 km/hr in the northern part, but during the spring tide, currents of about 9.5 km/hr+ are often met with; the swiftest current are formed by the combined ebb and stream of the rivers during the rains.

The current passes from west to east; consequently the change of tide is seen earlier in the west, i.e. along the rivers Hooghly and Matla than in further east. Velocity of the tidal current in the northern part increases owing to gradual constriction in river-width combined with the large spill area in surrounding swamp forests. The maximum rise and fall occur at places where the speed is highest. The current in large rivers continues to run, on its own momentum, an hour or so after the tidal wave has passed; consequently, at the end of ebb, the suction of current deepens the trough of tidal wave, and, by retarding the progress of oncoming wave, causes it to bank up and heighten its crest.

This speed is the ultimate driving force for nutrient distribution in the estuary. Water movement, caused by surface and bottom currents, carry plankton upstream and also maintain salinity gradients, which is important for maintaining populations of sessile or sedentary benthic organisms, majority of which have planktonic dispersal stages. Some of the islands do not experience regular inundation by tidal action due to turtle-back shape at the center, causing increase in salinity and formation of saline blanks.

Shallow shores

Mangrove plants require a huge quantity of water. At the primary stage, shallow water areas are suitable for establishment of the seeds and germination. The mangrove seedlings cannot anchor in deep water. Shore mangroves occupy a very narrow zone only.

Mud substrate and aridity

The most extensive mangroves are associated with mud and muddy soils. The high flow of the rivers, upstream water flow of rivers carry alluvial sands, mud to tidal flats thus, forming substrate for mangrove colonisation and succession.

Wind speed

The wind has a great influence on the tidal flow. While a south wind prolongs the period of flow, a north wind shortens the same. During cyclones, the wind causes storm waves. The flood tide was observed to continue rising long after the hour at which the ebb should have set in. Winds are generally light to moderate with a slight increase in summer and rainy season, but stronger in the coast; 5-30 km/hr; occasionally, 70 to 80 km/hr and rarely above; directions south-east and south-west during May-September; variable in October, west-north in cold season against south and south-west in March-April. Diurnal distribution of wind velocity shows two peaks, the first around 12:00 hr and the second around late night and is effective in transporting the temperature. The dry wind blowing from the landward direction helps advective transportation of different types of pollutants (e.g. trace gases) to the mangrove atmosphere.

Wind is one of the important factors in dispersing hydrochorous mangrove propagules and quantification of (meta-) population dynamics. Dispersal velocities significantly vary among different mangrove species based on buoyancy orientation of their propagules. For example, *Heritiera* propagules floating on the water surface are most influenced by prevailing wind conditions, yielding significantly higher velocities when the wind is equidirectional to the water current, but strongly limiting the dispersal range when the wind acts opposite or under a certain angle to the dominant water flow. Among elongated propagules, their density distribution must be taken into consideration, since it determines the propagule's floating orientation and, thus, indirectly the degree to which the fate of the propagule is influenced by the wind. Vertically oriented propagules are influenced significantly less than their horizontally floating counterparts.

The surface roughness becomes gradually more important as the body of a propagule protrudes above the water surface.

Nutrients

Growth of mangrove plants depends on availability of nutrients, which influences photosynthetic performance and resource utilisation of species and may alter species composition in the forests. A mangrove ecosystem rich in organic matter is linked to high levels of carbon allocation to roots along with litter fall and low rate of decomposition imposed by anoxic soils. Such decomposed organic materials are a major source of nutrients in the mangrove ecosystem and distributed via tidal flushing.

Vegetation

The vegetation of the Sundarban may be broadly classified as (a) the sea-face (beach forests), (b) the formative island flora, (c) the flora of the reclaimed low-lying cultivated tracts, and (d) the swamp forests. In the reclaimed areas, a complex flora of the original Sundarban species together with some plants from other parts. Besides the flora of the northern plains a number of littoral species also occur in this area along the embankments and edges of creeks.

Curtis (1933) divided the forests into three zones in terms of salinity-

(1) the fresh-water zone, which consists mainly of less dense, well stocked forests of first and second quality, including predominant *Sundari*, followed by *Genwa*;

(2) the moderately salt-water zone, which consists mostly of less densely stocked third quality forests; and

(3) the poorly stocked salt-water areas, which contain forests of either fourth quality, or poor third quality. These salinity zones are changing over a century because whereas Curtis located these zones in the north-south direction, at present the direction has shifted to the east-west. For example, in Bangladesh, the slightly saline zone consists of Chandpai-

Sarankhola range (21°50'-22°25'N, 89°45'-89°25'E), Khulna range (21°20'-22°30'N, 89°40'-89°20'E) falls under the moderately saline zone and Satkhira range (22°20'-21°40'N, 89°30'-89°25'E) is within the high saline zone.

Special type of the marsh vegetation composed of elements mainly of the Malay Peninsula and Polynesian regions, together with some Indo-Chinese, Ethiopian and a few of the New World is represented in these estuarine islands.

Nutrient concentrations

The nutrients are Phosphate, Nitrate, Nitrite, Silicate Ammonia and total nitrogen (TN). According to Chaudhuri *et al.* (2012, 5-6),

"Nutrient concentration i.e. total nitrogen (TN), ammonia-nitrogen, total phosphate (TP) and silicate showed higher concentration in post-monsoon and monsoon compared to pre-monsoon. TN and ammonia-nitrogen were estimated to be 34.14 µmol/L, 2.06 µmol/L in post-monsoon, 20.52 µmol/L, 1.24 µmol/L in pre-monsoon and 28.22 µmol/L, 1.36 µmol/L in monsoon (average) respectively. Highest TN and ammonia-nitrogen concentration were observed in the month of February (36.25 µmol/L) and January (2.3 µmol/L) respectively and lowest in the month of June (14.15 µmol/L, 0.97 µmol/L). Total phosphate concentration was estimated to be 2.09 µmol/L in post-monsoon, 1.44 µmol/L in pre-monsoon, 1.73 µmol/L in monsoon (average), being highest in February (2.15 µmol/L) and lowest in June (1.08 µmol/L). Silicate concentration was observed to be 24.23 µmol/L in post-monsoon, 14.58 µmol/L in pre-monsoon and 22.47 µmol/L in monsoon (average). Highest silicate concentration was recorded in October (29.58 µmol/L) and lowest in June (11.02 µmol/L). TN:TP ratio was greater than Redfield ratio (16:1) in post-monsoon (average 17.3) and less than Redfield ratio in pre-monsoon (average 14.5) and monsoon (average 14.7)".

The latest (2019) nutrient concentrations recorded from the estuaries are phosphate concentration 0.75-1 micro mol/lit, nitrate concentration 2.4-39.9 micro mol/lit, nitrite concentration 0.6-1.6 micro mol/lit, silicate concentration 4.8-49.1 micro mol/lit, ammonia concentration 0.2 -0.4 micro mol/lit and TN concentration 29.2-36.5 micro mol/lit.

Carbon storage (t/ha)

Ranking among the dominant species varies due to capacity- *Sonneratia apetala* (47.82)> *Avicennia alba* (32.01)>*Avicennia marina* (26.37)> *Excoecaria agallocha* (10.57)> *Avicennia officinalis* (4.47)(Ecosystem Acumen, Special Edition Newsletter, IUCN CEM South Asia, February 2, 2019).

Dissolved oxygen (DO) and biological oxygen demand (BOD)

In the Sundarbans, moderate to high DO concentration (6.5-9.8 mg/L) was observed throughout the year, highest in January (9.8 mg/L) and lowest in June (6.5 mg/L). High concentration of DO throughout the year indicates consistently high wave action and steady primary productivity in the estuary. BOD values ranged from 3.2 to 6.6 mg/l during high tide and 3.0 to 7.5 mg/l during low tide. It was higher during pre-monsoon. The DO level has exhibited two peaks-

(i) 2009 peak due to super cyclone Aila and (ii) 2020 peak during COVID-19 lockdown phase.

The percentage of DO increased during the months of COVID-19 lockdown than pre-COVID-19 years. Percentages of DO increased maximum at Diamond Harbour (38.54%) and minimum at Ajmalmari (12.40%). During pre-COVID-19 years, DO values were highest near Ajmalmari (5.79 ± 0.40   ppm).

During the lockdown phase, these values were comparatively higher than the predicted values, which may be due to negligible input of wastes from several anthropogenic sources that arise from industrial and domestic activities, complete stoppage of most of the industrial operations and movements of water transports that ultimately upgraded the estuarine water quality leading to improved fish resources of the estuarine system.

Chemical oxygen demand (COD)

The COD values ranged from 66-140 mg/l during high tide conditions and 70-149 mg/l during low tide conditions. Range of 101.29 and 114.8 mg/l was observed in the open sea islands.

Riverbed hydraulic conductivity of mangrove sediment

Estimated rate ranges between 1-10 m bulk of sediment/day based on void created due to innumerable animal burrows on horizontal and vertical scales of >1 m+ in SBR. Where these burrows are absent, the rate is many times more. Materials composing the sediment affect the effective porosity of the sediment and determine the hydraulic conductivity.

Suspended particulate matter

During the post-monsoon, SPM concentration was observed as 177.7-92.8 mg/l and in pre-monsoon 101.5-192.3 mg/l and during monsoon 233.7-255.6 mg/l.

Surface and groundwater

Surface water is saline, where EC ranges 41300-5460 μ mhos/cm at 25°C and Chloride (Cl) content varies from 1358 to 11067 mg/l. Ground water is fresh and potable in 915-4000 μ mhos/cm at 25°C and Chloride (Cl) content 64-1255 mg/l. Rain water stored in the Reservoir Pond has EC-611 μ mhos/cm at 25°C and Chloride content 99.3 mg/l.

Turbidity

Low aquatic nephelometric turbidity is 20–25 NTU and high aquatic turbidity is 57–125 NTU. The average range is <40 NTU. Maximum turbidity was observed in monsoon (60.8-131.3) followed by post-monsoon (61.2-31.3) and pre-monsoon (26.6-22.4). Somewhat near-stable conditions prevail during the period from February to June.

Bioturbation

Bioturbation has numerous effects on benthic community structure. In muddy sediments, particle reworking by deposit feeders appears to reduce densities of suspension feeders. Conveyor-belt feeding can displace surface-dwelling benthos. Bioturbation changes the depth-distribution of organic matter and can increase the inventory and quality of food for deposit feeders in sediments. The provision of both food and energetically favorable electron acceptors such as oxygen, nitrate, and iron oxide to depth in the sediment leads to higher abundance of bacteria relative to archaea in surficial sediments. Bio-irrigation increases microbial activity at depth and in the vicinity of burrows. Bioturbation can also increase nutrient fluxes leading to elevated rates of benthic primary production and increased microbial productivity as well. But bio-irrigation enhances rates of denitrification, reducing the efficiency of sedimentary nitrogen cycling.

Salinity of water

Parameters for saline water: Fresh water - < 0.5 parts per thousand (ppt) or less; Low < 15 ppt; Moderate 15-25 ppt; High >25 ppt. This is the toleration limit of the primary productivity and planktonic biomass, because beyond this threshold level their reduction was observed. In SBR, the salinity of water was found to be increased gradually from post-monsoon (17.3 PSU) to pre-monsoon period (24.5 PSU) and decreased to a lowest value in monsoon (12.6 PSU). Highest salinity was observed in June (24.5 PSU) and lowest in October (12.6 PSU).

Different kinds of mangrove species development depend on different levels of salinity. *Rhizophora* is an obligate halophyte with poor production because of the reduction rate of salt. *Heritiera fomes* has strong preference for low salinity and its growth rate depends on tidal inundation and fresh water supply and the production rate can be decreased with increased level of salinity. Increase in salinity also affects the species association and regular succession patterns in the Sundarbans as some non-woody shrubs and bushes replace the tree species, reducing the forest productivity and habitat quality for the wildlife.

pH of water

pH7 means that the water is neutral and within the average range of water pH; range 8.13-8.74 except Canning (7.51 ± 0.20). pH range of 6-9 is good for the fish and aquatic life.

Phytochemicals

Phytochemicals are biologically active, naturally occurring chemical compounds found in plants that generally help resist fungi, bacteria and plant virus infections and also consumption by insects and other animals. The major classes of phytochemicals like steroids, alkaloids, flavonoids, saponins, phenolics, terpenoids and tannins have been isolated from the mangroves, which have huge potential to prevent diseases.

Alkalinity

The alkalinity values ranged from 155 to 270 mg/l during high tide and 137 to 258 mg/l during low tide. Proximity to BoB and luxuriant mangrove vegetation is related to high alkalinity values which protects the fish and aquatic life at the time of fluctuating pH.

Sulphate

Sulphate values in Sundarbans ranged from 497 to 1200 mg/l during high tide due to higher sea water flow and 349 to 1106 mg/l during low tide due to mixture with the river water. It is higher downstream and lower upstream.

Chloride

Chloride values ranged from 1305 to 11065 mg/l during high tide and 809 to 10236 mg/l during low tide. The downstream values are higher because of proximity of BoB and salinity of soil and water.

Sodium

Sodium values ranged from 715 to 6619 mg/l during high tide and 436 to 6100 mg/l during low tide. Downstream values are higher as explained above.

Potassium

Potassium values ranged from 27 to 245 mg/l during high tide and 22 to 236 mg/l during low tide. Lower potassium levels in the upper estuary in comparison to downstream zones cause higher erosion in the areas close to BoB.

Nitrogen

Nitrogen values ranged from 21 to 57µgat/l during high tide and 27 to 68 µgat/l during low tide. This variation is caused by polluted materials from upstream, such as sewage, shrimp farms, agricultural run-off and benthic fluxes.

Others

Silicate, phosphate, phosphorus (PO4), etc. were found maximum during the monsoon.

Bank erosion

Bank erosion occurs commonly in rivers and upper reaches of estuaries. It takes place mainly due to combined activities of man and nature-discharge, flow diversion, formation of shoals, migration of ephemeral bars near the bank, channel scouring, cohesiveness of texture of the banks, soil humidity, bank configuration, bank failure, anthropogenic activities, construction of embankments and excavation of the bank side areas for different purposes, difference in seasonal sediment load, etc. The sediment load is higher in the rivers of Sundarbans rather than the marine tidal water. Two prominent eddies or vortexes are created due to the mixing of sluggish undercurrent with higher sediment load coming from north-west to southeast.

In Sundarbans, erosion and deposition go on simultaneously. Here, the process of erosion is done in four ways described below.

(i) Hydraulic action: Repeated flows of striking tidal waves enlarge the incipient joints, fracture patterns and lead to breaking the bank materials into smaller pieces. Moreover, the high tides, especially the spring tides, also exert enormous pressure on the air trapped hollows and crevices within the banks. The hollows, which are created due the continuous walking by the banks, also accelerate this process.

(ii) Attrition: In Sundarbans, particularly in the active zones, where the sea water has more accessibility, corrasion is an effective process at the mouth of the rivers like Matla. It is mainly done with the help of the natural tools of the tidal wave like coarse sand, pebbles etc.

(iii) Abrasion: Mechanical wear and tear as well as resultant breakdown of fragments due to mutual collision affected by the backwash, which remove the fragments from base of the banks, which are transported back to BoB during the ebb periods.

(iv) Corrosion or Solution: This chemical process dissolves different soluble rock materials consequent upon contact with the sea water. Owing to the upstream tidal inflow of sea water and the downward outflow of river water, different types of soluble material affect chemical composition of the bank materials, which leads to future modification of nature of the banks.

(v) Eddies originated due the mixing of the river water and the tidal water at the basement of banks: In the Sundarbans, the inflow and outflow of the tidal water do not follow the same route. Hence, small eddies originate regularly around a suitable area of base of the banks. Initially, the high tide hits base of the banks and, at the next stage, weakens basement of the banks and continuous wetting and drying twice a day make the banks deficient to bear the hydraulic pressure of the larger waves. Then in the third stage, an eddy is originated enlarging the hollows and the air trapped joints or crevices. Thus the hollow slowly enlarges due to the undercurrent at the basement. From the third stage, the erosion process actually starts by the low tide water level at basement of the banks and undercutting process continues even after the ebb periods. Scouring and enlarging of the hollows at toe of the banks goes on by the erosion action done by the low tide water at the hollows or bores. Ultimately, in the fourth stage, the upper portion of the bank collapses on the river and forms semi-point bar at the side of bank with the deposited material accumulated during the breakage of the banks. It is a continuous and most common process of bank failure in the Sundarbans.

Coastal erosion

Coastal erosion is a natural process by which land is lost or displaced due to the long-term removal of soil, sand and rocks along the coastline by the sea level rise, strong wave action, currents, tides, wind-driven water, coastal flooding, impact of storms, etc. Consequently, people living in the impact zone are forced to move inland, which squeezes the living space of humans, the beach biodiversity and ecological balance is destroyed and causes direct or indirect harm to human life and natural environment.

Coastal erosion is constantly reshaping the islands of Indian Sundarbans (Hazra *et al.* 2010). Continuous movement of sediment and water flow in the estuary expedites coastal erosion, which appears to be a product of change in sea level and tidal hydraulics. With continuing tidal and storm surges of higher intensity and height, sediments that were previously in equilibrium with the hydraulic system, are dislodged, eroded and are deposited within the channels or sub tidal areas leading to progressive reduction of land area with shallow channel floor or near-island bathymetry. The eroded materials are carried by the tidal currents to the north and re-deposited in the tidal regime on the north and east.

Geology

Geologically there are two major groups of deposits:

(i) Quaternary-Recent to sub-recent–Newer alluvium and
(ii) Pleistocene-Older Alluvium

Estuarine hydrology and habitats

Rivers

Total length of the channels and creeks in Indian Sundarbans is estimated to be about 6,000 km. There are 31 rivers in SBR in addition to many streams and rivulets (*Khal*) and forested Swamps and marshy wetlands (*Beel*). The estuaries are Hooghly (elevation 1-11 m; depth 5-20 m; width c.25 km at mouth), Muriganga (28.418 km, 3-4.8 km width, depth 6.05-13.5 m), Saptamukhi (10 km wide), Thakuran (62x10 km), Matla (length 125 km; width 26 km), Bidya (Length c.56 km; width c.0.12-5.31 km), Gosaba and Harinbhanga (length 79 km; width 6 km). In the Bangladesh Sundarbans, 13 rivers are important Passur, Sibsa, Chunnar, Kholpetua, Bal-Jhalia, Baleswar, Betmargang, Notabaki, Passakhali, Arpongasia, Kathka, Nilkamol and Malancha, which are classified into three different series i.e., Raimangal-Sibsa series, Passur-Sibsa series and Passur-Baleswar series. From east to west the river systems also comparise four estuaries viz., Bangra estuary, Kunga estuary, Malancha estuary and Raimangal estuary.

The degraded water quality during the last fifty years is damaging the other elements of ecosystem like soil, wildlife and vegetation due to high salinity, siltation and shortage of fresh water.

Creeks

This drainage system is divided into three types as follows:

(i) **Dead channel-**Those channels which have lost their connection with one of the main river systems on either side are called dead. Some of these channels are completely choked up with no tide water flowing except during monsoon. Channels like Giapati in Namkhana-Kakdwip, Piyali, Bidyadhari and Ghughudanga blind creeks are examples of this type.

(ii) **Decaying channel-** Those channels that are considerably silted up but having regularly recorded tidal fluctuations are called decaying. The upper part of Saptamukhi estuary, Hatania-Doania channel etc. are examples of this type.

(iii) **Active channel-** All those channels that are having pronounced tidal action are called active.

These wetland systems have a great ecological importance, especially in creating groundwater flow, aquaculture and these are the only routes in this inaccessible area.

Tidal/Mud flats

It is more muddy than sandy in nature. Three types of mud flats are found here- namely Saline water mud flat, Brackish water mud flat and Fresh water mud flat. In the sea-facing sectors, the mud is very saline and a major portion is inundated by daily tides, while mud flats of central and deep-inland sector get inundation by pushed back freshwater during spring tide and monsoon period only. Mud flats occur as elongated strips on both sides of the tidal rivers and creeks, as patches within mangrove swamp and along the periphery of the islands. A considerable part of this highly productive unit is now under aquaculture. For example, the vast mudflat of the Canning area has been reclaimed for traditional practice of brackish water aquaculture. At places, the mudflat area has been converted into a vegetable garden. In certain areas, linear patches of sand are seen lying within the mudflats. Mudflats are found near Phuldubi Gang (Naraharipur-Radhakrishnapur village), southern part of Agnimarichar, eastern coast of Jambudwip (now discontinued), Chemaguri, Banibishnupur and Baneswar chak, Lohachara and Suparibhanga island (subsided), peripheral part of Dalhousie island, along Piyali riverbed, Purba Gabtala, Haripur, LWLS, Patharpratima *Abad*, Mahabatnagar, Upendranagar, Kishorimohanpur etc. Mud flat near Goalpara along River Ichamati is extensively used for brick preparation.

Sand/Beach/Spit/Bar

Beaches of Sundarbans are more muddy than sandy. Vast extension of muddy beach is found on the east of Hooghly estuary, especially to the east of Bakkhali. These are mostly narrow linear ridges occurring on the narrow muddy beach or at the peripheral part of the islands, viz. Jambudwip, Sagar Island, Chuksardwip, Bakkhali beach ridge, Bulchery island, Halliday island, near Jagdal Gang, along the periphery of the island near Harinbhanga river mouth. Submerged bars and chars consisting of sand, silt and clay are found all along the coast and rivers namely, the Hooghly, the Matla, the Thakuran, the Saptamukhi, the Gosaba, the Raimangal, the Harinbhanga etc. Few of these are partly covered with marsh vegetation as well as mangrove swamps at places. These shoals are very mobile and change their position and shapes, especially during SW monsoon.

Littoral zonation

The littoral zone extends from the high water mark, which is rarely inundated, to shoreline areas that are permanently submerged. The littoral zone always includes the intertidal zone. However, the meaning of littoral zone can extend well beyond the intertidal zone. The erosive power of water results in particular types of landforms, such as sand dunes and estuaries. The littoral drift refers to natural movement of the littoral along the coast. Biologically, the ready availability of water enables a greater variety of plant and animal life and, particularly, formation of an extensive wetland. In addition, the excessive humidity due to evaporation usually creates a microclimate supporting unique types of organisms.

The intertidal zone close to BoB includes a variety of habitats such as sandy shores and beaches, soft sediments, salt marshes and mangroves. This region may be divided into four distinct physical zones on the basis of air exposure each gets at low tide and submergence at high tide, when the organisms inhabiting there have adapted to extremely harsh environments, high temperature and salinity. These environmental gradients cause apparent zones, which are detailed below along with the characteristic and distinct biological community.

1. Supra littoral or uppermost spray/splash zone: This part of the land is submerged only by rare extreme high water of spring tides or severe storms, but is repeatedly hit by wave splash and wind-blown spray. The organisms living here must cope with the downpour, deluge, storm, heat, predation by land animals and seabirds. This zone is dominated by the life forms like the lichens, snails and others. The terrestrial supra littoral zone is located above this part.

2. High littoral zone: This long stretch is flooded during the daily high tides, once or twice, and the organisms living here must be able to withstand the varying conditions of temperature, light, and salinity. This zone is dominated by the crabs, barnacles, seaweeds, etc.

3. Mid or medio littoral zone: This zone is generally submerged, except for a short period during the turn of the low tide. The upper sub-zonal belt dries out almost completely at low water, but the lower sub-zone hardly gets dried out completely and even at low waters remain moist by the action of waves or sprays. This zone is the domain of the anemones, algae, snail, sea urchins, etc.

4. Low littoral zone: This area is exposed only during the lowest spring tides. The anemones, red algae, crabs, sea urchins, kelp, etc.

Many more vertebrate (e.g., mammals, waterfowls) and invertebrate (e.g., snakes, insects, etc.) animals use the littoral zone as well as the terrestrial ecosystem for food and shelter. The littoral zone is more negatively affected by anthropogenic activities.

The extensive hinterland, historically formed by the littoral zone described above, but reclaimed during the colonial and post-independence periods, where the tidal effects are either minimal or nil now due to closure of the links or dying out of the tidal rivers flowing through them, is referred to as fringing freshwater wetlands

These wetlands are again subdivided into four categories- wooded wetland or swamps dominated by trees, wet meadow [open, groundwater-influenced (minerotrophic), sedge and grass dominated), marsh (frequently or continually inundated, usually treeless and dominated by grasses and other herbaceous plants) and aquatic vegetation, adapted to the unique hydric soil. Growing human settlements and anthropogenic activities have demonstrated a negative impact on the productivity and biodiversity of these wetlands.

Forest types based on salinity

Based on salinity distribution, the mangrove forest of Sundarbans can be classified into three types:

(i) Freshwater/less saline forest *(Ceriops decandra, Heritiera fomes, Xylocarpus mekongensis, Bruguiera gymnorhiza, Cynometra ramiflora, Amoora cucullata)*

It occurs on the north and the eastern portion of the Sundarbans; almost extinct.

(ii) Moderate salt water forest *(Excoecaria agallocha, Ceriops decandra, Xylocarpus mekongensis, Bruguiera gymnorhiza, Sonneratia apetala, etc.)*

It consists of the forest near the sea-face and where salinity is less in the monsoon and lean period.

(iii) Salt water forest *(Sonneratia apetala, Avicennia officinalis, Acanthus ilicifolius, Excoecaria agallocha, Xylocarpus spp, Aegialitis rotundifolia, Rhizophora conjugata)*

It occurs in the western part of the Sundarbans, where Garan and Genwa dominate the forest.

Present forest types

The mangrove forests of Sundarbans are now commonly categorised as Group 4, which is subdivided into-

4A. Littoral forests (L1), facing the sea: This coastal habitat is further divided into two nitches:

(i) Open frontal beach forests with *Ipomoea pes-caprae*, *Vigna marina* (vines), *Imperata cylindrica*, *Saccharum spontaneum*, *Chrysopogon zizanioides*, *Hemarthria compressa* (grass); *Canavalia gladiata* (legume), *Cyperus arenarius*, *Fimbristylis dichotoma* (sedges), *Aristolochia indica* (creeper), *Oldenlandia biflora*, *Rothia indica*, *Launaea sarmentosa* (herbs), *Rivea hypocrateriformis* (climbing shrub), *Gisekia pharmacodes*, etc; and

(ii) Open back beach forest Sandy dunes, back dunes, sandbar vegetation with perennial herbs and grasses such as *Alternanthera paronychiodes*, *Alysocarpus vaginalis* and others.

4B. Tidal Swamp forests with following subgroups:

TS1: Mangrove scrub: *Ceriops, Avicennia alba, Aegialitis rotundifolia, Excoecaria agallocha, Phoenix paludosa* (drier ground) etc. Along the edge of tidal waterways and sheltered muddy coast. Dense forest with average height 3-6 m. Few species are markedly gregarious, all evergreen with leathery leaves. Vivipary seen. Common in western Sundarban.

TS2: Mangrove: *Rhizophora, Kandelia, Avicennia, Excoecaria, Ceriops, Bruguiera, Xylocarpus, Sonneratia*- Along the edge of tidal waterways and sheltered by a muddy coast at slightly higher level. An evergreen forest of moderate height. Tidal mud flats are permanently wet with salt water and submerged with every tide. Stilt roots and vivipary seen.

TS3: Salt water (mixed)- *Heritiera fomes* (Sterculiaceae), and *Excoecaria agallocha, Ceriops decandra, Xylocarpus mekongensis, Avicennia officinalis, Aegialitis rotundifolia* (near sea face) etc. *Nypa fruticans* are relatively uncommon. Fairly dense forest, more than the freshwater type, but not so high. Rarely over 20 m. Trees do not attain girth. Ground flooded in every tide with brackish water. Less silt deposition than fresh water type. Less humus, soil stiffer, clayey liable to crack extensively when exposed. Bigger river deltas.

TS4: Brackish water (mixed)- *Heritiera fomes, Sonneratia apetala, Acanthus ilicifolius, Xylocarpus mekongensis, Bruguiera* sp, *Sonneratia caseolaris, Excoecaria agallocha, Ceriops decandra, Phoenix paludosa* (high land), *Acanthus ilicifolius, Hibiscus tiliaceus, Nypa fruticans* (fringing banks). In the larger deltas, notably High forests over 33 m, stilt roots are rarely met but pneumatophores are present. Forest is flooded for some portion each day, the water is never very salty and very fresh during the rainy season or slightly brackish. Good amounts of fresh silt deposition.

4B/E1: Palm swamp- *Phoenix paludosa* seen on drier areas within saltwater mangrove scrub or forest.

This habitat is also sub-divided into the following niches:

(i)   Beach–bank transitional (tidal/fluvial tidal) zone at the mouth of estuaries and tidal inlets;

(ii) Interior islands-

(a) Estuary bank (tidal/fluvial tidal) along the sides of estuaries and tidal channels;

(b) Interdistributary estuarine swamp (biotidal), i.e. inner parts of all tidal islands characterised by thriving mangroves/mangrove marshes;

(c) Inter-creek reclamation (bio-tidal/anthropogenic), i.e. almost all of the supratidal interior areas with abundant old mangrove roots and pneumatophores including anthropogenic modifications like agriculture and aquaculture.

Estuarine system

Hooghly mouth at mouth at Sagar Island (21.56°–21.88°N, 88.08°– 88.16°E) 25 km wide.

Saptamukhi (21.604°N, 88.352°E), width 10.0 km, has two arms- (i) West Gulley length 41.0 km, (ii) East Gulley length 64.0 km= 105.0 km.

Thakuran (21.658°N, 88.492°E), length 62.0 km, width 10.0 km.

Matla (21.603°N, 88.655°E), length 125.0 km, width 26.0 km.

Harinbhanga (21.718°N, 89.081°E), length 79.0 km, width 6.0 km.

Raimangal (21.781°N, 89.138°E), length 114.0 km, width 8.0 km.

The Bidya falls into the Matla around 21.917°N, 88.667°E; length 83.5 km, width 4.5 km.

The Gomdi (Gomdi khal or Gomor) falls into the Matla around 22.069°N, 88.747°E; length 67.0 km, width 1.0 km.

Tidal range

The tidal range of the river Hooghly varies between 4.24 and 5.19 m at Sagar (21.65°N, 88.05°E) and increases up to 5.3–6.1 m at Diamond Harbour (22.192°N–88.185°E) and the tidal amplitude ranging between 3.88 and 4.78 m at Garden Reach within the Kolkata Port area, 145 km north of Sagar, having at least a dozen of shifting sandbars in this stretch of 70 km.

Bangladesh Sundarbans

Sundarbans Forest Biogeographic Zone (SFBZ) encompasses the mouths, deltas, alluvial pans and coastal tributaries of important rivers such as Baleswar river on the east, and the Sela-Gang Bangra rivers, the Pasur-Shibsa-Kunga rivers, the Arpangasia-Manalcha rivers, and the Jamuna-Raimongal-Harinbhanga rivers on the west.

The SRF floor rises from 0.9 m to 2.11 m above the mean sea level.

During the late 20th century, major forest types according to the predominant species in forest area of 399,471 ha were- Sundari 74,992 ha, Sundari-Genwa 105,973 ha, Sundari-Pasur-Kankra 9,556 ha, Genwa 21,520, Genwa-Sundari 75,703 ha, Genwa-Garan 34,604 ha, Garan/Garan-Genwa 64,807 ha, Pasur-Kankra-Baen 4,030 ha and Keora 8,286 ha.

The floral composition varies in the protected areas depending on the salinity status. Sundarban East, a fresh or slightly saline habitat, mainly harbours Sundari mixed with varying quantities of Genwa followed by Pasur frequently associated with kankra, beneath the Sundari stands. Singra is found on comparatively dry soils and Amur on moist soil. Goran is rare but Golpata is common. The salty Sundarban West habitat is having sparsely spaced, short-boled Genwa over dense Garan, interspersed with dense patch of Hental palm on the drier soils. Dhundal, Pasur and Kankra occur sporadically throughout the area. Sundari does not thrive well and Golpata is very scarce. The Sundarban South is moderately salty where Genwa predominates mixed with Sundari in varying proportion, growing over a very dense jungle of Garan. Pasur is associated with Kankra and Baen. Golpata is abundant.

Species distribution

The northern part, new depositions and intertidal mud flats are characterized by *Avicennia alba, A. marina, A. officinalis* and *Sonneratia* spp, flanked by foreshore grassland (*Oryza coarctata*). *Avicennia* has gradually been replaced by *Excoecaria agalocha* and *Ceriops decundra* in most of the areas. The southern and eastern associates are composed of *Rhizophora apiculata, R. mucronata, Bruguiera sexangula, B. gymnorhiza* and a few patches of *Heritiera fomes, E. agallocha, Ceriops decandra*. Pure forests of *Phoenix paludosa* exist on relatively high lands. Occurrence of *Phoenix, Xylocarpus* and *Nypa* is extremely limited. Nypa palm swamps are common on central, eastern and southern portions, along creeks and rivers. The sea-facing areas have a long line of *Saccharum* grassland ideally suited for tigers and prey species.

Zoogeographical region

The Sundarban is a part of the Oriental Realm, where two Zoo-geographic Regions are merged, namely the Indian Region and the Indo-Chinese region. Ethiopian and Palaearctic elements are also found.

Eco-regions

The Sundarbans is divided into two ecoregions-

1. Sundarbans freshwater swamp forests (IM0162)

This ecoregion is characterised by *Heritiera* sp., *Xylocarpus* sp., *Bruguiera* sp., *Sonneratia* spp. and *Avicennia* sp., with *Pandanus tectorius, Hibiscus tiliaceus*, and *Nipa fruticans* along the fringing banks. ecoregion lies behind the latter comprising of a tropical moist broadleaf forest and representing the slightly brackish water region washed by the upstream freshwater during the rainy season along with a deposit of silt. Originally, the region covered 14,600 km², extending from the northern part of Khulna District up to BoB with scattered portions extending into the Indian counterpart.

This eco-region is nearly extinct. Most of the natural habitat has long been converted to agriculture, making it almost impossible to even surmise the original composition of the eco-region's biodiversity. Centuries-old human habitation and random forest exploitation have resulted in degraded and fragmented habitat as well as loss of biodiversity.

There are two PAs in Bangladesh-

Narendrapur (110 km²) and

Ata Danga Baor (20 km²) that cover a mere 130 km².

There are two PAs in India-

Bibhutibhushan WLS (0.64 km²) (23°-11'23.14''N and 23°11'36.84''N, 88°46'23.94''E and 88°46'25.00''E)

Floral biodiversity of the PA comprises about 209 species of angiosperms including 59 trees, 98 herbs, 34 shrubs, 15 climbers and 3 creepers. A thin line of swamps of Hijal (*Barringtonia acutangula*) representing Tropical Seasonal Littoral and Swamp forest is seen along the bank of river Icchamati. This forest is known to harbour more than 200 deer, birds, rabbits and a large number of langurs. 0.31 km² (almost 50%) of the WLS as the "Eco Sensitive Zone" (ESZ) on 1st October, 2018.

Chintamani Kar WLS (0.06956 km²)(22°42'N, 88°40'E)

Being surrounded by temporary and permanent small water bodies as well as terrestrial vegetation covering a wide variety of orchids, ferns, epiphytes, plants and trees like mango, jackfruit, coconut, tamarind, guava, dumur or ficus, bamboo yards, safeda, chatun etc, this sanctuary harbours diverse invertebrates, insects, molluscs, reptiles, birds, and small mammals like Civet Cat (*Viverridae* spp.), Fishing Cat (*Felis viverrina*), Mongoose (*Herpestes* spp.), Water Monitor Lizard (*Varanus salvator*), Jackal (*Canis aureus*), Squirrel (*Funambulus pennanti*), Common Fox (*Vulpes bengalensis*), etc. It is a paradise for birds such as Red throated flycatcher, Oriole, Paradise fly catcher, Drongo, Jungle Babbler, Asian Koel, spotted Dove, Common Kingfisher, Indian cuckoo, Rufous Woodpecker, Little Cormorant, Lineated Barbet, etc. 0.105 km² of the WLS, including the surrounding 100 m belt, was declared an ESZ on 12th October, 2017.

East Kolkata (Calcutta) Wetlands (EKWs)(22°27'N, 88°27'E)

It is a habitat for waterfowls and home for a large number (1,925 species) of flora and fauna, was declared a Ramsar site (No.1208) in 2002 over 125 km² to save the wetlands from continued deterioration. The solid waste dumping area on the western periphery of the wetlands were converted to horticulture since 1876. The first attempt to freshwater aquaculture is reported in 1918. The large scale usage of sewage for fish culture began in 1930s. EKW consists of 264 operating fish farms. The post independence expansion of the urban area prompted reclaiming of nearly 1,000 hectare of the northern portion of the wetland and hundreds of fish ponds for establishment of the Salt Lake City. In 1969, redistribution of land through land reforms led to further filling up of approximately 2,500 hectare of water bodies for conversion into paddy fields.

16 of mammalian species recorded from the wetland area represent 8 Carnivore species, 2 bat species, 6 species of Squirrel, Rat, and Mouse. Out of these 16 species, 9 are common, 1 is sporadic and rest 6 are rare. One mammalian species is endemic to EKW i.e. Marsh Mongoose (*Herpestes palustris*). 10 amphibians and 29 reptiles (9 lizard, 1 turtle and 19 snake) species were recorded from EKW (EKWMA, 2020), as against 19 reptilian species including 2 species of snake, 2 of monitor lizard, 3 of common lizard and 1 species of fresh water tortoise during a survey between 1980-1995.

Comparison of few inventories of birds available shows a rapid change in bird biodiversity in recent years in the area. A study between 1960 and 1965 has recorded 248 species (90 aquatic birds and 158 land birds) in EKW. Among these 50 per cent of the aquatic birds were reported to be migratory. Prakriti Samsad recorded 123 species of birds from Salt Lakes during 1978-83 (Roy Chowdhury, 1984). At least 16 species have not been recorded after 1978-83 survey. Of the 271 species of birds recorded from the wetlands, only 162 species have been variably noted during the last 30 years. It is assessed that 109 species of birds have become locally extinct, majority being aquatic birds. From the annual bird population estimation under Asian Water-bird Census in Nalban bheri it was observed that still a handful of Gadwals, Garganeys, Snipes, gulls, Terns, Egrets and Cormorants occur throughout the year though showing declining trends in population.

Similarly, there has been significant loss of vegetation diversity, particularly those of mangroves and other brackish-water species.

The wetland which in early twentieth century teemed with a large spectrum of brackish-water and freshwater fishes, only supports cultivable freshwater species. The commercially important fin and shellfish in the EKW include 79 species of fish, 11 species of prawns, 3 species of crabs and 20 species of molluscs. Among the fish species, 17 are cultured species and 41 are wild species. Occurrence of 80 species of fishes was reported from the Salt Lake. The low-lying region with saltwater lakes acting as spill reservoirs for the Bidyadhari was utilised for farming of brackish water fish such as Bhetki (*Lates calcarifer*), Parse (*Mugil parsia*), Bhangar (*Mugil tada*) and Prawns (*Macrobrachium rosenbergii*), etc. Occurrence of 40 fish species was also reported from EKW. 27 ornamental fish species of 21 families have also been recorded from EKW. CIFRI has observed entry of four exotic species (Crocodile fish) in EKW in recent years. The presence of invasive exotic fish species such as *Clarius guripinus* and *Pangasius sutchi* in the aqua-culture farms poses great threat to the native diversity.

2. Sundarbans brackish water mangroves (IM1406)

It was estimated that during a period of 238 years (1776-2014), the mangrove forests in Indian Sundarbans slashed down by approximately 71.88% [1776: 6588; 1873: 6068 (-7.89%); 1968: 2307 (-61.98%); 1989: 1983 (-14.04); 2001: 1926 (-2.87%); 2014: 1852 (-3.84)]. Therefore, maximum reclamation took place between the rivers Hooghly and Matla during the period 1873-1968 and the rate of deforestation has progressively reduced thereafter due to strict implementation of the legal provisions and management actions including the forcible eviction of the refugees from

Marichjhapi for unauthorized occupation of the reserve forests in January, 1979.

During the first two decades of 21st century, land loss in nine sea-facing islands is shown below in km²:

Dalhousi: 11.6, Bhangaduani: 10.9, Chulkati: 5.6, Bulcherry: 4.9, Baghmara: 4.7, Dhanchi: 4.1, Chamta: 3.0, Mechua: 2.6 and Jambu: 1.3.

Land loss in nine islands during these two decades was about 49 km², all of which (except Jambu Island) are tiger habitats.

This ecoregion is characterised mainly by *Heritiera fomes* and other tree species including *Avicennia, Xylocarpus mekongensis, Xylocarpus granatum, Sonneratia apetala, Bruguiera gymnorhiza, Ceriops decandra, Aegiceras corniculatum, Rhizophora mucronata,* and *Nypa fruticans*. The commonly identifiable vegetation types in the dense Sundarbans mangrove forests are salt water mixed forest, mangrove scrub, brackish-water mixed forest, littoral forest, wet forest and wet alluvial grass forests.

During the 21st century, percentage loss of areas of mangrove genera in nine erosion-prone sea-facing islands is maximum in case of *Ceriops* sp. (26.3) followed by *Avicennia* sp. (22.3), *Excoecaria* sp. and *Ceriops* sp. (20.9), *Excoecaria* sp. (13.6), *Aegialitis* sp. (7.1), *Sonneratia* sp. and *Heritiera* sp. (3.9), *Phoenix* sp. (3.3), Mixed mangrove (2.7).

Causes of change in and impact on the mangrove forests

1. Declining sediment supply

   Avulsion of river courses eastward with a consequent reduction in freshwater flow and sediment supply- Reduced resilience of mangrove to relative sea-level rise and increased coastal erosion.

2. Salinisation

   Avulsion of river courses eastward with a consequent reduction in freshwater inflow and increasing marine influence due to sea-level rise- Salinity stress, loss/decline of low salinity mangroves, stunted growth and mangrove health deterioration.

3. Relative sea-level rise: Climate-induced sea-level rise and deltaic subsidence- Loss of mangroves due to erosion and inundation.

4. Temperature rise

   Continuing warming of land and sea- Stress on mangrove germination and propagation, with potential adverse impacts on ecosystem functions.

5. Change in rainfall

   Seasonal rainfall change and variability including in the monsoon- Stress on germination and propagation, and potential for stress on established mangrove forest.

6. Cyclones and Storm surges

   High wind speeds and extreme water levels- Abrupt loss of mangrove canopy cover, reduced leaf area, death along river margins due to storm/surge thrust.

Present reserved and protected areas

Most of the Reserved Forests has been designated as Protected Areas under the highest protection and management regime with NP and WLS with some parts set aside for forest produce extraction and eco-tourism. These areas act as 'ecological islands' within the mangrove forests.

Indian Sundarban

STR was declared in May, 1973 [1948: 2,584.89 km^2; rounded off= 2,585 km^2 (Land= 1,680.13 km^2 and water= 904.76 km^2); 2003: 2,626.36 km^2 (Land= 1,624.18 km^2 and water= 1,002.18 km^2)].

The forest area in South 24-Parganas Forest Division was 1,635.11 (Land: 755.56 km^2, water: 879.55 km^2) in 1948 has been increased to 1,699.85 km^2 (Land: 585.74, water: 1,114.11 km^2); total increase (55 years) being 64.74 km^2.

Five protected areas have been set up in the Western Sundarbans-

Tiger Reserve

One NP 1,330.12 km^2 (1984) as well as Critical Tiger Habitat (CTH)= 1,699.62 km^2, including primitive area (113 km^2) to act as gene pool, which is kept free from any human activity and resultant disturbances;

Management Zones-

(a) Core Zone: 1,692 km^2,

(b) Buffer zone (2,233 km^2) with recuperation zone (Pirkhali and Panchamukhani), open only for eco-tourism purpose (fishing, honey-collection etc prohibited), multiple-use zone (Jhilla, Arbesi, Harinbhanga and Khatuajhuri), for regulated fishing, honey and beeswax collection;

(c) Transition zone for cooperation (5,705 km^2) comprises some mangrove areas, mostly non-forest and reclaimed habitation areas for agriculture, where all work together to manage and develop the resources in addition to research and eco-tourism;

Four Wildlife Sanctuaries-

Tiger Reserve

Sajnekhali 362.33 km^2 (1960 and 1976),

South 24-Parganas Forest Division

Halliday Island 5.95 km^2 (1976),

Lothian Island 38 km^2 (1948, 1976) and

West Sundarban 556.45 km^2 (2013) with limited human activities compatible with sound ecological practices are allowed in addition to research, education and eco-tourism.

Bangladesh

Three Wildlife Sanctuaries (Sundarbans West 715 km², East 312.27 km² and South 369.70 km²) were constituted in on 6th April, 1996 and declared on 29th June, 2017 under the Bangladesh Wildlife Act 1974.

In 2012, three river areas (Chandpai, Dudhmukhi and Dhangmari) covering a total area of 10.7 km² were declared as WLS, mainly for the protection of dolphins. The rest of the Sundarban remained as Reserved Forest, which was declared in the colonial era.

Swatch of No Ground Marine Protected (SoNG) Area was declared in 2014 in a 120-km wide belt of estuarine, coastal and deep-sea waters across the SRF and offshore to a 900+ meter deep under-sea canyon for conservation of marine cetaceans and birds.

Four more WLS for the dolphins were declared recently.

CRZ

Categorically, SBR falls under CRZ-I (i) and (ii) categories including the ecologically sensitive conservation areas like Dunes/Runnels, Fraserganj, Gangasagar Island, etc. Coastal stretches of BoB, estuaries, backwaters, creeks, etc. influenced by tidal action up to 500 m from HTL and the wetlands between(LT) and the HTL have been declared as CRZ. Maximum habitation with vegetation (545.191 km²) is grown on both sides of tidal river/creeks and roads.

Out of 2,227.19 km² (mostly in Sundarbans), mudflats cover 223.94 km². Marsh vegetation occupies 22.59 km². Built-up land is 1,303.60 km². Reclaimed areas are mainly converted for settlement and agriculture. About 347.46 km² is used for saline and brackish water aquaculture. However, the landscape has been changed since 1991. Some islands have been completely subsided and some new islands formed.

SBR covers two distinct physiographic sectors of this L-shaped CRZ- (i) along the river Hooghly (Central Sector) and (ii) on the east of river Hooghly (Eastern Sector).

Regulations as per CRZ notification

The following activities are prohibited within CRZ, namely:

(i)   Setting up of new industries and expansion of existing industries, except those directly related to waterfront or directly needing foreshore facilities;

(ii)   Manufacture or handling or storage or disposal of hazardous substances.

(iii)   Setting up and expansion of fish processing units including warehousing (excluding hatchery and natural fish drying in permitted areas);

(iv)   Setting up and expansion of units/mechanism for disposal of waste and effluents, except facilities required for discharging treated effluents into the water course with approval under the water (Prevention and Control of Pollution) act, 1974; and except for storm water drains;

(v)   Discharge of untreated waters and effluents from industries, cities or towns and other human settlements;

(vi)   Dumping of urban waste for the purpose of land filling or otherwise;

(vii)   Dumping of ash or any waste from thermal power stations;

(viii) Land reclamation, bunding or disturbing the natural course of sea water with similar obstructions, except those required for control of coastal erosion and maintenance or clearing of waterways, channels and ports and for prevention of sand bars and also except for tidal regulators, storm water drains and structures for prevention of salinity ingress and for sweet water recharge;

(ix)   Mining of sand, rocks and other substrata materials, except rare minerals;

(x)   Harvesting or drawl of ground water and construction of mechanisms within 200 m of HTL; in the 200 m to 500 m zone it shall be permitted only when done manually through ordinary wells for drinking, horticulture, agriculture and fisheries;

(xi)   Construction activities in ecologically sensitive areas;

(xii)   Any construction activities in between LTL and HTL except facilities for carrying treated effluents and waste water discharges into the sea, facilities for carrying seawater for cooling purpose, oil, gas and similar pipelines and facilities essential for permitted activities; and

(xiii) Dressing or altering of sand dunes, natural features including landscape changes for beautification, recreational and other such permissible purposes.

Forest management

The management variations in both the countries have been pointed out by Nishat *et al.* (2019):

- Management boundaries differ between the two countries. In Bangladesh, while a 10-km wide band was declared ecologically critical area (ECA), forest conservation essentially focuses on the reserve forest. There are a few uncoordinated activities by government and non-government agencies, no real initiative has been taken to manage and develop the surrounding areas which house the population dependant on the Sundarban so far. On the other hand, SBR consisting of uninhabited National Park as well as inhabited areas and there are quite a few programmes and initiatives by government, non-government and research agencies focusing on the entire SBR.

- As a result, in Bangladesh planning and management has typically been focused on the physical resource, its status and extent. In West Bengal, India, socio-economic issues affecting forest change, dependency and interlinkages on forest resources are better addressed.

- In Bangladesh only one agency, the Forest Department is solely responsible for the conservation of the Sundarban. While in India there are quite a few agencies at the state and national level involved in conservation activities. These include, West Bengal Forest Department, National Tiger Conservation Authority, Sundarban Biosphere Reserve Authority. There is often overlapping roles and responsibilities creating confusion.

- The geographic location of Sundarban in both countries mean that some of the issues are also different. The Sundarban is situated a 100km from Kolkata, the state capital, and is accessible by road and railway. This means more tourists have access to the mangrove forest. Tourism infrastructure are also more developed. Whereas, in Bangladesh, the Sundarban is quite remote and can be accessed only by rivers. This has restricted the number of tourists in Bangladesh, however, tourists are on the rise.

- Despite human habitations and economic exploitation of the forest, Sundarban retained a forest or canopy closure of about 70%. The forest-village interface in Bangladesh is limited to the northern boundary of the Sundarban, limiting human-wildlife interaction. In India, many human habitations share boundaries with forested areas, causing wildlife, especially tigers to stray into the habited areas. In order to prevent straying of tiger into villages, Nylon net as well as well as Goran fencing have been erected along the forest-village interface. Tranquilisation and capture of the straying animal and their subsequent release into the forest, is also frequently resorted to.

Natural hazards

Triangular shape of BoB produces concentration and funnelling of storm energy in a northerly direction towards the apex, with landfall typically in the Sundarbans.

Storms and cyclones

Tropical cyclones are measured by the wind speeds and are divided into five categories as follows:

- Category 1: Very dangerous winds ranging from 74 to 95 mph will produce some damage.
- Category 2: Extremely dangerous winds ranging from 96 to 110 mph will cause extensive damage.
- Category 3: Devastating damage by winds ranging from 111 to 129 mph will occur.
- Category 4: Catastrophic damage by winds ranging from 130 to 156 mph will occur.
- Category 5: Unprecedented catastrophic damage by winds ranging from 157 or higher mph will occur.

Storms are common in the Sundarbans during pre-monsoon (May), monsoon (June-September) and post-monsoon (October-December). Pre- and post-monsoon storms are more violent than those of the monsoon season. Life span of a severe cyclonic storm in the Indian seas averages about 4 days from the time it forms until the time it enters the land. Super

Cyclonic Storms over BoB registered 26% increase over last 120 years, intensifying in post monsoon.

Cyclones bring strong wind, heavy rainfall and flooding causing severe coastal erosion and embankment failure. Some of these often develop into cyclones of varying intensity usually accompanied by tidal waves and cause much loss of life and damage to property and forests. Since the reclamation was started before the completion of siltation process therefore, most of such reclaimed lands are below the high tide level thus making them most vulnerable during the cyclones and there are frequent breaches in the embankments thus causing enormous damages to the villages.

The region had experienced severe cyclonic storms as listed below.

1582, 1688, 1699, 1707, October 7–12, 1737, 1742, 1830, 1832, 1833, 1839, 1840, 1844, 1848, 1850, 1852, 1858, 1859 (twice), 1862, 1864, 1867, 1869, 1872, 1877, 1878, 1880, 1881, 1882, 1883, 1884, 1887 (twice), 1888 (twice), 1889 (twice), 1893, 1894, 1896 (quintuple), 1898 (quadruple), 1899, 1901 (quadruple), 1904 (twice), 1909, 1913 (twice), 1916 (twice), 1917, 1919, 1922, 1927, 1928, 1932 (severe), 1934 (severe), 1935 (severe), 1936 (severe), 1937 (severe), 1940, 1941, 1942, 1943, 1948, 1956, 1960, 1962, 1965, 1968 (severe), 1970, 1973, 1976, 1981, 1982, 1985, 1988 (severe), 1991, 1994, 1997, 1999, 2001, 2002, 2004, 2006, 2009, Fani in May 2019, Bulbul in November 2019, and Amphan May 2020.

The records of storms and cyclones have revealed an increasing trend in the degree of their intensity while showing a decrease in the frequency of occurrence as a result of the warming of trend. This change has significant bearing on the extent of coastal flooding, soil erosion in twelve sea-facing islands of the western Sundarbans from Sagar on the west up to Bhangaduani on the east and saline water intrusion due to storm surges. The increasing intensity also implies increasing precipitation pattern in the Sundarbans.

On the night of November 1, 1867, a violent cyclone passed over Port Canning. A wave nearly 2 m height carried a portion of the riverbank jetties. The port was abandoned. The rails and station were also destroyed. Earlier in May 1833, the Sagar Island was struck by a severe cyclone and five settlements were almost lost. When subsequent storms struck in June 1842, October 1848 and again in June 1852, the settlers abandoned the island. After resettlement, another cyclone struck on 5 October 1864 and caused great havoc. Thereafter another storm struck in November 1867 that reduced the cultivated area on the island to about 440 ha.

Amphan was the worst cyclone to hit SBR in nearly 300 years. With wind speeds after landfall of 155-165 kmph, it was categorised as a Category 5 supercyclone- a cyclone attaining wind speeds of 260 kmph at sea before landfall. 28 percent of the region had been completely destroyed.

These mangroves have also reduced the intensity of Amphan- from Category 5 to Category 3 through what is called the "mangrove reduction effect", saving the mainland from high scale devastation. In addition to destruction of mangrove forests, agricultural land in the non-forest areas shrank from 2,149 km^2 to 1,691 km^2 between 2002 and 2009, mainly because of frequent storms, primitive agricultural methods, and a rising population. In May 2009, cyclone Aila destroyed 778 km of the embankments, though local people estimate the loss to be much higher. Immediately after the cyclone, a tiger that had been swept along by strong currents took refuge in a hut in Jamespur. Since the area was still chest-high in water, the Forest Department staff had to cut a hole in the roof of the house to tranquilise the tiger. It was caged, removed from the village, checked for any injuries and then released in the forest the following day. Three cheetal deer were also rescued on the same day. On the night of 30th June, a young tigress strayed onto an inhabited island at Adivasi para, on the eastern edge of STR. It was trapped, tranquilised and released on 3rd July. On the night of 2nd July, another young tigress entered Lahiripur village on Satjelia Island. It was successfully trapped the next night and released on 5th July. A fourth tiger strayed into Chargheri village on the night of 6th July and was trapped the following night and released in the forests. The tiger had some scratch marks on its front legs, possibly due to a territorial fight.

Cyclones significantly affect the morphology and ecosystem of the Sundarbans. The winds and waves generated by a low-level storm can cause extensive defoliation of mangrove trees bordering the sea-face, while a major cyclone can displace vast swaths of beach sand and dune fields and convert a long stretch of forest into mounds of organic rubble.

Cyclones cause tremendous disruption to wildlife. The loss of wildlife is either due to being washed away by the cyclone and tidal surges or falling under the broken trees. The dead animals also cause environmental hazards to the remaining herbivores and other wildlife. The cyclone damaged broken trees restrict the movement of wildlife and cause scarcity of fodder due to loss of regeneration in the forest floor. As the existing sweet water ponds in the forests have been infested with the salt water, safe drinking water for the animals is not available.

Impact of Yaas on 26th May, 2021

Since the landfall coincided with high tide on Wednesday morning, it led to major damage to Bhagabatpur crocodile breeding and Jharkhali rescue centres and inundation of at least 18 forest department camps and three range offices.

The biggest impact of flooding due to the cyclone Yaas was all the ponds of captive bred *Batagur baska* in the Sundarbans were inundated by salt water. Carcass of a full grown river terrapin was found in one of the isolation ponds in Jhingakhali camp. Before the cyclone, 35 hatchlings were kept in safe locations and could be saved. Four-five were also rescued. There were 200 captive bred terrapins in stock and would be searched out after the flood water receded.

There were more than 300 crocodiles were kept the Bhagabatpur crocodile breeding centre, but the hatchlings and sub-adults were shifted to safe place before arrival of Yaash, but a few adults were still there in the ponds. The boundary wall of the centre was completely damaged and storm water entered the compound through a breached embankment. A female crocodile around 9.6 ft, which was displaced at the time of flooding by Cyclone Yaas was rescued by a team of South 24-Parganas Forest Division on 28th May, 2021 at Patharpratima. The crocodile had entered into a pond owned by a local resident. After rescue, she was released at Lothian Wildlife Sanctuary.

At least five cheetal deer and one wild boar were also rescued by the villagers and forest staff at the time of flooding from both forest areas like Jhingakhali and Dayapur and Dulki villages on 26th May, 2021 and would be released in the wild. The police also rescued a deer near river Gomor and handed over to the forest department.

Above all, a full grown male tiger, aged about 11-12 years was found lying on the bank of pond of Harikhali camp under STR on 30th May, 2021. It was very weak and could not even be fed any drink. When it was being brought to Sajnekhali, it expired on the way. This a great loss to STR.

240 m of nylon net fencing, 850 m of boundary wall and a jetty were also damaged.

Earthquakes

1737, 1762, 1842, 1895, 1897, 1981: Magnitude 4.9 on Richter scale shook north of Sundarban on 26th March (Epicenter Latitude 22.3, Longitude 89.1).

Famine

1770, 1791, 1866, 1874, 1921, 1942

Floods

1823, 1834, 1838, 1856, 1864, 1868, 1871, 1885, 1890, 1900, 1904, 1907 (twice), 1966, 1978.

CHAPTER 4
HABITATS AND HABITANTS: ADAPTION AND ECOLOGICAL ROLE

Ecosystem services (ES)

The most productive Sundarban mangroves and salt marshes, often called "Blue Forests", provide many valuable ES for human well-being.

1. *Provisioning services*: Timber, fuel wood, fibre, food- fish, prawn, crab, honey, grains, fruits, etc. and thatch materials, fresh water, biochemicals (pharmacological resources), natural medicines, genetic materials (for resistance to plant pathogens and ornamental resources- shells and flowers), rivers (for transport);

2. *Regulating services*: Climate- temperature, precipitation and other processes, ground water recharge or discharge, coastal stability and protection, bioremediation, control of erosion and flood, protection from cyclones and storm surges, prevention of saltwater intrusion, retention, recovery and removal of excess nutrients and other pollutants, pollination, biological control, etc.

3. *Cultural services*: Eco-tourism, river cruising, beach relaxation, canopy-walking, wildlife watching, participation in ethnic festivals and worship, recreation, aesthetic, heritage, education and research.

4. *Habitat or supporting services*: Preservation of natural and semi-natural habitats for genetic diversity of communities and individual species, transportation and accumulation of inorganic and organic matters in the surface through tides, litter fall, etc. to provide nursery grounds of fish and other aquatic species, retention of soil and sediment, carbon sequestration, storage, recycling, processing and acquisition of nutrients.

Mangroves are being alarmingly degraded and lost owing to conversion (agriculture, aquaculture, coastal/upstream development and infrastructure), over-exploitation by the communities and, finally, inadequately restored (very much lesser than loss) due to weak institutional arrangements, policies and management systems.

Biodiversity values of Sundarbans

- Sundarban is the largest mangrove forest in the world.
- It is the largest estuarine delta in the world shared with Bangladesh.
- It is a significant example of on-going ecological processes of delta formation and subsequent colonization of the newly formed islands and associated mangrove communities.
- Indian Sundarbans mangrove forest is considered threatened (range: Vulnerable to Critically Endangered) under Red List of Ecosystems.
- It harbours exceptional biodiversity in terrestrial, aquatic including benthic and marine habitats, ranging from micro to macro flora and fauna.
- It is the only remaining habitat in the lower Bengal Basin for a wide variety of faunal species.
- Globally, it is the only Mangrove Tiger land.
- It is the habitat for the largest number of mangrove species in a single area.
- It is the habitat of the world's largest species of Estuarine Crocodiles.
- It has core breeding areas for a number of globally endangered species including *Panthera tigris*, *Platanista gangetica*, *Orcaella brevirostris*, *Crocodylus porosus* and the critically endangered endemic river terrapin *Batagur baska*.
- Two of the four species of highly primitive horseshoe crab (*Tachypleus gigas* and *Carcinoscorpius rotundicauda*) are found here.
- It forms a good number of monsoon heronries and winter swamp home for Trans-Himalayan and other migratory birds.
- It is the last Great Coastal Wetland left in the world, where the terrestrial animals including the tiger and prey species have adapted to a semi-amphibious life, being capable of swimming for long distances and feeding on the fish, crab, etc.
- Mangrove flora has many adaptations for salinity resistance like extensive lateral root system, presence of pneumatophores in the roots, thick and waxy leaves and vivipary germination.

- 53.78% of Indian Sundarbans are designated as PAs under the highest protection and management regime.
- The islands are also of great socio-economic importance as a storm barrier, shore stabiliser, nutrient and sediment trap, a source of timber and non-timber forest products (NTFPs).
- Biodiversity has direct consumptive value in food, agriculture, medicine, industry, etc.
- It also has aesthetic and recreational value.
- Biodiversity maintains ecological balance and continues evolutionary processes.
- The indirect ecosystem services provided through biodiversity are photosynthesis, pollination, transpiration, chemical and nutrient cycling, soil maintenance, climate regulation, air and water system management, waste treatment and pest control.

Biological diversity (flora and fauna) and habitat preferences

Biodiversity of the estuary consists of terrestrial, freshwater and marine communities.

Mangrove habitats

A collection of mangrove trees in an area makes up an important mangrove habitat by providing food, shelter and nursery areas for fish, birds, crustaceans and other lifeforms and a source of livelihood for many humans, including fuelwood, timber, NTFPs and areas for fishing as well as forming a buffer that defends coastlines from flooding and erosion.

Genus-wise habitats

1. *Acanthus*- Regularly inundated intertidal inner estuarine areas.
2. *Acrostichum*- Disturbed habitats with irregular inundation.
3. *Aegialitis*- Coarse-textured soil in saline intertidal areas.
4. *Aegiceras*- Less saline and newly silted inner estuarine areas with regular inundation.
5. *Aglaia*- Upstream mid-intertidal estuaries and swamps.
6. *Avicennia*- Intertidal river flats and slopes with normal salinity and newly accreted areas.

7. *Bruguiera*- Riverbanks and ridges in the middle estuarine region with frequent inundation.
8. *Ceriops*- Moderate to high-saline middle estuarine areas with regular inundation in general.
9. *Excoecaria*- Slightly elevated and well-drained inner estuarine areas with moderate to low salinity.
10. *Heritiera*- Less saline inner estuarine area along rivers that flood only during spring tide and that receive sufficient freshwater from upstream.
11. *Kandelia*- Soft mud along inland river banks.
12. *Lumnitzera*- Higher intertidal zone, beaches, bank of creeks.
13. *Nypa*- Less saline wet and flat river banks in the inner estuarine region receiving regular supply of freshwater.
14. *Phoenix*- Inner estuarine areas on mature dry soil along rivers that flood only during spring tides.
15. *Rhizophora*- Mouths of creeks, creeklets and small rivers in the middle part of the estuary with regular inundation and moderate to high salinity.
16. *Sonneratia*- Seaward fringes of the outer estuary, and best adapted in areas with regular inundation and normal salinity. Also found in newly accreted areas.
17. *Suaeda*- Slightly raised and less inundated areas, sometimes on denuded surfaces where soil salinity is particularly high.
18. *Xylocarpus*- well-drained areas in the middle estuarine region with fewer than 10 inundations per month.

Review

Geomorphological, environmental and hydrological factors are responsible for distribution of the mangrove species. There are ecological conditions like soil type (sandy clay, silty clay or silt), rainfall, tidal inundation and fluctuations, salinity gradients [polyhaline (>4 dsm^{-1}), mesohaline (2-4 dsm^{-1}) and oligohaline (<2 dsm^{-1}), limnetic than euhaine]. Salinity levels vary due to daily tidal fluctuations, changing moon phases between spring and neap tides and variation in freshwater inflow. The biotic factors play the real role in distribution of the floral species. For example, the succession of mangrove genera is variable at five tide levels (MLWP= Mean Low Water Spring, MTLL= Mid Tide Lower Limit, ML= Mean Level, MLWN= Mean Low Water Neap, MLWS= Mean Low Water Spring) as follows: *Avicennia, Sonneratia, Aegiceros* (MLWP 0-1 m),

Heritiera, Lumnitzera, Schyphiphora, Brownlowia (MLWS 1-2 m), *Xylocarpus* (MLWN 2-3 m), *Excoecaria, Aegialitis* (MLWN 2-4 m), *Ceriops* (ML 3-5 m), *Bruguiera, Kandelia, Nypa* (MTLL 3-6 m) and *Rhizophora* (MTLL 5-6 m).

The salinity tolerance limit also varies significantly from species to species, which is shown (range in ppt) against each mangrove species- *Avicennia* (9-30), *Sonneratia* (4-15), *Aegiceros* (10-14), *Heritiera* (5-15), *Lumnitzera* (5-12), *Scyphiphora* (5-9), *Brownlowia* (7-12), *Xylocarpus* (15-20), *Excoecaria* (3-18), *Aegialitis* (5-7), *Ceriops* (6-15), *Bruguiera* (6-20), *Kandelia* (10-25), *Nypa* (5-18) and *Rhizophora* (9-25).

In the river dominated mangroves, the species distribution also depends on the landforms or contours of the forest, such as river flat (minimum value)- *Avicennia, Sonneratia, Aegiceros, Ceriops, Excoecaria, Aegialitis*; river slope (bed and flow)- *Rhizophora, Kandelia*; river flat-slope- *Bruguiera*, river mouth- *Nypa* or ridge (elevated) forest- *Xylocarpus, Heritiera, Lumnitzera, Schyphiphora* and *Brownlowia*.

Moreover, the mangrove species occupy their respective positions in the intertidal areas depending on their requirements for substrate, water replenishment, topography, etc. For example, some of the mangrove species, such as *Avicennia* spp, prefer firm exposed substratum towards the landward side, whereas *Sonneratia* sp. grows in soft substrate in the sheltered estuarine mouth. *Bruguiera* and *Rhizophora* spp. are found in sandy, firm bottoms and also in the low tide areas.

If the distribution pattern of the mangrove species is considered on the basis of combined forces, such as substratum and salinity, it may be seen that there is no record of such growth in the rocky sandy-cum-euhaine (>30 ppt) zone; maximum diversity in sandy clay-cum-polyhaline (18.0 to 30.0 ppt) zone (*Sonneratia, Rhizophora, Avicennia, Bruguiera, Acanthus*), moderate diversity in silty clay-cum-mesohaline (5.0-18.0 ppt) zone (*Kandelia, Avicennia, Rhizophora, Aegiceros* and *Sonneratia*) and low diversity in silt-cum-oligohaline (0.5-5.0 ppt) zone (*Sonneratia, Acrostichum* and *Cyperus* spp).

Accordingly, in the Indian Sundarbans, the restricted species were found to be *Acanthus volubilis, Aglaia cuculala, Bruguiera cylindrica, Bruguiera parviflora, B. sexangula, Caesalpinia spp, Cerbera odollam, Ceriops tagal, Cyanometra ramiflora, Dodonaea viscosa, Hibiscus tiliaceus, Kandelia candel,*

Premna corymbosa, Rhizophora apiculata, Salicornia brachiata, Scyphiphora hydrophyllacea, Sonneratia caseolaris, Stenoclaena palustris, Thespesia populnea, T. populnoides and *Volkameria inermis.*

The Sundarban represents all the true mangrove species, of which the dominant ones were found to be *Rhizophora mucronata, Bruguiera gymnorhiza, Ceriops decandra, Avicennia alba, A. marina, Sonneratia apetala, Xylocarpus* spp, *Excoecaria agallocha* and *Phoenix paludosa.* Of them, the family Rhizophoraceae is dominant compared to other true mangrove families. Considering the palm species, *Nypa fruticans* are distributed better in the eastern part, particularly Chamta, Haringbhanga, Khatuajhuri and Arbesi forest blocks, located far from BoB because of supply of sweet headwaters of Raimangal-Harinbhanga than the more saline western region, i.e. Matla estuary. But the other species, i.e. *Phoenix paludosa,* is abundant in most of the forest blocks with large patches, particularly in the intertidal zone. In the sea-facing forest blocks of Mayadwip and Goasaba the diversity was also good like that of Bagmara, and together they protected the region from the onslaught of frequent cyclonic storms.

Salinity has two correlations with the mangrove species. Salinity was found negatively correlated with the species richness and abundance of *Aegiceras corniculatum, Rhizophora apiculata, Bruguiera* spp, *Nypa fruticans, Heritiera fomes,* but *Sonneratia* spp., *Avicennia officinalis,* etc. have positive correlations with salinity. However, *Heritiera fomes,* the glory of Sundarbans, was found to be abundant in a few blocks, such as Chotohardi, Mayadwip, Goasaba, Chamta and Chandkhali, all are within the core area of STR and Chamta is also designated as the primitive area.

It was observed that the diversity is highest in the remote forest blocks not suffering from any anthropogenic interventions, whereas those forest blocks having great impact of such interventions are less diverse. Hence, the diversity of the mangrove species largely depends on human interventions. For example in Bagmara the mangrove density is highest but the density of primitive conservation blocks not as remote as Bagmara is comparatively low due to human interventions.

Ecological change

Since Curtis (1933) described the Sundarban forests in the early twentieth century, there has been a great change in the status of individual crops, forest exploitation vis-à-vis natural regeneration in SBR at the end of the century (Chaudhuri and Choudhury, 1994):

- *Ceriops decandra* and *Excoecaria agallocha* association forms the overall matrix of the vegetation. But the dynamic locality factors often impart metastable shape of the vegetation.
- *Heritiera fomes* is sprinkled all over the matrix of *Excoecaria-Ceriops*.
- Forests to the east of Matla have mixed mangrove vegetation of *Rhizophora apiculata, R. mucronata, Bruguiera gymnorhiza, B. sexangula, B. parviflora, B. cylindrical, Kandelia candel, Aegialitis rotundifolia, Aegiceras corniculatum, Lumnitzera racemosa* and sprinklings of *H. fomes.*
- On the sea shore occurs *Thespesia populnea.*
- *Xylocarpus granatum* and *X. mekongensis*, which were scattered all over the blocks east of the River Matla, are now few and far between. The generation status is poor.
- *Ceriops* spp have a degenerated form, which occupies a wide area. Similarly, *Avicinnia* has a degenerated form in *A. alba*, which occurs in areas where the anthropogenic stress is heavy and inundation level has gone high.
- *Nypa fruticans* has best development in Champta, Harinbhanga, Khatuajhuri and Arbesi blocks, but the height growth is much less than those growing in less saline areas. This species has been over-exploited and the regeneration status is poor.
- *Avicennia alba, A. marina* and *A. officinalis* have profuse regeneration in newly formed delta.
- *Phoenix paludosa* occupy uninundated fringe of the forests. In the forests west of Matla it occurs over wide areas as biotic sub-climax vegetation. Hental area has increased to a great extent.
- Beach forests have *Excoecaria-Ceriops* thickets and several freshwater species occur on the sandy areas.
- Saline blank areas have been extended considerably.
- *Bruguiera* sp. is not common. The regeneration status is poor.
- *Rhizophora* sp. has been over-exploited and is not common. The regeneration status is poor.
- *Sonneretia apetala* dominates in the southern half while *Avicennia marina, A. alba* in the north forest shores. *Sonneretia caseolaris* and *S. griffithii*

occur in large numbers near the sea face. *S. apetala* has been over-exploited and the regeneration status is poor.

- *Excoecaria agallocha* is extending and the regeneration status is poor.
- *Ceriops* sp is extending.
- The status of *Aegiceras, Aegialitis, Amoora* is discouraging. *Aegiceras* suffers from pilferage of timber and fuelwood.
- The effects of erosion by ebb tides and function of muddy channels extending to the interior of the islands is very high in Gosaba block (1–3). Gosaba-3 is the most affected compartment, which contains large patches of mud flats at the heads of these muddy channels. Considerable parts of the block is either devoid of trees or is an open type of forest with sporadic growth of bushes or dwarf trees like *Ceriops decandra* (Jhanti Garan).
- The entire crop is now low, dense forests of monotonous crop of *Ceriops* and *Excoecaria*. The higher diameter class of *Excoecaria* hardly exceeds 10– 15 cm, that of *Ceriops* hardly exceeds 7–10 cm. Heterogeneous distribution of *Avicennia* sp and *Excoecaria* sp occurs in large areas of land both in the core and buffer area and in different parts of the intertidal flats and they generally occur in the following combinations namely (i) *Excoecaria-Avicennia-Phoenix,* (ii) *Phoenix-Excoecaria-Ceriops,* (iii) *Ceriops-Rhizophora-Excoecaria–Avicennia,* (vi) *Avicennia-Excoecaria-Sonneratia,* (v) *Nypa-Ceriops-Excoecaria,* (vi) *Avicennia-Rhizophora* and (vii) *Rhizophora-Avicennia-Ceriops.*

Zonation and habitat preference

The zonation of mangrove vegetation in the eastern and western side of the study area is quite distinct and shows specific patterns. While the western side is dominated by *Avicennia* sp, the eastern side is clearly dominated by Excoecaria sp. The zonation of mangroves in the core region, as identified in the intertidal river flood plains, includes both overlapping and non-overlapping nature of forests. *Avicennia, Excoecaria, Rhizophora* and *Ceriops* are the widely dispersed taxa in the intertidal flats. Mangroves of different genera and species in a forest generally overlap with each other. However, *Sonneratia* sp. maintains a unique identity of its own by appearing in the lowest intertidal zone. It exhibits distinct non-overlapping characters from the major cluster of other mangroves that occur in the middle to upper intertidal zones. With time, there is a gradual filling up of the intermediate space by the appearance of other species of mangroves.

The study area is also classified into three salinity- low, medium and high zones in the northern, central and southern parts respectively. Although water salinity tends to increase as one proceeds downwards from the northern zone to the central zone and then to the southern zone, a similar pattern is also noticed so far as soil salinity index (starting from 5 to 6 PPT in the northern zone, viz. Sajnekhali, Sudhannyakhali, Pirkhali, Panchamukhani and Netidhopani) is concerned, but the frequency of human casualties does not vary according to such zoning, which indicates that the man-eaters of Sundarbans are concentrated in some habitat formations only, which also incidentally records significantly more salinity (Chaudhuri, 2007).

Three ecological units of mangrove vegetation are found within the above zones:

(a) The eastern patch, lying east of the River Harinbhanga in STR, where some sweet water flows from Bangladesh.

(b) The western patch, lying west of the River Thakuran, where a tickle of sweet water reaches from the River Hooghly in South 24-Parganas Forest Division.

(c) The central true mangrove patch, lying between the Rivers Harinbhanga and Thakuran in STR, is almost completely cut off from the upstream flow and fed by backwaters of the BoB.

The western side of Sundarbans such as LWLS in Bhagabatpur range under South 24-Parganas Division shows that the seaward edge of the forest is endowed with saline grass *Porterasia coaretata*, which are followed towards landward side by *Avicennia officinalis*, *A. alba*, *Acanthus ilicifolius*, *Excoecaria*, *Bruguiera gymnorhiza*, *Sonneratia apetala*, *Ceriops* sp and *Phoenix* sp.

The generic and specific diversity of the Sundarban mangroves varies in three identified zones. The tiger mostly uses the low diversity areas above the general tidal level (Zone A). At Zone C (southern), both the generic and specific diversity are the highest amongst the three zones, but the generic diversity of Zone A (northern)(above the tide level) is more than that of Zone B (central), which is frequently inundated, while the reverse relationship exists between Zones A and B so far as specific diversity is concerned. But, according to timber and fuel resources, Zone A is richer than Zone B, which, in turn, is richer than Zone C. So, the richness of resource is inversely related to the generic and specific diversity of the vegetation. Again, it may be seen that Zone C, which records the highest generic and specific diversity, does not form an ideal and permanent abode of the tigers as the tiger possibly avoids the constant tidal fluctuations and, hence, the increased salinity. Salinity indexes are the most in Zone C followed by Zone B and then by Zone A. A unique example of a very dense but closed vegetation matrix is evidenced from the low figures of both generic and specific diversity. Hence, Zone A is the mostly used habitat of the tigers and its prey animals.

Littoral or supralittoral forest fauna

Mangrove forest is inhabited by terrestrial animal communities. They may occupy trees or ground or both. Some have the power of flight. Most of the mangrove animal communities have distinct zonation in relation to tidal height, but the tree fauna exhibit vertical zonation in the vegetation.

Tree community

Animals under this community include both aerial and arboreal forms. The upper canopy of mangrove trees is the home of birds, bats, monkeys and insects. But such as *Pipistrellus mimus* can be found flying on the onset of evening inside the TR areas. *Macaca mulatta*, the only species of primate occurring in the Sundarban, is well distributed in the entire forest. They are often found feeding on *Sonneratia apetala* but are also well adapted to crab-eating.

Many species of birds build their nests in the mangrove trees. Herons, egrets, cormorants and darters enjoy roosting in colonies on the tall trees. Several species of birds use trunk, branches and aerial roots of mangrove as observation posts for feeding on fishes, molluscs, crustaceans and aquatic insects.

Flowers of some species of mangals attract honeybees and offer opportunities for the production of honey. The Rock Bee, *Apis dorsata*, is an important species of commercial importance. They are known to build their honeycomb inside the forest in large numbers. About 39% of honey is produced from *Excoecaria agallocha*, 16% from *Avicennia* species, 11%

from *Ceriops* species, 10% from *Rhizophora* species and only 24% from the rest of the plants. They have considered that *Phoenix, Excoearia* association is the ideal site for comb formation.

Ground dwellers

The forest floor of Sundarban is the domain of the Royal Bengal Tiger, *Panthera tigris*. This 'majestic creature', a conservation-dependent species, was conferred the status of national animal of India on November 18, 1972 because of its elegance, strength, agility, and colossal power. It is also the national animal of Bangladesh. It has been placed under the endangered category by IUCN in 2006 and is listed under Schedule-I of the Wildlife (Protection) Act in 1972 in India and Appendix-I of CITES. The Bengal tiger is categorised as 'Critically Endangered' (IUCN-Bangladesh, 2000). It is listed in the third schedule of the Bangladesh Wildlife Act of 1974, implying its full protection by interdicting killing and capturing (MoEF-Bangladesh, 2004).

The Sundarbans tigers exhibit strong ecological and genetic differences from other Bengal tiger populations. However, there was evidence of recent historical connectivity with the peninsular Indian tiger population. There is no knowledge of the current gene flow between Sundarbans populations in India and Bangladesh, although at individual level movement in between the two countries is ascertained first in recent times.

Osteological studies in the Indian Sundarbans revealed that an adult male tiger aged about 14 years was 2.5 m in length of which the tail was 0.8 m and the chest-girth was 1.2 m, whereas another adult male appeared to be 1.95 m in length from the nose-tip to the tail-tip and the tail was 0.8 m long. A sub-adult tiger, aged about 3 years, was measured 2.2 m in length and 1.1 m of chest-girth, which weighed about 8 kg (dilapidated condition). The individual skulls of the available samples (adult males) varied from 26.1 cm to 28.3 cm in length and 18.1-19.1 cm in width; the scapula varied from 22 cm to 22.6 cm in length, in width 11.9 to 12.9 cm; humerus in length 29 cm to 29.2 cm, girth at deltoid tuberosity 10 cm to 10.5 cm; ulna in length 28.6 to 30.9 cm, girth below coronoid process 9.3-9.8; radius in length 23.4- 24.6, girth at the base of radial tuberosity 6-6.1 cm; pelvic girdle in length 26.3-26.8 cm, distance below two wings 12.8-14 cm; distance below two tuber 9.5-10.8 cm; transverse diameter 7-7.3 cm;

conjugate diameter 11-11.2 cm; acetabular depth 2.7-2.8 cm, acetabular diameter 4 cm; femur length 32.5-33 cm; girth below lesser 9.5-9.6 cm and tibia in length 27.9-28.6 cm, girth below tubial 11.9-13 cm.

After chemical analysis of the body odour collected from three tranquilised wild tigers (*Panthera tigris tigris*) from SBR, the electron micrograph of hairs from two tigers has been elucidated. A few saturated, unbranched fatty acid methyl esters like methyl tetradecanoate, methyl pentadecanoate, methyl hexadecanoate, methyl stearate, methyl eicosanoate and methyl eruciate were identified. A few benzenoid compounds were also detected.

The Sundarban tigers represent one of the top five largest global populations of the species and along with the unique adaptations of the Sundarban tigers for a struggling life in the mangrove forests, this population is also extremely important for global recovery of the species and a pride for both Bangladesh and India. The mangrove forest is also the home of many grounds dwelling wildlife species.

Among mammals, ground fauna comprises *Felis viverrina, Axis, Sus scrofa, Lutra lutra*, etc. The wild boars feed on underground tubers but also relish dead fishes, prawns, crabs, molluscs and sea turtle eggs. The Spotted Deer preferably browse on leaves, twigs and fruits of *Sonneratia apetala, Avicennia officinalis* and *Excoecaria agallocha*. Herds of deer follow troops of *Macaca mulatta* from one *Sonneratia* tree to another to pick up what the monkeys drop from the treetops while feeding. Two or three deer are seen in a herd inside the forest but in the open meadow the number increases to 40-50.

Among the existing mammals the magnificent tiger has got the pride place as the top carnivorous animal occurring in the hazardous environment of the Sundarban. It leads an almost amphibious life and is quite capable of crossing wide rivers. It has well adapted itself in this difficult terrain which is characterized by sharp pneumatophores, muddy substratum, innumerable rivers and creeks with tidal rhythm, variable salinity and lack of freshwater source. It feeds on pigs, deer even on fishes, crabs and water monitors. The wild pig constitutes a substantial proportion of its diet as the majority of tiger scats are predominated by the presence of pig hairs. This is in contrast to the common belief that the tiger mainly preys on the Axis deer. The tiger's hunting tactics include concealment, stalking, sudden rush and then dispatch of the prey. These tactics allow tiger to have a wide range of prey types and sizes from few hundred grams of fish and crabs to a wild boar or deer of about 50 kg. It is generally believed that the tigers in this mangrove forest do not have territories due to the obliteration of urination marks by the tidal waters.

The Sundarban tiger is not an obligate man-eater, for which various reasons have been given. The study on tiger straying showed the remarkable feature. It was reported that in the last 12 years out of 120 cases of tiger straying, the tiger has attacked human beings only in six cases whereas most of the deaths have occurred within the forests. This peculiarity in the tiger behaviour perhaps has only one hypothesis that within the forest area i.e. their habitat, they consider all moving objects as part of their prey. It has been reported that the tiger also preys upon the killing of crocodiles.

Three secretive species of Lesser Cats- Jungle Cat (*Felis chaus*), Leopard Car (*Prionailurus bengalensis*) and the Fishing Cat (*Prionailurus viverrina*) are mainly found in the northern forests of the Sundarbans. The jungle Cat prefers the peripheral forest areas and the fishing Cat prefers swampy areas and are also reported to stray into the fringe villages. Later these strayed animals are relocated in the wild.

The Fishing Cat is a robust, deep-chest feline, and the largest of the lineage, with length of about 70 cm, standing over 35 cm at the shoulder level and a marked sexual dimorphism. Males are significantly larger and more robust. The adult male weighs 8-14 kg, where the female is measured 5-9 kg. It has short and dull coat, which is camouflaging grizzled grey in colour and rather coarse in texture. The coat is often tinged with olive brown, and has distinctive spots and stripes. Six to eight black lines run from the forehead to the neck, and breaks up into shorter lines and longitudinal spots on the shoulder. Fishing cat has two distinctive dark 'collars' on its throat. Its short, low-set, muscular tail, marked with about six or seven incomplete dark bands, distinguishes it from the leopard cat. Its double coated fur is crucial for its adaptation to immersion in water. When viewed from the front, the back of its small ears appear black with round whitish spot at the centre. Although the exact function of the spot is still unknown, it is often said that the ear-spot serves as the 'follow me' signals for the kittens, especially in low-light conditions. Each ear is controlled by more than 20 muscles, and is independent of the other ear. Thus, when one ear is pointed forward, the other may be pointed backward enabling to hear sounds from all the directions, including those from behind. They are even sensitive to ultra-sonic noises made by small preys, such as the rodents.

The head resembles a broad modified wedge with rounded contours. It has vertical markings above the eyes; at the back of the skull horizontal white and black markings (mascara) emerge from the edge of the eyes and extends up to the grayish white cheeks. The whiskers, called vibrissae, are short. These are deeply embedded in the skin and connected to nerve endings that transmit information to the brain, so that the cat can feel its way as it moves. It can read air currents and locations of the obstacles and items around in order to stalk almost silently. When capturing prey, all of these whiskers point forward like a net, to detect exactly where the prey is and the direction of its movement. The flattened nose skin is deep-brick in colour. Its legs are short with two distinct elbow bars in forelimbs, long phalanges and claws in the forepaws. The hindfoot is 13-15 cm long. The unique feature of the fishing cat is that their hind feet are webbed, but the webbing beneath the toes is not very developed. Without full sheaths its claws are partially visible even when retracted, perhaps an adaptation for a greater hold on the slippery surface while moving. This is also to prevent the wear on the claws while walking on the rough surface. Its pugmarks are bigger than those of other smaller cats residing in the same habitat. In the impression on the soil its middle toe is seemed to be prominent with the claw-marks. The partial membrane between its toes on the forefeet helps gain better traction through the slippery mud and water in the wetlands. The cat's double fur is also very crucial for its adaptation to immersion in the water. Next to the skin there is a layer of dense hair, as a result of which the water cannot penetrate it. This waterproof coat also keep the cat warm and dry in the coldest days. Emerging from the first layer of fur is another layer made up of long 'guard hairs', which gives the cat its glossy pattern of the shine.

It is a secretive ground-dweller. The beds of *Typha* spp., *Phragmites* sp. and *Saccharum* spp. are very good niches for this predator and many of its prey species. It often swims skillfully even in the deep water or long distances similar to the tigers, Axis deer and Wild boars in the Sundarbans. It catches fish by pursuit and using its long claws as fish-hooks. It was also observed fishing at the edge of waterbody, sometimes from overhanging branch of a tree above the shallow water and even scooping the prey from the depth of water. The cat lightly taps the water-surface with its paw for creating ripples, as if an insect has landed the water. When a fish approaches to investigate the source, the cat pounces on and batters it. It swims in the shallow water to grab the fish with its mouth. It was observed to hunt in the moonlit nights.

The fishing cats are often trapped by the villagers for depredation. Although a few animals are handed over to the forest department for rehabilitation in suitable habitats, many are killed on the spot in retaliation. An adult female frequently raiding on the poultry birds at Badu (North 24-Parganas) was chased by a huge mob. She could only save one of her kitten in her mouth and fortunately the other one could be rescued. The orphans are often rescued and temporarily housed in the Salt Lake Rehabilitation Centre. In the Sundarbans during early 21st century, many fishing cats were killed- two kittens at Sajnekhali on 12th November, 2003; one at Jamespur village on 30th September, 2004; again three at Sajnekhali on 1st to 3rd November, 2005; one at Namkhana on 1st February, 2005; one at Sajnekhali on 1st March, 2006 and one at Sonakhali on 1st March, 2007. In STR, 12 fishing cats were rescued during this period. The Lothian Island WLS was the common releasing site of these rescued fishing cats.

The fishing cat was common in the Sundarbans during the early 20th century, but became rare in various negotiable islands. Presently, it is distributed in the outer estuarine zone (within 20 km from the sea-face) and the inner estuarine zone (beyond 60 km from the sea-face). But it is strangely found absent in the central estuarine tract. It is usually seen in Sudhanyakhali, Jhingekhali, Jhilla and Haldi compounds close to the sweet water ponds. It is rarely found around brackish water wetlands of Kakdwip and Kultali areas of South 24-Parganas district. In fact, the fishing cat population is found in the fragmented habitats. In 1990, its population in the Indian Sundarbans was estimated to be about 500. The population trend is decreasing rapidly in the human-dominated landscape due to poaching.

In the aqua-terrestrial habitat of the Sundarbans, the fishing cat is a unique example of the great abilities and diversities of the exotic cat family. It is an ecological adaptation known as 'meso-carnivore release', as this felid plays a significant role in controlling the complex food chain called 'tropic cascade', its status in the wild is indicative of the health and functioning efficacy of the ecosystem it dwells (Mallick, 2010c).

The fishing cat is the focal and charismatic species to conserve the important wetlands vis-à-vis ecological communities. Hence, it has been declared as the solemn State Animal of West Bengal. Though protected by the national legislation, conservation of the species depends on adequate protection of the remaining wetlands in human-dominated landscape.

In the Sundarbans, where the Hog deer and the Swamp deer have become extinct long back and the Barking deer is on the verge of extinction, only the Axis deer could survive because of its adaptability including, high breeding capacity, while the fawns are seen throughout the year. Their feeding in the Sundarbans depends on the tidal timings. They are not normally seen along the fringe areas during the high tide. They are mainly seen during the low tide feeding under *Sonneratia* or *Excoecaria* forest patches, by standing erect on their hind legs, particularly in the early mornings and afternoons. This species does not usually graze on the newly regenerated grasses like *Porteresia* sp. on the mud flats or patches of *Saccharum* sp. on the sea face. It also browses on the palatable small leaves and fruits of *Sonmratia* and *Ceriops* as well as the fallen leaves of *Excoecaria*. In the Sundarbans, the Chital can cross the channels by swimming and even eats crabs and other molluscs. They are found to be very cautious in their approach to sweet water pond in all the observation sites at Sajnakhali, Sudhanyakhali, Dobanki, Chamta, Haldi and Jhingekhali. They observe the ground situation all around carefully before drinking water. They drink deeply and continuously. It is interesting that they do not visit the sweet water pond during the midday when the tiger is supposed to visit the pond. But the wild boars are seen during the daytime. The composition of deer herd is unstable in the Sundarbans as was observed in Sajnakhali and Haldi, because the large herds often split into a number of fraction groups, which again reunite sometimes in part or whole or they even form new herds. However, the large herds are seen only in comparatively open highlands, which is not usually submerged in the floods. Their mortality or predation by the tiger or poaching are common.

In the Sundarbans the wild boars found inside the forest areas appears to be of bigger in size than those seen in the fringe areas. Since domesticated pigs were offered to the tigers as bait till 1980s mainly in the tourist places located in the fringes, their descendants may be of smaller size. So, it is very difficult to distinguish between a feral and wild boar. The wild boar is also a swimmer as was evident in an incident when a wild boar strayed into the village Saterkona from Pirkhali forest area by crossing the Gomdi river and the animal was captured by fishing net and relocated in Gajikhali after providing necessary veterinary care.

Among the reptiles, the king Cobra, the common cobra, Banded Krait, Russell's viper comprise the community of venomous reptiles, while the Python, Chequered Keelback, Dhaman, Green Whip Snake and several other species constitute the non-venomous snakes. The tidal creeks harbour Homalopsid snakes adapted to living in water, the common being *Cereberus rynchops*.

The snakes live in mangrove forests and coastal lowland forests, usually near water. The carnivorous snakes feed on reptiles, birds, and small mammals and play a very important ecological role in their environment, both as predator and prey species in Sundarban. They help control populations of small mammals, birds and reptiles they prey on. Apart from this, the poisonous snakes have immense medicinal values for their venom.

The snake populations in Sundarbans are declining due to habitat loss caused by tremendous population pressure, netting by fishermen and retaliatory killing to avoid snakebite because thousands of people are bitten and die from snakebite every year. The common poisonous snakes are Common krait (*B. caeruleus*) and Monocled cobra (*Naja kaouthia*). Nearly 66% of the snake bite deaths were due to Common krait bite and about 34% by Monocled cobra, most of them occurring on the floor bed in the months of June to September. Only one case of death was due to sea snake bite, but there is no reported death due to Russell's viper (*D. russelli*) bite. The most affected areas are Basanti Block followed by Gosaba, Kultali, Patharpratima, Jaynagar, Mathurapur, Namkhana and Sagar Blocks. Hospital cases are better treated. Still the villagers go for the local quacks due to ignorance and traditional belief system (superstition). Awareness campaigns are often arranged by the NGOs at the grass root level. Whenever rescued, the species are handed over to the forest department for release in the wild.

The lizards are mainly the Varanus salvator (*V. flavescens*) which is a rare monitor reaching about 2.4 mtr. in length. This can be very frequently found now within the Reserve. The most commonly sighted Monitor Lizard in the Indian Sundarbans is the Salvator Lizard, which also forms an important prey base for the tiger as is evident from the scat analysis reports. The Salvator Lizard has been frequently sighted in a wide variety of habitats, which includes seashores, sweet water ponds, mud flats of different forest types and various water bodies. This species is highly mobile and an excellent swimmer. It has been observed to swim and remain under water for over half an hour in the sweet water ponds. When they swim, they keep their limbs to the side of their body and propel themselves through sinuous undulations of the flattened tail. With its powerful legs, it can run fast and climb swiftly up the trees, chasing after its prey, using their strong curved claws. They are found to live in burrows built along riverbanks. Ecologically, they serve as scavengers in the mangrove forests and were found to scavenge on soft visceral parts of a dead tigers and prey upon juvenile birds like Little Cormorants and eggs of birds, turtles and even crocodiles. They are omnivorous. They are hunted for food by the larger carnivores such as crocodiles and birds of prey. Their juveniles are particularly vulnerable to the large birds such as herons.

Olive ridley (*Lepidochelys olivacea*) turtle is the most common visitor on the eastern coast from November till the first week of April, while, historically, there have been records of green and hawksbill turtles as well in the Chhaimari and Kalas islands. The nesting season lasts between December and March. Sea turtles are also known to occur in the sea-facing islands of the Sundarbans, mostly in the uninhabited regions.

During late 20th century, nesting of Olive Ridley turtle's was occasionally recorded from the newly developed Nayachar Island in the Hooghly estuary. A Government of India and UNDP Sea Turtle Project was launched in SBR during April-May, 2000. Nesting grounds of the sea turtles were located in the extreme southern part of Sundarbans during November, 2000 to April, 2001. Part of it falls under TR and the rest is under 24-Parganas (South) Division.

Turtles prefer sandy beaches. The entire area falls under estuarine deltaic region of Sundarban. Under the former area, turtle habitats are located at Mechua and Chhaimari Islands and under the latter area, the islands are

Jambudwip, Kalash and Bijeara. Kanak Island is also a known habitat of these turtles, but the area is often submerged under the sea. An Olive Ridley was recorded to crawl on the Kalash Island at 11.30 pm, just two days before full moon. Similarly, a Ridley laid eggs on Bijeara beach on a full moon night.

It has been observed that maximum numbers of nest are found during February and March. The turtles come out of the sea water and crawl up the sandy beaches during the night, lay eggs and go back to sea again during the same night. It has also been observed that the turtles come for laying eggs from one day earlier to 3 days after new moon and full moon. The atmosphere during the period remains foggy and there is a southern wind. The temperature during February is recorded as 24°C (Air), 25°C (Surface) and 25°C (Nest). The corresponding figures in March are 32°C (Air), 34°C (Surface) and 31°C (Nest). Moisture content in the nest is recorded as 50% in mid-February and 62% in mid-March. The nests are destroyed by wild animals, erosion, deposition of fresh sand, fierce wind and cyclone.

There are records of turtle exploitation from the islands of Sundarbans as well. The water monitors are the greatest predators of their eggs and hatchlings along with Wild Pigs, Terns and Seagulls. Tigers, water monitor lizards and wild boars have been the natural predators of the nesting females and eggs in the Sundarban region.

The endangered River Terrapin (*Batagur baska*) also uses the beaches as their nesting ground. The Mechua beach in Bagmara block is an important nesting ground for such terrapins. Other coastal soft-shell turtles (*Pelochelys bibroni*), Sengal eyed terrapin (*Morenia ocellata*) and three keeled terrapin (*Geomada tricarinata*) are also recorded.

The dolphins are the top predator of the food pyramid in a water body and play a significant role by keeping the ecosystem healthy and balanced. They feed mainly on old and ill fish that are weaker. Thus, they fight against the infectious diseases among the fish populations and ensure reproduction of healthy fish resources. Possessed with a long life, whales and dolphins are slow producing predators and sensitive to environmental disturbances and negative human impacts. They are, therefore, ideal indicator species for both fresh water and marine ecosystems' health. The river dolphins are becoming increasingly vulnerable as their habitat is restricted and those rivers are now becoming polluted as a result of irresponsible and aggressive industrialization (Alam *et al.*, 2020).

Two genus of cetaceans *Platanista* and *Orcaella* are found in the Sundarbans waterways. The Gangetic Dolphin (*Platanista gangetica*) are frequently found in the eastern side particularly in the Raimangal river, the upper part of which is fed by some freshwater flow. It descends during the monsoon season when salinity is low but appears to decrease in number during the summer (Mallick, 2011). This migration also seems to be associated with the migration and dispersal of fishes, which are their main prey. The Irrawaddy dolphins are not found in the more saline Matla-Hooghly estuary. Seventeen individuals were seen in the pre-monsoon and twenty two in the post-monsoon months.

The Irrawaddy dolphin (*Orcaella brevirostris*) is identified by a bulging forehead, a very short beak, triangular pectoral fin and a small dorsal fin on the back. It was recorded mostly in the lower reaches of the river Raimangal. A few were also observed at the Jhilla and Amlamati. Two sightings are recorded- one in the river Hatania-Dotania (21.7500°N, 88.2160°E) between Narayanpur and Namkhana during 2012-2013 and another in the river Bidya (21.9330°N & 88.7000°E) in 2016 (Mitra and Chowdhury, 2018). There was no sighting record of this species in the river Matla, which is a habitat of the Ganges dolphin. Previously, the species was once reported in the river Hooghly, at the steamer *ghat* of Kakdwip (21.8500°N, 88.1660°E) during 2012-2013. A single individual of this species was sighted four times in the upstream covering a distance of about 100 km (51-98 km) from the sea-face during one year's systematic double observer boat-based surveys at four acoustic monitoring locations (Chowdhury *et al.*, 2020), viz. Raichak (22.201°N, 88.108°E; 51 km) at 11.07 h on 28th June, 2018, Falta (22.271°N, 88.087°E; 65 km) at 16.54 h on 24th March, 2018, Burul (22.349°N, 88.097°E; 73 km) at 10.15 h on 21st July, 2018 and Batanagar (22.508°N, 88.202°E; 98 km) at 12.20 h on 09th January, 2019, all in South 24-Parganas district. Since the sightings encompass both wet and dry seasons (monsoon as well as winter) and the number of observations has been small taking into account the considerable time spent on the river (March 2018–March 2019), it was believed that a resident but small population of the Irrawaddy Dolphin was present in this stretch of the river (*ibid*).

A survey of a nearly 100-km stretch of SBR adjoining Bangladesh, has also confirmed presence of the dolphin populations only in the westernmost segment, in the lower reaches of the river Hooghly, where the salinity is lower than that of natural seawater. This endangered species stayed away from the central Sundarbans, where siltation has disrupted freshwater flow leading to high salinity levels. The easternmost part of SBR having freshwater connectivity and is moderately saline but the salinity level increases downstream and not usually visited by this dolphin.

The Indo-Pacific finless porpoise (*Neomeris phoceaneides*) is also found in rivers near the estuary.

The marshes and the rivers offer asylum to the Estuarine crocodile (*Crocodylus porosus*) one of the rarest and the largest of crocodiles and Olive Ridley are artificially hatched, reared and subsequently released within the Reserve. Systematic monitoring of the released crocodiles tagged with location specific colour is done and a drift up to 6.5 kms from the released location has been noted during the year.

A wide and varied assortment of Fishes, Molluscs, Crabs and Prawns inhabit the estuaries. The mangrove leaves, which decompose slowly, offer food for the larval shrimps and they migrate from the sea to the mangrove estuary for spending up to their adult stage. Even the snappers or mullets depend very much on the mangroves. Mullets are mainly edible fish in the area like Bhetki and Bhangon. *Pangasius* i.e. Pangas fish is the primary heterotroph which often swallows full keora fruit. The sharks also lurk in the creek water like the marine wood borers like ship-worm or Teredo often cause concern to the watercraft.

The anatomical and behavioural adaptations of the 10-30 cm long mudskippers allow them to live effectively on land as well as in the water. Among the different species of mudskippers Boleophthalmus is the largest, measuring 25-30 cm. The heads are blunt having large, movable, close-set, and protuberant eyes on the top for an all-round view and mouth faces downwards as they feed on the mud surface. Their eyes are well-adapted to vision in air and, when they are laying waiting for their prey, only their eyes stick up out of the muddy water. In order to keep the eyes wet, they have little cups underneath the eyes and, when blink, the eyes roll down into the skull and get remoistened by the water held in these little cups.

The mudskippers have multiple modes of breathing. They can respire through their skin, mucosal lining in their mouth and throat (the pharynx). But they need their body to remain always wet since they can take oxygen by diffusion process only. Due to this habit, the mudskipper population is limited to the humid environment only, which is exactly a similar habit adopted by the amphibians, known as cutaneous air breathers. An important adaptation to respiratory system is their enlarged gill chambers, where they are able to retain a bubble of air. The large gill chambers closes tightly when it needs to keep the gills moist, which can allow them to function and supply oxygen for respiration, when they remain on the land. When their burrow is submerged, they can also maintain an air pocket inside, which make them able to breathe even in very low oxygen supply.

The anatomical and physiological adaptation of the mudskippers to aerial respiration varies from species to species. They are poorly adapted to respire aquatically in hypoxic conditions, but respire aerially to avoid low oxygen-stress. They are essentially ammoniotelic, excrete mainly through their gills and are highly tolerant of ammonia concentrations in their environment. They are very much capable of active excretion of ammonia, even at pH 9.0. When the mudskippers are on land, they detoxify ammonia through partial amino acid catabolism. They are euryhaline and can withstand rapid and drastic changes in salinity. Osmoregulation in hypersaline waters is done by accumulation of free amino acids (FAA) and ammonia in their muscles and through the rapid activation of gills. In many species of the mudskipper, regulation in hyposmotic conditions is partially behavioural evapo-transpiration through their skin. Hot and humid is their most suitable environment for survival as they need their body to be moist in order to breathe through their skin. In very hot summers, the mudskipper can remain active for several minutes, when they are out of water. For thermo-regulation, they dig in deep burrows in soft sediments. They have variable annual ranges of body temperature between 14°C and 35°C and that of air temperature between 10°C and 42°C. The mudskippers swim by side to side movement in the water. But, on the land, they have two main ways for their mobility- walking on their pectoral fins and skipping or jumping, for which their name is mudskipper. They are much capable to climb, walk and skip out of water. Their very strong and well-muscled pectoral fins move down the body and allow them to swing, inhabiting them between the tides and the rear fin acts as a stabiliser.

The mudskipper used to live in their own territory. They maintain polygonal mud-walled territories provide which is considered an excellent example of the elastic disc concept of territories. They also construct mud walls around their territories to avoid aggression of the neighbouring predators. The mud walls are considered to play a secondary and indirect role in maintaining their populations within the territories by maintaining diatom population which is the primary source of food for the mudskippers. This territorial behaviour helps them maintain high population densities within a confined habitat full of all elements for their survival.

The common mudskipper (*Periophthalmus kalolo*) and barred mudskipper (*Periophthalmus argentilineatus*) are found exclusively in mangrove habitats, ambon rock skipper (*Paralticus amboinensis*), lined rock skipper (*Istiblennius lineatus*) and streaky rock skipper (*Istiblennius dussumieri*) are more widespread in the pools along the coastal margins and mangrove zones. The adults inhabit the upper subtidal down to the high intertidal zone, including tidal reaches of rivers, supratidal eco-tones to freshwater swamps and diverse sympatric assemblages.

The amphibious species, such as *Periopthalmus* and *Boleophthalmus*, are of considerable interest. The former creeps up the trees with the rising water level. They use their pectoral fins to walk on land. The fish adapted to intertidal habitats (both high and low tide) and are very active when out of water for feeding, interaction with another and for defense they dig his own deep burrow to keep them away from any disturbances in their habitats. Some species emerge only at night and graze on algae, to escape from predators, to avoid hypoxic conditions that develop in pools when there is very low tide. The mudskippers inhabiting in the mangrove swamps are directly influenced by the tidal fluctuations and, at the time of any danger, they can jump into the open water or can move rapidly onto muddy land using their strong pectoral fins. The major prey organism of the otter species is the mudskippers in the Sundarbans.

In polluted coastal areas, the mudskippers are the potential biological indicators for environmental monitoring and assessment of coastal waters (Ansari *et al.*, 2014). Regularly discharged pollutants released into the coastal environment by the industrial, agricultural, domestic and transportation activities have imminent detrimental effect on the flora and fauna of the coastal ecosystems, especially the mangrove ecosystems and tropical mudflats. They play an important role in benthic ecology as they prey small crustaceans and graze diatoms and algae from mudflats.

Abundance and distribution of mudskippers on land as well as in coastal waters may be considered as a direct indicator of habitat health. While protecting and improving the state of coastal waters and mangrove forests ecosystems, which are the natural habitat for mudskippers, the mudskipper populations are also protected. The physiological, histological, and embryological changes in the mudskippers are considered as strong indicators of water quality parameters.

Among the crustaceans the One Armed Fiddler Crab (*Uca* species) often shows off to his mate with the colourful arm. They have diurnal clocks inside which regulates their colour change along with tides. Another interesting crab is the *Clibarnius padavensis* i.e. Hermit crabs occupy gastropod shells of the genus *Telescopium, Nerita, Cerithidea* or *Semifusus*, apart from the edible crab *Scylla serrata* there are a number of species of crabs found within the creek waters. Amongst them, Ghost Crab and Patal Chingri (*Thalassina anomala*) are important ones.

There are two species of trilobite viz. *Tachepleus gigas* and *Carcinoscorpius rotundicauda* commonly known as Horseshoe crab or king crab. King crabs, which are also called living fossils, are now protected owing to its medicinal value for catalyzing other medicine. They have hardly changed in 400 million years and are found to be an ideal medium for detecting poison in vaccines. Laboratories have been taking them by the ton. Since they are exploited as animal fodder the numbers are declining.

They live primarily in and around shallow coastal waters on soft, sandy or muddy bottoms at about 30 m depth. Though they are generally considered to be entirely marine species, they are also found entering into rivers through its mouth or confluence. They tend to spawn in the intertidal zone at spring high tides. They are usually seen when they move into shallow water to spawn and lay eggs on muddy/sandy beaches in the protected areas. In the estuarine ecosystem of Sundarbans, the horseshoe crabs are found in the mangrove creeks and mudflats or *chars*.

Habitat segregation of two species of horseshoe crabs found in the Sundarbans was observed, i.e. the smaller species live close to the mangrove forests, the large species prefer relatively sandy coasts. While

Carcinoscorpius rotundicauda (Latreille, 1802) occurs on the muddy beaches or intertidal flats of the estuarine Sundarbans delta (Hooghly-Matla estuary) associated with mangrove vegetations and adjoining coastal areas, *Tachypleus gigas* (O. F. Müller, 1785) prefers soft sandy beaches in the intertidal zone, having potentially low wave action. It was found that the Horseshoe crabs in Sundarbans lay eggs on the beaches, which constitute a mixture of silt (mud) and sand. But the larvae after hatching live in the silty mudflat of the mangroves for about seven to nine years until a total body length of 15 cm has been reached before they migrate to deeper waters in the sea. When they live buried in the mud of seafloor, they hunt for small animals, worms, crustaceans, mollusks (live young clams) and even small fish which they eat.

Occurrence of *Carcinoscorpius rotundicauda* (February to August or September) on the muddy shore is not uniform in all zones and is localised in certain pockets (particularly narrow channels or sub-creeks in the islands) during different seasons. Their occurrence in the mudflats during low tide is more in the 24-Parganas (South) Division than TR areas. Moreover their occurrence is comparatively higher in the Saptamukhi-Thakuran zone than the Matla-Gosaba zone. The population is fluctuating in different seasons, i.e. more abundant during the pre-monsoon period depending on availability of more food items during this season. With gradually receding water marks during low tides, animals are seen slowly coming out of water and crawling on soft mud.

Occurrence of *Tachypleus gigas* (late February or March to July) is comparatively less in SBR. It is recorded from the shores of Sagar Island, Prentice Island and Fraserganj during summer and monsoon period. *Tachypleus gigas* come to the beaches in pairs to lay eggs at the time of highest high tide of full moon and new moon days. The larvae after hatching migrate a little offshore and live there.

From ecological point of view, amongst the two Asiatic species found in Sundarbans, only *Carcinoscorpius rotundicauda* is found more adapted to a wide range of salinity than *Tachypleus gigas*. This is evidenced by the occurrence of *Tachypleus gigas* mostly at the sea shores directly facing the high saline sea-water regime and *Carcinoscorpius rotundicauda* in the brackish water estuaries and river mouths.

Habitat loss and the non-targeted catch of horseshoe crabs during traditional, semi-mechanised and mechanised fishing, are undoing the creature's abiding presence.

Insects abound in the forests amongst which the Honey Bee (*Apis dorsata*) is a source of considerable revenue and a lure for men to enter forests. About 500 quintals of honey are produced annually from the Reserve.

Avifauna

The birds are in abundance, including a large number of migrants from the higher latitudes that visit the Sundarbans in winter. All of its terrestrial, arboreal and hydrological habitats are conveniently utilised by the bird communities, a mixture of resident, breeding birds, summer and winter visitors (mostly aquatic) for their survival. The birds make up their nutritional needs from Annelida (74), Arthropoda (1,326), Mollusca (173), Crustacea (334), Amphibia (10), Testudines (11), Lizards (13), Snakes (30), Chordata (823) including fish (350) as per the field records. These resources serve the biological needs of the aves. Hence, they prefer the mangrove forests and surroundings for their survival.

Like the rest of the world, the highest number of species of birds in the study area belongs to a single Order named Passeriformes- passerine (203). The least number is in the Order Bucerotiformes - hoopoe with a single species, Common Hoopoe (*Upupa epops*), but no hornbill species under Bucerotiformes was recorded in Greater Sundarbans, though they are found in the mangroves of Malaysia. Worse than that, no species of the Order Otidiformes (Bustards) exists in the Sundarbans today, although one species thrived in the grassland of Sundarban till the beginning of the 20th century. This extirpated bird was Bengal Florican (*Houbaropsis bengalensis*), which is a globally critically endangered species.

The avifaunal diversity in Sundarban is dynamic and depends on various factors such as the flowing water level, seasonal variations and changes in tide levels twice (two high and two low) a day. At low tide, vast rivers turn into tiny streams that twist along sandy islands, exposing a rich ecosystem, the birds were seen foraging in the open mudflats and resting in the mangroves during a high tide, when large tracts of the forest disappear underwater. During monsoon, many water birds were seen foraging in puddles in the adjoining grasslands.

The post-monsoon period is most productive in Sundarban, when the species diversity and congregation is increased, particularly with the arrival of long-distance migratory (stopover and wintering) birds since September from three flyways- the Central Asian Flyway, the East Asian-Australasian Flyway and the Asian-East African Flyway. Here, they need to feed voraciously, increasing their body mass by 20-60% and energy reserve to undertake the return journey in summer.

Again, low diversity was observed during the monsoon owing to heavy rains [average for pre-monsoon (March-May): 239.4 mm; monsoon (June-September): 1,355 mm; and post-monsoon (October-February): 226.8 mm], increased flow of water, extensive and prolonged inundation, when availability of food and space is limited. Whereas a huge number of migrants depart Sundarban during March-April, vacating the habitat for limited summer (from March onwards) migrants from their wintering grounds to the mangrove habitat, no severe food crisis takes place.

Overall bird density is often correlated to insect abundance, which is varied between seasons and mangrove patches. Insect abundance is highest, when mangroves are flowering. For example, occurrence of flowering for *Excoecaria agallocha* (Euphorbiaceae) in July, for *Bruguiera sexangula* (Rhizophoraceae) in April-June, for *Ceriops decandra* (Rhizophoraceae) in March-April, but variable for *Heritiera fomes* (Malvaceae) from March to April in the less saline zone, from March to early May in the moderately saline zone and from March to early June in the strongly saline zone. In general, partitioning of the available foraging niches is limited in the extensive habitats, resulting in dominance of the bird assemblage by a few species that are generalists, in terms of feeding, compared to the specialists.

Sundarban now harbours some sympatric species like kingfishers (9) and cuckoos (16); several raptors such as, eagles, falcons, vultures, kites, harriers, etc; some terrestrial birds including doves, woodpeckers, pigeons, fly-catchers, oriental magpie robin, red jungle fowl, owls, rose-winged parakeet, etc; aquatic birds like storks, herons, egrets, adjutants, little cormorant, etc; semi-aquatic plovers, red-wattled lapwing, avocet, stint, curlew, sandpiper, common greenshank, gulls, terns, etc.

A few species are recognized as mangrove specialists like Mangrove Pitta (*Pitta megarhyncha*), Mangrove Whistler (*Pachycephala grisola*), Brown-winged Kingfisher (*Halcyon amauroptera*) and Collared Kingfisher (*Todiramphus chloris*). Those species that prefer mangroves, but are not mangrove specialists, include Masked Finfoot (*Heliopais personata*), Lesser Adjutant (*Leptoptilos javanicus*), Buffy Fish Owl (*Ketupa ketupu*), Great Thick-knee (*Esacus recurvirostris*), Streak-breasted Woodpecker (*Picus viridanus*), White-browed Scimitar Babbler (*Pomatorhinus schisticeps*) and Whitebellied Sea Eagle (*Haliaetus leucogaster*). Many species of Sundarban are known to prefer grasslands, viz. Blue-breasted Quail (*Coturnix chinensis*), Red-wattled Lapwing (*Vanellus indicus*), Paddyfield Pipit (*Anthus rufulus*), Zitting Cisticola (*Cisticola juncidis*), Bengal Bushlark (*Mirafra assamica*), Grey Wagtail (*Motacilla cinerea*), etc.

According to the occurrence rate of the extant species during the study period, about 35% was considered rare. These birds include Ferruginous Pochard (*Aythya nyroca*), Ruddy Kingfisher (*Halcyon coromanda*), Chestnut–winged Cuckoo (*Clamator coromandus*), Slaty-legged Crake (*Rallina eurizonoides*), Pintail Snipe (*Gallinago stenura*), Eurasian Thick-knee (*Burhinus oedicnemus*), Grey Plover (*Pluvialis squatarola*), Darter (*Anhinga melanogaster*), Cinnamon Bittern (*Ixobrychus cinnamomeus*), Great White Pelican (*Pelecanus onocrotalus*), Indian Pitta (*Pitta brachyuran*) and so on.

Threatened animals are grouped into four basic categories of conservation concern, as per IUCN Red List- Critically Endangered, Endangered, Vulnerable and Near threatened. Among the globally Critically Endangered birds, Sundarban harbours two resident species- White-rumped Vulture (*Gyps bengalensis*) and Slender-billed Vulture (*Gyps tenuirostris*). Two globally critically endangered species, Spoon-billed Sandpiper (*Eurynorhynchus pygmaea*) and Baer's Pochard (*Aythya baeri*), are winter migrants to Sundarban. Two resident globally endangered species are Masked Finfoot (*Heliopais personata*) and Black-bellied Tern (*Sterna acuticauda*). Globally Endangered species also include winter migrants to Sundarban like Great Knot (*Calidris tenuirostris*), Egyptian Vulture (*Neophron percnopterus*), Steppe Eagle (*Aquila nipalensis*) and Yellow-breasted Bunting (*Schoeniclus aureoles*). Four resident globally Vulnerable species of Sundarban are Lesser Adjutant (*Leptoptilos javanicus*) and Indian Spotted Eagle (*Clanga hastata*), whereas two globally Vulnerable species are breeding migrants to Sundarban. They are Pallas's Fish Eagle (*Haliaeetus leucoryphus*) and Bristled Grassbird (*Chaetornis striata*). The other globally vulnerable winter migrants to Sundarban are Common Pochard (*Aythya ferina*), Woolly-necked Stork (*Ciconia episcopus*),

Indian Skimmer (*Rynchops albicollis*), Wood snipe (*Gallinago nemoricola*) and Greater Spotted Eagle (*Clanga clanga*). Among the Globally Near Threatened birds of Sundarbans, familiar residents are Oriental Darter (*Anhinga melanogaster*), Red-necked Falcon (*Falco chicquera*), Brown-winged Kingfisher (*Pelargopsis amauroptera*), Great Thick-knee (*Esacus recurvirostris*), River Tern (*Sterna aurantia*), Mangrove Pitta (*Pitta megarhyncha*), Red-breasted Parakeet (*Psittacula alexandri*), Alexandrine Parakeet (*Psittacula eupatria*), Blossom-headed Parakeet (*Psittacula roseata*) and Spot-billed Pelican (*Pelecanus philippensis*). Globally Near Threatened winter migrants to Sundarban are two ducks, namely Falcated Duck (*Mareca falcata*) and Ferruginous Duck (*Aythya nyroca*), shorebirds like Eurasian Oystercatcher (*Haematopus ostralegus*), Eurasian Curlew (*Numenius arquata*), Bar-tailed Godwit (*Limosa lapponica*), Blacktailed Godwit (*Limosa limosa*), Red Knot (*Calidris canutus*), Curlew Sandpiper (*Calidris ferruginea*), Red-necked Stint (*Calidris ruficollis*), and Asian Dowitcher (*Limnodromus semipalmatus*).

It was observed that Masked Finfoot, the rare breeding resident, mainly occurs in the eastern half of the Sundarban. Its population appears to have declined after the devastating cyclones of the 21st century. There are recent sightings of the Spoon-billed Sandpiper on the shore of Sundarban, east or west. Both White-rumped Vulture and Pallas's Fish Eagle are rarely found in the northern Sundarban. The Greater Spotted Eagle is rarely found on the riverbanks or shores of the Sundarban.

About 65% of the species found during the study was common, (not threatened or Least concern category), mostly belonging to Phasianidae (Galliformes), Cuculidae (Cuculiformes), Psittaculidae (Psittaciformes), Caprimulgidae (Caprimulgiformes), Columbidae (Columbiformes), Ardeidae (Pelecaniformes), Scolopacidae, Charadriidae and Laridae (Charadriiformes), Dicruridae, Sturnidae, Hirundinidae, Pycnonotidae and Muscicapidae (Passeriformes), Ciconiidae (Ciconiiformes), Accipitridae (Accipitriformes) and Alcedinidae (Coraciiformes). These include Red Jungle fowl (*Gallus gallus*), Greater coucal (*Centropus sinensis*), Rose–ringed parakeet (*Psittacula krameri*), Large-tailed Nightjar (*Caprimulgus macrurus*), Eurasian Collared Dove (*Streptopelia decaocto*), Whimbrel (*Numenius phaeopus*), Indian Pond Heron (*Ardeola grayii*), Bronzed Drongo (*Dicrurus aeneus*), Oriental Magpie Robin (*Copsychus saularis*), Jungle Myna (*Acridotheres fuscus*), Barn Swallow (*Hirundo rustica*), Red-vented Bulbul (*Pycnonotus cafer*), etc.

The birds under the Orders Charadriiformes, Procellariiformes and Suliformes like Noddies, terns like Sooty and Roseate, gulls (Larus spp), Shearwaters, Boobies etc. use the coastal habitat for breeding purpose, when the sea food provide them nutrition.

About 40% of Sundarban is under deep water in the form of estuaries and large rivers. This aquatic habitat is used by the birds belonging to six Orders- Anseriformes, Podicipediformes, Pelecaniformes, Ciconiiformes, Gruiformes and Charadriiformes.

The new depositions and intertidal mudflats are characterized by *Avicennia* (Avicenniaceae) and *Sonneratia* (Sonneratiaceae), flanked by foreshore grassland, gradually replaced by the mangrove species. The tidal mudflats provide the much needed benthic food supply for the birds. The inland and foreshore grasslands are also used by many small and large bird species and many make their nest in this habitat. Specifically, the birds such as larks, pipits and buntings use the grass and shrub areas. The birds like Red-vented Bulbul (*Pycnonotus cafer*), Red-whiskered Bulbul (*Pycnonotus jocosus*), Cattle Egret (*Bubulcus ibis*), Indian Pond Heron (*Ardeola grayii*), Little Egret (*Egretta garzetta*) and Purple Sunbird (*Cinnyris asiaticus*) are often seen moving from grasslands to the mangrove areas because they use the grasslands for foraging and the mangroves for perching.

High species richness in the Sundarban mangroves is associated with plant species richness, the density of the understory and food resource distribution. The immensely productive Sundarban estuarine wetlands are constantly fed by nutrients brought in by the north-south flowing freshwater channel and flushed by the ebb and flow of the tides from the BoB to support a diverse plant and animal communities including the terrestrial, arboreal and aquatic birds. Whereas the phyto- and zoo-planktons feed the fish, crabs, prawns, shrimp and mollusks in the shallow intertidal network, these in turn support the wading migratory and resident birds. Mangrove bird species with larger bills comprise both arboreal and ground foragers that feed primarily on crabs or insects, and larger bills are most prominent among passerine species that feed primarily on the ground.

Mangrove birds may, therefore, be divided into three categories-terrestrial, semi-aquatic and aquatic. Since the terrestrial type of habitat provides ideal foraging and breeding sites and also shelter for the birds, it is utilized by a large number of terrestrial species, grouped according to their favourite food and technique of capture.

The availability of trunks, limbs, and foliage comprising the tree canopy enables a variety of passerines and non-passerine birds, which are not found in other wetland areas, to use mangrove swamps. It also allows extensive breeding activity by a number of tree nesting or ground-dwelling birds. In fact, Sunderbans is a potential breeding ground of immense variety of birds like Heron, Egret, Cormorant, Fishing Eagle, White Bellied Sea Eagle, Seagull, Tern, Kingfisher, Black-tailed Godwit, Little Stint, Eastern Knot, Sandpiper, Golden Plover, Whistling Teal, Pintail, and White-Eyed Pochard etc. because the higher density of mangroves provides them a secure habitat for a nest and the fledglings. Clamorous Reed Warbler (*Acrocephalus stentoreus*), Ashy Prinia (*Prinia socialis*), Plain Prinia (*Prinia inornata*), Scaly-breasted Munia (*Lonchura punctulata*), Baya Weaver (*Ploceus philippinus*) and many others are often seen breeding in the mangroves and associated plants, especially in the monsoon. During monsoons heronaries develop in eastern part of Sundarban. The key breeding areas of the threatened waterbirds are the Haor (backswamp) areas, new accretion in river systems and mangroves of offshore islands.

The mangroves, mudflats, estuaries and adjacent areas are rich in food resources, which include fishes, crabs, turtles, snake, amphibians, small mammals and invertebrates such as gastropods (more dominant), bivalves (less dominant), prawn, nekton and insects. Most of the mangrove-restricted bird species feed primarily on insects in the canopy, on muddy substrate, in water, on trunks and limbs of trees and from air. There are other species which feed on crabs, nectar and fish.

In terms of preference, these mangrove birds are divided into four feeding guilds-

(i) Aerial hunters or sallying birds like swifts (Apodidae), wood swallows (Artamidae), swallows (Hirundinidae), fish eagles and kites (Accipitridae), bee-eaters (Meropidae), kingfishers (Alcedinidae), which used to hover on mudflats and mangrove areas in search of their prey. The raptors (Falconiformes and Strigiformes) prey on fishes, small birds, small mammals, reptiles, amphibians and large invertebrates. They also roost in the mangrove areas.

(ii) Foliage and bark gleaners: These birds, mostly terrestrial species, prefer using mangrove vegetation, i.e. trees, shrubs, palms, and ferns for foraging, perching, nesting and roosting. The users are-woodpeckers (Picidae), Tailorbirds, Warblers, Flyeaters (Sylviidae), Flycatchers (Muscipcapidae), Thrush, Shyama and Robins (Turdidae), Nuthatch (Sittidae), Sunbirds, Spiderhunters (Nectariniidae), Pigeons (Columbidae), Owls (Strigidae), Cuckoos and Malkohas (Cuculidae), Parrots (Psittacidae), Tits (Paridae), Orioles (Oriolidae), Drongos (Dicruridae), Ioras (Chloropseidae), Cuckooshrikes (Campephagidae) and Pittas (Pittidae). This category includes the frugivores (pigeons and parrots), insectivorous eating the harmful caterpillars, beetles, bugs, and aphids (tailorbirds, shrikes, flycatchers, ioras, woodpeckers, robins, warblers, tits, etc.), nectarivores or pollinators (sunbirds, white-eyes and spider-hunters).

(iii) Surface and diving foragers: The birds like Pelicans (Pelecanidae), Ducks and Goose (Anatidae) mostly swim on the surface of water in search of small fishes, amphibians, aquatic invertebrates and vegetable matter, while Cormorants (Phalacrocoracidae), Darters (Anhingidae), and Grebes (Podicipedidae) dive into deep water, particularly river beds, in search of food, mainly fishes, amphibians and aquatic invertebrates such as mollusks as well as vegetable matter. Asian Openbill is a molluscivore.

(iv) Waders: The birds like Egrets, Herons, Bitterns (Ardeidae), Finfoots (Heliornithidae), Plovers (Charadriidae), Oystercatchers (Haematopodidae), Sandpipers, Curlews, Shanks, Stints, Ruffs, Godwits, Knots, Dowitchers, Turnstones, Whimbrel, Snipes, Oystercatchers (Scolopacidae), Stilts and Avocets (Recurvirostridae), Phalaropes (Phalaropidae), Gulls, Terns and Noddys (Laridae), Spoonbills, Ibis, and Storks (Ciconiidae) and Frigate birds (Fregatidae) wade in shallow water to prey on fishes, prawns, mollusks, crustaceans, polychaetes and other invertebrates of the mud-flats during the low tide. These bird species utilize the mangrove areas for foraging, roosting, nesting for breeding and shelter from inclement weather.

The carnivore is a generalised group composed of Piscivorous, Insectivorous, Avivorous, Crustacivorous and Molluscivorous species. Carrion is also a popular food source for carnivorous birds, particularly vultures.

In terms of diet, the herons (e.g. striated and pond), egrets (cattle, great, intermediate, little), terns (little), kingfishers and cormorants (little) are piscivores. Waterhen (white-breasted), rails (slaty breasted), lapwings (red-wattled), sandpipers (common, green) are examples of crustacivore. The species belonging to Falconiformes and Accipitriformes are carnivores. The gulls, mynas and starlings are omnivorous species. Graminivores include Jungle fowl, sparrow, weaver, silverbill and munia, whereas the columbids are both frugivores and graminivores. The parakeets are frugivores, so also cuculids and barbets. The swifts, picids, most of the passerins and bee eaters are insectivores. Hoopoe is both insectivore and graminivore, whereas bulbuls are both insectivores and frugivores.

A few passerins (family Nectariniidae) are nectarivores, whereas the flowerpeckers are both carnivores and insectivores.

Diet of most of the birds in Sundarbans is insects, which in terms of nutritional value, is adequate because of rich and easily digestible protein and fat. Predominantly, the insects preferred by the birds in Sundarbans were identified as belonging to the Orders Coleoptera, Hemiptera, Hymenoptera, Orthoptera, Odonata, Lepidoptera and Diptera.

There are six more orders with a low (2-8) number of species and not preferred by the birds. Beetles are most frequently preyed upon. There seems to be a direct correlation between habitat resource availability and utilization, i.e. the birds choose food opportunistically.

There was a monsoon nesting ground at Sajnekhali covering 1.5 km^2 area of mixed heronry from middle of June to end of September. This nesting ground is however, no more, attractive to the visitors and the site for nesting was subsequently shifted to Jhilla Block and recently to Arbesi Block.

Honey bee

There are three species of honey producing bees in the Sundarbans. They are *Apis dorsata, Apis cerena and Apis florea*. But, honey and wax collection is done from the hives of *Apis dorsata*, which migrate to Sundarbans in January and leave after June. *Aegiceras, Avicennia, Bruguiera, Ceriops, Excoecaria, Rhizophora, Sonneratia, Xylocarpus, Acanthus, Heritiera*, etc. are considered as a good honey and nectar producing plants for *Apis dorsata*. The honey flow in the Sundarbans continues for a period of four months from March to June. The nectar yielding plant is also divided into two categories- early honey and nectar producing plant (e.g. *Khalsi, Keora, Chaila, Passur, Kankra, Jhana Gorjan*, etc.) and late honey and nectar producing species (*Kankra, Gewa, Baen, Hargoja*).

The bees move through the forest in different ways. When searching for flowers, they move indirectly back and forth until they reach a flower. However, when the bees travel from the flower to the hive, they travel in a straight line. In the forest, the honey harvesters follow the returning bees until they reach the hive. When one person finds a hive, he calls for the others. The group prepares smoke to drive off the bees by burning *hental* leaves wrapped in a bundle. Finally, they cut down part of the beehive for the honey and beeswax. The harvest time is from the first of April to the thirtieth of June.

Other than human beings, honey is consumed by wild animals like mammals (tiger, rhesus monkey, wild pig, spotted deer). Crabs are found to cling to the combs. Some lizards and snakes are also honey consumers.

Snake

Snake diversity is higher in Sundarbans than other mangrove habitats. 42 species (including the sea snakes) have been recorded from Sundarbans. The two most commonly observed poisonous species are *Bungarus caeruleus* (51%) and *Naja keutia* (40 %) and that of non-poisonous varieties are *Ptyas mucosus* (41%), *Typhlina bramina* (34%), *Xenochrophis piscator* (12%) and *Amphiesma stolata* (10%). Snake biting and human mortality rate is much higher in the Sundarbans.

Aquatic habitat

Aquatic habitat has not yet been studied in full detail. The most interesting is the formation of Phytoplankton in the shallow clear water of the tidal creeks receiving enough sunlight for a luxuriant growth. Phytoplankton production in the tidal creeks in Sundarbans estuary averaged 151.07 g/cm2 Y-1 designating this ecosystem to be moderately productive compared to other nutrient rich estuarine ecosystems of the world.

Primary productivity increases steadily from November to February, despite the fact that salinity of water is increased. Thus, primary production is independent of salinity as long as the salinity regime does not exceed a certain tolerance level. But this pattern breaks down at high salinity values in the month of April. Thus, the annual cycle of phytoplankton biomass (chlorophyll concentration) and abundance (phytoplankton standing crop) evidenced seasonal variation being highest in February (43.80 µg L-1, 1.37 x 1011 cells m-3) and lowest in June (4.85 µg L-1, 1.07 x 108 cells m-3). Phytoplankton primary productivity also followed a seasonal cycle being highest (597.3 mgC m-2 d-1) in the month of February (post-monsoon) and lowest (311.0 mgC m -2 d-1) in the month of August (monsoon).

This estuary remains autotrophic for five months (November to March) of the year, when the primary production is greater than community respiration resulting in export or burial of organic matter through conversion of inorganic matter and carbon dioxide. The estuary remains heterotrophic during the remaining seven months (April to October). Community respiration is greater than primary production during that period and allochthonous materials are re-mineralized leading to production of inorganic nutrients and carbon dioxide. Thus, Sundarbans estuary is a net sink of CO_2 for five months of the year and net source of CO_2 for the remaining seven months. If net ecosystem metabolism of the entire year is taken as a whole, the estuary can be designated as a net source of CO_2.

The available phytoplanktons are the sources of augmentation of oxygen content in the water. This influx, however, is checked by the zoo-plankton particularly by the shrimp population which invade mangrove estuaries during the semi larval stage to adult stage. The zoo-planktons consume the phytoplankton and diminish the oxygen content and the whole equilibrium is also controlled by the seasonal salinity of the creeks. The total catch of fish diminishes to a minimum during the highest salinity. The microorganisms like Noctuluca, Din-oflagellates produce bioluminescence during winter night particularly near the sea face and the entire atmosphere turns into a fairy land.

This estuary shows chlorophyll concentration greater than 10 µg/L throughout the year except for the months of May (9.07 µg/L) and June (4.85 µg/L). Thus, the estuary is identified as meso-trophic only in the months of May and June but eutrophic in the remaining ten months.

Estuarine crocodile

Crocodiles porosus lives in the coastal brackish mangrove swamps and tidal rivers, creeks and river deltas, where salinity changes with different seasons; muddy banks of the rivers during low tides. It is mostly found in the Saptamukhi, Thakuran, Choto Noubanki, and Mechua Khal areas of Indian Sundarbans.

It is the top predator of the estuarine habitat, i.e. the rivers and creeks. Its food is mainly the fish occurring in its habitat but it preys upon large animals i.e. spotted deer, wild boar, turtles, monkeys, birds and even the tiger. The young prey upon small creatures like crustaceans, arthropods, amphibians, small fish and reptiles. Human-crocodile conflict often takes place in the fringe forest area. During the breeding season, it makes nests in the form of mounds made with mud and leaves of Hental (*Phoenix paludosa*). These nests are protected by females with aggression against any intruders.

The first census was conducted for four days in January, 2012 in all the creeks and rivers covering about 1,160 km area. This operation was conducted by the Wildlife Institute of India, Dehradun and the Wildlife Wing of the forest department of West Bengal. The field exercise for the census was carried out over three days in January 2012 by 35 groups of enumerators who traveled, along a straight path, for 613 hours, covering 1,164 km. The enumerators found evidence of 240 crocodiles; 141 direct sightings of the reptile and river bank trails of 99.

In all, 240 crocodile were spotted during the period. Presence of 0.12 crocs/km was found including the breeding population. Five or six of the crocodiles sighted were more than 20 feet long. The eastern part of the Sunderbans National Park, which has lower salinity, has the highest density. The juveniles are concentrated in Ram Ganga area of the estuarine region. A good number of sightings were recorded from the East (28 direct and 15 indirect), West (40 direct and 51 indirect) and Sajnekhali (30 direct and 14 indirect) ranges in the Tiger Reserve, whereas the Ramganga (15 direct and 3 indirect) Range in South 24-Parganas Division. The census for the first time has set a benchmark for the crocodile population in the Sunderbans. This healthy population is an indication that the crocodiles have the capacity to withstand the increasing salinity and surface water temperature in the region. Rapid destruction of breeding grounds, poaching and fishermen's indiscriminate access to rivers and canals of the Sundarbans contribute to rapid fall in the crocodile population in the forest areas.

In 2014-2015, 42 crocodiles were tracked using GPS across a riverine area of about 351 km in the Bangladesh's Sundarbans. The number has decreased at an alarming rate compared to a population between 150 and 200 during the previous census in 1985.

Marsh crocodile

Crocodilus palustris once occurred in the Sundarbans. It prefers freshwater habitats including rivers, lakes and marshes. But it became extinct during the 20th century.

Gharial

Gavialis gangeticus of deep and fast-flowing freshwater rivers is critically endangered in the Sundarbans and scarcely found due to salinity. It was photographed on the riverbank on 19th July, 2018 in STR. A sub-adult female gharial of about 7 feet was also rescued from a canal of Rajnagar in Namkhana at about 11 am on 24th May, 2022. It has been transported to and kept at the Bhagabatpur Crocodile Centre.

Shark and ray

There are 28 species of sharks and rays in the Sundarbans. 10 shark species have been identified in the lower, middle and upper reaches of the Indian Sundarbans (caught in Kakdwip, Fraserganj, Raidighi, Namkhana and Sagar islands), of which the important species of the lower habitats are Black-tip shark (*Carcharhinus limbatus*), White-cheek shark or Wide-mouth black-spot shark (*Carcharhinus dussumieri*), Bull shark (*Carcharhinus leucas*), Wing-head shark (*Eusphyra blochii*), Smooth hammer-head shark (*Sphyrna zygaena*) and Tiger shark (*Galeocerdo cuvier*). One species inhabiting little upstream are Spade-nose shark (*Scoliodon laticaudus*) and three main species in the upstream are the Ganges shark (*Glyphis gangeticus*), the Spear-tooth shark (*Glyphis glyphis*) and the Irrawaddy river shark (*Glyphis siamensis*).

A study in the Sundarbans has shown that the distribution and number of sharks has decreased significantly. The rate at which the sharks are declining indicates that these sharks may disappear completely from the mangrove wetlands within the next few decades. Water pollution due to presence of heavy metals (Zn, Cu, Cd and Pb) in the sharks' body is the main threat in the aquatic habitats of the Sundarbans. It is seen that the pollution level in the forested areas is the least because the area between Bidya-Matala rivers has less industrial waste materials, but its quantity is highest in the Nayachar Island of Hooghly River, which is located just opposite the industrial area of Haldia. The samples indicate that all the aquatic animals, including sharks, living in the Sundarbans area, and the animals they eat are affected by this type of pollution and have become increasingly endangered.

Fish

The mangroves and associated extensive mudflats are ideal nursery grounds of the breeding population of fishes. Elasmobranchs comprise only 10% known fishes. The rest are bony fishes. Gobiidae is the most diverse family followed by Sciaenidae and Engraulidae. Gray sharks (Carcharhinidae) dominate the elasmobranchs followed by Stingrays (Dasyatidae). About 20% of the fish diversity may be termed non-commercial and 4.6% species are not edible. More than 43% are commercially important and contribute significantly in the region. However, more than 20% of fish have ornamental values for aquariums in the market. 4.85% of species are threatened and 6.3% are near threatened. The FD imposes restrictions on catching fish in the core and buffer areas of BR, but illegal fishing is a great management problem here. It appears that 82 species under 24 families lose their juveniles in a significant number regularly due to wild harvest of lucrative prawn seeds.

Rare and threatened species and endemism

Sundarbans harbour a good number of rare and globally threatened animals including 'Bengal Tiger' and 'Estuarine Crocodile'. The prominent among them are Fishing Cat, Salvatore Lizard (Water Monitor), Gangetic Dolphin, Batagur Baska (River Terrapin), Marine Turtles like Olive Ridley, ground Turtle and Hawksbill turtle, King Crab (Horse Shoes), etc.

Batagur baska is a large turtle with a carapace (upper shell) that grows up to 60 cm in length. The male is quickly recognised by its jet black head and neck, and a wrinkled, orange/peach-coloured body and forelimbs, that turn hot neon red during the breeding season. The female, however, has a creamish-grey coloured head and body. River terrapin is one of the important species of estuarine ecosystem, not only it takes green plants but also it takes rotten food items and clear the water, hence it is called ' water vulture'. The species is found in parts of Bangladesh and Indian Sundarban. It is an aquatic species but uses terrestrial nesting grounds, frequenting the tidal zones of estuaries large rivers and mangroves. It prefers nesting in colonies on sand banks. The species prefers freshwater habitats and moves to the brackish river mouths in the breeding season (mostly in September-November). During January-March, it nests in the sea coast and lays egg in clutches comprising of 19-37 numbers, measuring 40-60 mm; the incubation period being 60-66 days.

Predator-prey dynamics

Tiger niche

There is adequate forest cover for its hunting, hiding and procreation. However, it is generally observed that the forest of *Phoenix* forms a favoured tiger's den and its presence is commonly felt in the mangrove forests having plant consociations of mainly *Rhizophora-Bruguiera,*

Avicennia-Oryzha, Excoecaria-Phoenix, Exoecaria-Ceriops, pure *Ceriops* and pure *Oryzha.*

The uniqueness of the habitat is said to have contributed to certain behavioural traits, which are the characteristic of Sundarban tigers only. The man-eating trait of Sundarban tigers has become almost a legend in the Sundarbans. It is considered that man-eating propensity of tigers in this area is hereditary acquired over a period of generations.

Due to typical terrain conditions, where land features are broken by rivers and creeks, which form a substantial part of the total surface area, Sundarban tiger has become an excellent swimmer. Bahuguna and Mallick (2010) cited a strong case of such an arduous exercise of the tiger:

An old tiger was trapped in July 2008 on the fringe of Sundarbans. It was released in the forest. Although apparently it was healthy, it had only one canine intact. It was again captured in October at the same place. This time it was released deep inside the forest. In November, it again appeared in same area from where it was captured. This time it was released at the southernmost tip of Sundarbans. But, this time also it came back to the same place, covering a distance of about 100 km by crossing many rivers, one of which is 4-5 km wide. Finally, it was trapped in December and handed over to the Kolkata Zoo.

Other prey animals also cross the streams. Prey animals occupying the Sundarban Tiger Niche are the *Axis axis, Sus scrofa, Macaca mulatta* among other large animals. The secondary predators are mainly *Prionailurus viverrinus* and to a small extent *Felis chaus.* The fodder grass is almost absent except *Porteresia coarctata.* There is, however, no dearth of fodder for *Axis axis,* which substitute mainly on the leaves, twigs and fruits of *Sonneratia apetala* and leaves of *Avecennia officinalis* and *Bruguiera gymnorhiza.*

In STR, maximum tigers were sighted in Pirkhali block, but the frequency was highest here during the four months from January to April and then the sightings started reducing from May onwards recovering only in December. This may indicate a periodical fluctuation of population in this block. Another remarkable feature is that Netidhopani is the second important sighting area in the region, but here also the sighting records were not uniform throughout the year, but fluctuates during the rainy season and winter months. Arbesi block is the third important sighting area. Here, most sightings were recorded in December. On the contrary, the sighting record in the adjacent Panchamukhani block is low throughout the year. Matla block may be termed as very low in terms of sighting, where from August to December no sighting record was available. Experience is almost similar in Chamta, Chhotohardi, Chandkhali and Gosaba blocks. But in the southern blocks of Gona, and Bagmara, sighting was almost negligible and in Mayadwip it was nil. Jhilla, Khatuajhuri and Harinbhanga blocks were not much important in terms of sighting. So, the presence of tigers was mostly felt in the northern belt of forest blocks, whereas only one block in the central zone, i.e. Netidhopani, holds most of its residents throughout the year.

However, since 2007, according to the national prescription for the tiger monitoring, a range of techniques including camera traps, DNA sampling, pug mark surveys and the assessment of tiger claw marks on trees have been followed in the Indian Sundarban to get a reliable estimate of tiger numbers. The estimate showed that the Indian Sundarbans has >50 (but <100) adult tigers. A density of 4.3 tigers/100 km^2 (taking into account about 1,600 km^2 land area) was found with the range being 64–90 tigers. However, that was not a total count but only a tidal channel search and the inner mangrove forests were excluded due to lack of proper animal trails and fear of tiger attacks.

While the male tigers change their territory more, the females tend to hold a territory for comparatively longer periods. The territory of a male also overlaps the territories of more than one female. The cubs are reared exclusively by the mother and they remain attached to her for up to two and half years. The female cubs often occupy territories or home ranges adjacent to that of their mother.

In the tidal estuarine habitat the marsh tigers of Sundarban spray marking fluid (MF) on the tree branches to mark their territories and communicate via biochemical messengers. The marking is subject to twice-daily tidal inundation, and is thus likely to be washed away. Perhaps together with the fixative lipids of MF, the waxy surface of certain mangroves further help slow down the release of the odor molecules. Of three mangroves *Excoecaria sp.*, *Sonneratia sp.*, and *Heritiera sp.* studied by Poddar-Sarkar and Brahmachary (2014), *Heritiera* sp. best retained the smell. A tranquillized mangrove–marsh Bengal tiger of Sundarban was released back to nature after collection of sample for GCMS for the analysis of chemical compounds of body odor. It was observed by them that presence of 2 acetyl-1-pyrroline (2AP) in the marsh tiger, a recent adaptation in the Sundarban mangrove swamp is not surprising. The wildlifers in the Sunderbans are divided on the issue of MF smell surviving twice-daily tidal inundations; some hold the view that marking does not last at all, while others are of the opinion that tigers spray MF only on trees above the high-water mark. Preliminary findings on the body odor of three tranquilized mangrove-marsh Bengal tigers of Sundarban were reported (Poddar-Sarkar *et al.*, 2013).

Dust bathing to protect the body from insect-infestation is done by the Sundarban tiger (Kalyan Chakrabarti, pers. com). During the daily activity, every tiger records its presence while walking on soil amenable to receiving its pugmarks. While the tiger pugmarks were found along the banks of the creeks and rivers, their scats (wet as well as dry) were found at Harinbhanga 3, Jhilla 1, 2 and 3, Arbesi 5, Chandkhali 2 and 3, Bagmara 3 and 8, Chamta 4, Matla 3, Chhotohardi 3, Panchamukhani 5, Khatuajhuri 2, Mayadwip 1, Pirkhali 2 and also near the sweet water ponds visited by the tigers like Dobanki camp (Pirkhali 5), Sajnekhali camp (Pirkhali 1), Choragazi (Panchamukhani 3), Jhingakhali (Arbesi 1), Tushkhali (Khatuajhuri 1), etc. The nail marks (scratching) of some tigers on the trees like *X. granatum* (males), *S. apetala* and *Avicennia* spp. (females) were also observed. The tiger-roars were heard mostly between the pre-monsoon and the pre-winter, but rarely thereafter. These calls are generally related to the hunting, mating and sometimes while the mother moving with her cubs encounters an enemy.

The size of a tiger's home range depends on prey abundance and in case of the males, access to the estrus females. In the study area, the mean distance moved by the radio-collared tiger/tigress per day was found to be 4.64 km^2, whereas the maximum distance covered/day was 13.5 km^2. While following a tiger on the bank of the Nabanki khal, it was observed to move a distance of 1 km in ten minutes. A radio-collared tigress moved over a radius of c. 30 km^2 during a week. However, it could move much faster while chasing its prey.

Migration of the tigers between the two segments of India and Bangladesh was anticipated, but there was no proof in the past. But, during 2010, evidence of such territorial shifting was known. The straying

Khatuajhuri male, blind on the right eye with signs of territorial fights, which was captured on 20th May, radio-collared and released in the nearby forests on 22nd May, crossed the river Harinbhanga and strayed into the Bangladesh Sundarban, where it remained in the island of Talpati till his collar stopped functioning on 5th August. Hence, no further information regarding its movement was available.

Of late, another male tiger, who has mostly stayed in the Bangladesh Sunderbans, was captured from Harinbhanga forest, just opposite the Harikhali camp under Basirhat range of STR. This tiger was not clicked by the camera traps in the Indian Sundarbans. It was radio-collared on 27th December, 2020, after which he travelled 100 km in four months to reach the Bangladesh counterpart. During its long journey, the big cat crossed a few rivers, some of them wider than a kilometer. During this period, the tiger did not venture into any human habitats. After initial movements for a few days on the Indian side, it started venturing into the Talpatti island within the Bangladesh Sunderbans and crossed rivers, such as Choto; Harikhali, Boro Harikhali and even the Raimangal. Between December 27, 2019 and May 11, 2020 the tiger moved across three islands: Harinbhanga and Khatuajhuri in the Indian Sunderbans, and Talpatti island in Bangladesh. After May 11, the radio-collar stopped giving signals. The tiger's last recorded location was Talpatti island in Bangladesh. The gadget also had a mortality sensor, which gives signals in case of the tiger's death. But that didn't happen. No static signal was received from the collar, which pointed out that the tiger was safe. It is anticipated that the collar might have slipped off the tiger's neck. In the Sunderbans, salinity in the water can also damage the radio-collar.

In January 2017, a tigress was radio-collared and released in the South 24 Parganas division, the buffer area of the forest. This tiger, too, travelled a linear distance of over 100 km in four months to reach the tip of the BoB before finally settling in its territory. Before that, five other tigers- one of which had also ventured into Bangladesh's Talpatti island (described previously) were also radio-collared in the Indian Sunderbans.

Both the countries had undertaken the camera-trap exercise together in 2018. On both the sides, one camera trap station was placed for every 2 km². The tiger status report, 2018, has also revealed that three tigers of Bangladesh matched with one tiger caught in cameras at National Park East range and two at Basirhat range, both in Indian Sunderbans. These three tigers were also photographed in the Khulna range of Bangladesh Sunderbans and counted during the compilation of Bangladesh data. Since Khulna falls in the western part of Bangladesh Sunderbans, tiger movement from that side to the Indian Sunderbans is natural, like what had happened earlier with the Talpatti tiger in 2010.

Tiger enumeration

Studies on the tiger in the Indian Sundarbans date back to 1972 after its inclusion in the list of endangered species in 1969 and also the Red Data Book of the International Union for the Conservation of Nature and Natural Resources (IUCN) due to an alarming decrease in numbers of free living tigers. On the recommendation of Guy Reginald Mountfort, who led an expedition in 1967, IUCN and World Wildlife Fund (WWF) decided to sponsor an ecological and ethological survey of the situation in Sundarbans to assess the requirements of a viable tiger population without undermining the human utilisation of forest. The then DFO of undivided 24-Parganas Forest Division, Amal Bhusan Chaudhuri carried out an ecological study in part of the Indian Sundarbans and estimated the tiger population to be 112. Later in 1973, after creation of STR, the tiger population was estimated to be 181 in 1976, which increased to 205 in 1979 and 264 in 1983. This estimation was mostly based on the pugmark methods, attacks on humans and interviews with the local communities, but was criticised as an unreliable total count and prone to human errors.

Karanth and Nicholas (2000) carried out the first camera trap survey in the Indian Sundarbans and estimated fewer than 100 tigers. This study had limitations as the camera traps were set up around the sweet water ponds because no trails or roads were available due to the thick mangrove vegetation. Other studies in the Indian Sundarbans have been based on extrapolation from radio telemetry studies carried out by Jhala *et al.* (2011) in STR, estimating the tiger population to be around 70 in 2010 by Wildlife Institute of India (WII). As a part of the Phase-IV tiger estimation in Sundarbans, WWF-India conducted a camera trapping exercise in 24-Parganas (South) Forest Division, which is also a tiger habitat, for a more holistic estimate outside the PAs. As per the report (Das *et al.*, 2012), the population was estimated to be 8.0±0.2 (N-hat±SE) individuals for Ramganga range and 13±3.5 (N-hat±SE) individuals for Raidighi range. Tiger density was estimated as 4.3 individuals/100 km^2 at Ramganga range with an effective trapping area of 184.5 km^2 and 7.08 individuals

/100 km^2 at Raidighi range with an effective trapping area of 141.3 km^2. In MLSECR (Maximum Likelihood Spatially Explicit Capture Recapture) analysis, estimated tiger density was 3.8 (±SE 1.5) individuals/100 km^2 for Ramganga range and 5.2 (±SE 1.7) individuals/100 km^2 for Raidighi range.

Thereafter, the tiger abundance was estimated to be 76 (62-96) in 2014 in SBR by National Tiger Conservation Authority and WII, 87 in 2016-2017, 88 (86-90) in 2018-2019 and 96 in 2019-2020. Despite two cyclones Amphan and Yaas, triggering massive destruction and inundation of low-lying areas, the tiger population, as of 2020-21, also stood at 96 including 30 males, 52 females, 14 unidentified sex and four cubs (cubs are not included in the total estimation as per NTCA guidelines), over the age of one year. During the current year, it is expected from the preliminary data that the number might be increased by 27.

Over the years there has been a positive trend in tiger abundance and, therefore, the population can be considered stable. The low coefficient of variation (high precision) in tiger abundance was largely due to the extensive camera trap coverage and high recaptures of tiger individuals. It is known that in the hostile terrain, the carrying capacity is three to five tigers per 100 km^2 and in multiple blocks the density is more than that. As a result, young big cats might be forced to move out in search of new territories and weaker ones might look for a new home with easier hunting skills. This may be the cause of growing human-tiger conflict in the delta. As these tigers have strayed into the fringe villages, all of them were captured and released into the wild. The Wildlife Institute of India (WII) in a recent report stated that the delta has already saturated its tiger carrying capacity.

According to the Tiger Census 2015 and 2018, tiger population was estimated to be 106 and 114 tigers respectively in the Bangladesh part. After 2017-18 no tiger census was conducted in the forest which is the only habitat of Royal Bengal Tigers in Bangladesh. Forest Department data shows at least 50 tigers were killed in the last 15 years due to illegal poaching of wildlife in the forest.

Recently, an increase in lonely male tigers from the Bangladeshi side of the Sundarbans entering the Indian side during the mating season in winter, looking for female companions, was suspected in early 2022. They also tend to enter the fringe villages. In one such instance, a male tiger entered Mathurakhand village near the forest compartments of Pirkhali II and III on the fringes of the tiger reserve area late on a January night and killed a couple of cows and buffalos. One of the reasons for such tiger straying from Bangladesh might be the male-to-male fights over a breeding female. Generally, the male tigers who lose the battle of possession have to leave the area after the fight is over. So they enter the villages on the fringes of the forest. Moreover, a defeated tiger, who has been weakened by a fight, is not capable of hunting in the dense forest, which requires physical strength, so it enters the villages to find easy prey.

Camera trapping and tiger-prey density

The pugmark method followed during the twentieth century was field-friendly, but due to some drawbacks, the Project Tiger authorities developed a new methodology for monitoring of tigers, co-predators, prey and habitat. In fact, the camera trapping technique is more reliable than the traditional method of counting pugmarks. As the Sundarbans ecosystem is subjected to tides twice a day with varying tide levels, there is a high risk of the camera traps being inundated. Suitable camera trap locations are selected near the brackish water holes, in elevated places, river bends, regular channel crossing paths frequented by the tigers based on local knowledge of the frontline forest staff and sign surveys to maximise photo-capture and minimise chances of lethal encounters. Trap stations are regularly monitored and constantly supplied with baits to minimize the spatio-temporal variation in photo captures between traps.

Reconnaissance surveys were carried out in different grids for selecting the potential camera trap locations. During the 2016 study period 3,090 km^2 forest area was covered in SBR with break up along with the number of camera trap grids against 24-Parganas (South) Forest Division and four ranges of STR as follows:

1. 24-Parganas (South) Forest Division: 454 km^2 (50)

2. NP (East) Range: 850 km^2 (60)

3. NP (West) Range: 890 km^2 (60)

4. Sajnekhali WLS: 430 km^2 (60)

5. Basirhat Range: 466 km^2 (44)

Most of the trapping grids were standardised across the years. For the newly sampled areas and grids with no tiger capture for years, the trapping grids were selected based on the following criteria: (i) tiger pugmarks, (ii) comparatively high elevation areas unlikely to get submerged even during the high tides, and (iii) exclusion of the sample grids with excessive human disturbance to minimize losses of camera traps and data. Identification of tigers was done visually with the images from camera traps. As per the radio telemetry data from the Sundarbans, tigers rarely cross channels wider than 1 km in width within a short span of time. Therefore, channels wider than 1 km and forest fringe villages were masked in a GIS platform and classified as 'non-habitat'. All mangrove patches and channels <1 km wide were classified as 'habitat'.

In addition, Khal survey is carried out across the entire Sundarban when direct sightings and signs of tiger, fishing cat, otters, estuarine crocodiles, monitor lizard, wild pig, spotted deer and human disturbance along with vegetation covariates are collected.

During the period January 2012 - April 2013, to estimate tiger numbers, WWF-India and SBR collaboratively carried out camera trapping sampling in the western Sundarbans to establish a population baseline. In 24-Parganas (South) Forest Division, the exercise was carried out from January 2012 to March 2012. Twenty unique individuals were identified from this division and range-wise density was calculated. Tiger density was estimated to be 3.8 (±SE1.5)/100 km². and 5.2 (±SE1.7)/100 km². for Raidighi Range.

In NP (East) range, the exercise was carried out from November 2012 - January 2013. 21 unique individuals were identified from this range. Tiger density was estimated to be 3.69 (±SE0.82)/100 km².

In Sajnekhali WLS, the exercise was carried out from January - March 2013. 13 unique individuals were identified from this range. Tiger density was estimated to be 2.36 (±SE 0.64)/100 km².

In Basirhat Range, the exercise was carried out from March 2013 to April 2013. Thirteen unique individuals were identified from this range. Tiger density was estimated to be 2.57 (±SE 0.76)/100 km².

In session 2013-2014, camera traps were installed in Ramganga Range of 24 Parganas (South) Forest Division and in National Park (East) Range of the Sundarban Tiger Reserve. The same dataset was used in the All-India Tiger Census. Five unique individuals were identified from Ramganga Range. The estimated tiger density was 1.57 (±SE 0.74) individuals/100 km^2 for Ramganga Range. Twenty unique individuals were identified from National Park (East) Range. The estimated tiger density was 3.77 (±SE 1.03) individuals/100 km^2 for NP (East) Range. Camera traps could only be deployed in 20 locations in Basirhat Range for a short period of time due to rough weather conditions. The data from this exercise was not taken into account.

In 2014-2015, camera traps were again placed in Basirhat Range of STR and in the entire tiger habitat of 24- Parganas (South) Forest Division comprising three ranges (Ramganga, Raidighi and Herobhanga Range). Seventeen unique individuals along with four cubs were identified from 24-Parganas (South) Forest Division. The estimated tiger density was 3.42 (±SE 0.74) individuals/100 km^2 for 24 Parganas (South) Forest Division. Sixteen unique individuals along with three cubs were identified from Basirhat Range. The estimated tiger density was 3.33 (±SE 0.09) individuals/100 km^2 for Basirhat Range.

During the tiger count for 2015-2016 in SBR showed 81 tigers. No individual was found common between the ranges. A new individual means it has never been encountered in previous camera trapping sessions, although some crossed from one range to another over the period. Capture of four cubs (>1 year old) and four sub-adults (>2 years old) were not incorporated into density estimation.

24 Parganas (South) Forest Division

A total sampling effort of 2000 trap days (50 camera trap stations, each operating on 40 occasions) at 24 Parganas (South) Forest Division yielded 265 photographs (both flanks combined) of tigers. A total of 37 out of 50 camera trap stations recorded the photographs. There were no tiger captures on 22.5% occasions. A total of 27 unique tigers were identified. Out of these 27 individuals, 3 individuals— SB110 (Sub Adult Male), SB111 (Sub Adult Male) and SB118 (a cub) were not used for analysis as all of them were less than two years old.

There were 86 records of tiger captures of 24 tigers. This included fifteen captures of SB90, thirteen captures of SB83, seven captures of SB92, six captures of SB10, four captures of SB7, SB13, SB18 and SB114, three captures of SB14, SB91 and SB94, two captures of SB20, SB93, SB112, SB113, SB115, SB116 and SB117 and single captures of SB3, SB9, SB15, SB96, SB97 and SB98.

NP (East) Range

A total sampling effort of 1140 trap days (60 camera trap stations, each operating for 19 occasions) at National Park (East) Range yielded 130 photographs (both flanks combined) of tigers. A total of 23 out of 60 camera trap stations recorded the photographs. There were no tiger captures on 46.7% occasions. A total of 14 tigers were individually identified. The capture history file used in data analysis had 35 records. This included six captures of SB21, five captures of SB44, four captures of SB123, three captures of SB38 and SB120, two captures of SB35, SB40, SB86, SB87 and SB119 and single captures of S24, SB84, SB121 and SB122.

NP (West) Range

A total sampling effort of 2280 trap days (60 camera trap stations, each operating for 38 occasions) at National Park (West) Range yielded 319 photographs (both flanks) of tigers. A total of 38 out of 60 camera trap stations recorded the photographs. There were no tiger captures on 10.5% occasions. A total of 22 tigers were individually identified. Out of these 22 individuals, 3 were cubs (SB153, SB154 and SB155) and not used for analysis as all of them were less than one year old. The capture history file used for data analysis had 93 captures and recaptures. This included thirteen captures of SB134 and SB140, eleven captures of SB139, eight captures of SB141, seven captures of SB138, six captures of SB135 and SB142, five captures of SB144 and SB152, four captures of SB137 and SB 143, three captures of SB 148, two captures of SB 136 and single captures of SB145, SB146, SB147, SB149, SB 150 and SB151.

Sajnekhali WLS

A total sampling effort of 1800 trap days (60 camera trap stations, each operating for 30 occasions) at Sajnekhali Wildlife Sanctuary yielded 69 photographs (both flanks) of tigers. A total of 27 out of 60 camera trap stations recorded the photographs. There were no tiger captures on 36.7% occasions. A total of 11 tigers were individually identified. Out of these 11 individuals, 2 individuals- SB132 (Sub Adult Female), and SB133 (Sub Adult Female) were not used for analysis as all of them were less than two years old. The capture history file used for data analysis had 37 captures and recaptures. This included ten captures of SB48, seven captures of SB57, six captures of SB124, five captures of SB64, three captures of SB 129, two captures of SB49 and SB 131, and single captures of SB50 and SB130.

Basirhat Range

A total sampling effort of 1760 trap days (44 traps station for each operation on 40 occasions) at Basirhat Range yielded 75 photographs (both flanks) of tigers. A total of 16 out of 44 camera trap stations recorded the photographs. There were no tiger captures on 50% occasions. A total of 11 tigers were individually identified. The capture history file used for data analysis had 31 captures and recaptures. This included six captures of SB61, four captures of SB124, SB125 and SB126, two captures of SB69, SB72, SB104, SB105, SB106 and SB 127 and single capture of SB101.

Estimated tiger density in 24 Parganas (South) Forest Division, National Park (East) Range, National Park (West) Range, Sajnekhali WLS and Basirhat Ranges were 5.46 ± 1.27 individuals/100 km², 2.09 ± 0.66 individuals/100 km², 3.07 ± 0.8 individuals/100 km², 1.79 ± 0.63 individuals/100 km², 3.56 ± 1.35 individuals/100 km² respectively.

Estimation of population size and distribution pattern is an integral component of monitoring population status, demographic trends and management efficacy. Systematic monitoring of tiger population in the Indian Sundarbans over a period of four years suggests that the population may not be stable as portrayed by the capture history of individuals. The overall female: male ratio in the study area for the current study (2.73:1) is strongly female biased.

The capture pattern indicates that spatial overlap is minimal among the males whereas between females as well as males and females it is comparatively higher, which is consistent with the territorial behaviour of tigers. Density estimates have remained relatively stable over the years. It is worth noting that in spite of being reserve forests (with greater human use), both 24 Parganas (South) Forest Division and Basirhat Range support a higher density of tigers than the adjacent NP.

A need was felt among the stakeholders to monitor the tiger population of the entire tiger habitat in the Indian Sundarbans using camera traps within a season in a single calendar year. Keeping this in mind, the Sundarban Biosphere Reserve in active collaboration with WWF-India decided to monitor the tiger population of the entire tiger habitat in the Indian Sundarbans using camera traps in a season within a single calendar year (2015-2016).

Accordingly, camera trapping was carried out in eight blocks spread across the Sundarbans of India and Bangladesh (Jhala *et al.,* 2016). The absolute tiger density was estimated through spatially explicit capture mark-recapture using camera traps at three representative sample blocks (Sarankhola, Satkhira and Khulna ranges) covering an area of 1,264 km^2 in the Bangladesh Sundarbans and five blocks (covering five ranges-Basirhat, Sajnekhali, East and West NP in STR and Ramganga in South 24-Parganas Forest Division) comprising an area of 1,649 km^2 of SBR. According to this report, considering density of tigers on the Indian Sundarbans was about four tigers per 100 km^2, the Bangladesh population was below the potential carrying capacity of the habitat.

Tiger density was negatively correlated with human disturbance which was relatively high in Bangladesh Sundarban. Suitable camera trap locations were selected near brackish water holes, in elevated places, river bends, regular channel crossing paths frequented by tigers. The highest tiger density was recorded in Block VIII (Sajnekhali) while the lowest density was recorded from Block III (Khulna).

The tiger density in all the eight camera trapped blocks was 2.85 (2.10-3.61) tigers/100 km^2. Tiger density was found to have a positive relation with tiger sign intensity (r= 0.464; P= 0.008); and encounter rate of prey (r= 0.447, P= 0.012), while it had a negative relationship with encounter rate of human disturbance (r= -0.554, P= 0.0012). Tiger numbers, estimated by camera trapping using SECR within the minimum bounding camera trapped polygons and the multiple regression model for the remaining part, was 182 (145 to 226) in the total tiger occupied area of 6,724 km^2; with 106 (83 to 130) tigers on the Bangladesh side and 76 (62 to 96) tigers on the Indian side in 2014, where the other cycles showed 74 (64-90) in

2010 and 88 (86-90 in 2018) and the tiger density was estimated at 3.6 (SE 0.38) tigers per 100 km^2. The detection at the activity centre for females was 0.13 (SE 0.01) while for males it was 0.09 (SE 0.009). The movement/scale parameter for females was 3.2 (SE 0.1) km while for males it was 4.6 (SE 0.18) km. The female to male detection ratio was 0.64:0.36. A total of 6 young tigers were photo-captured. Parts of the SBR, Sajnekhali and West ranges of STR had the highest density of tigers.

The tiger count for 2019-2020 in SBR rose to 96 (male 23, female 43 and unsexed 30), identified on examination of their stripe patterns. 11 tiger cubs have been identified. 73 tigers were recorded in STR during December 16-January 13 [total of 1,156 Cuddeback camera traps (578 pairs)] and 23 tigers were recorded during January 22-February 19, in the 24 Parganas (South) Division. Tiger sign and sighting encounter rates were higher on the Indian side especially in the core zone of the national park as compared to other areas.

The tiger population in the Indian Sunderbans has remained stable in the past one year. the mangroves is home to 96 tigers, as per the camera trap exercise undertaken by the state forest department between December 2020 and February 2021, which is also known as the Phase IV estimation. As per this latest report, out of the 74 big cats in STR area, 12 were found in Sajnekhali WLS area, 17 in Basirhat Range (Buffer area), 21 in NP East and 24 in NP West ranges. Of the 22 tigers in South 24 Parganas Forest Division, four were found in Matla range, 11 in Raidighi and seven clicked in Ramganga range.

This time, 1,395 cameras (close to 700 pairs) were placed in SBR compared to 1,200 used during the same exercise in 2019-2020. Of this, WWF-India gave 400 cameras. The entire mangrove forest was divided into 2 km^2 grids, each with two camera traps. Four cubs clicked in Basirhat and NP East ranges were not taken into account while arriving at the final number as all of them were less than one year old. This is the minimum number that have been photographed and the actual number may be higher because some individuals might not be camera-trapped. In order to get more accurate details the entire mangrove forest of 4,500 km^2 should be divided into 1 km^2 grid and two cameras be placed at each grid. For this, installation of an additional 800 trap camera is essential.

During the tiger estimate (2018) in the Indian Sundarbans the tiger density was estimated at 3.6 (SE 0.38) tigers per 100 km^2. The g$_o$ (detection at activity centre) for females was 0.13 (SE 0.01) while for males it was 0.09 (SE 0.009). The movement/scale parameter for females was 3.2 (SE 0.1) km while for males it was 4.6 (SE 0.18) km. The female to male detection ratio was 0.64:0.36. A total of 6 young tigers were photo-captured. Parts of SBR and Sajnekhali and West ranges of STR had the highest density of tigers.

During this estimate the encounter rate of the prey species per km was 0.053 in case of Axis deer (total number of detection 64), 0.02 for the wild boar (23), 0.03 for the rhesus macaque (35), 0.05 for the monitor lizard (67).

The tiger population in the Sundarban landscape seems to be at stable density. Together with Bangladesh (Aziz *et al.* 2018), the Sundarban holds about 200 tigers, which have uniquely adapted to the mangrove forests and are, therefore, of global importance. It is important that this transboundary population is managed as a single population.

The density of the prey species is variable in both the countries. The Axis deer was captured throughout Sundarban landscape in India with highest concentration of photo-captures in Sajnekhali Wildlife Sanctuary during the tiger estimation in 2018. Two of the capture hotspots in this sanctuary coincide with the location (Dobanki camp) where chital had been released in the late 2000s. Spotted deer sign and sighting encounter rate was more or less uniform across the landscape in Bangladesh with pockets of higher encounter rates in the Sarankhola and Khulna ranges.

The wild pig's presence was recorded throughout the landscape with higher encounter rates on the Indian side with highest concentration of photo-captures in SBR. Prevalence of more capture hotspots in SBR concurs with higher sightings of wild pig during boat transects in this area.

Similarly, Rhesus macaque was captured more or less throughout the Indian Sundarban landscape with highest concentration of photocaptures in Sajnekhali beat of Sajnekhali Wildlife Sanctuary. The capture hotspot in this beat coincide with the location of the beat and range offices, where there is maximum instantaneous human presence inside the tiger reserve on any given day as these offices issue tourist and fishing permits. The Rhesus macaque's sign and sighting encounter rate was also higher in the

East range of SNP as well as on the buffer zone of the Basirhat range in the Indian side.

Monitor lizard distribution was primarily in STR with very few photo-captures in SBR. The capture hotspots in STR were in the Sajnekhali WLS, the buffer zone of Basirhat range and the core area of NP (West range), whereas in the Bangladesh landscape the monitor lizard's sign and sighting encounter rate was highest in the Chandpai, Sarankhola range, while its presence was not recorded in many areas.

The fishing cat was captured throughout the Indian landscape; however, there were fewer photo-captures in the East range of NP in the core area. No such record is available in the Bangladesh Sundarbans.

Similarly, the leopard cat was captured throughout the Indian landscape and its abundance is indicated by the numerous capture hotspots both in STR as well as SBR.

In the Indian landscape the Jungle cat was captured only at the fringe areas of the mangrove forests, with its highest photocaptures in the buffer zone of Basirhat range.

The golden jackal was captured for the first time in the Indian Sundarban and seems to be a recent coloniser restricted to the mangrove forest edges of Sajnekhali WLS and islands of SNR adjacent to the human habited islands.

On the contrary, the Lesser adjutant stork seemed to have more presence on in the Indian side rather than the Bangladesh Sundarban.

Sundarban is home to three species of otter, viz. smooth coated otter (*Lutrogale perspicillata*), Eurasian otter (*Lutra lutra*) and oriental small-clawed otter (*Aonyx cinereus*). However, it was not possible to identify the individual species from the camera trap photographs recorded during 2018 in the Indian landscape. Hence, the map depicts the varying intensity of photo-captures of all three species combined together. The capture hotspots were primarily in the core zone of NP beside wide channels.

The otters in general had a higher presence in the Bangladesh Sundarban as compared to the Indian side with ranges of Chandpai-Sarankhola and Khulna recording highest encounters. A survey was conducted over 351 km of water courses across four sample areas- 97 km in Satkhira block, 78 km in West WS, 101 km in East WS, and 75 km in Chandpai block, when two groups of otters were observed for several hours in two different locations in the south-eastern part of the Sundarbans during morning (0915-1040 h) and evening hours (1505-1755 h) and in all, 53 small-clawed otters in 13 locations, alongside signs of footprints and spraints were recorded (Aziz, 2018). Mean group size of this otter species was estimated to be 4.08±SE 1.13 (range = 1-12, n=13), whereas the mean encounter rate of sighting and signs (footprint and spraint) was 0.06/km, with higher in the East WLS areas (0.09/km) and lower in the Satkhira block (0.02). Mean encounter rate of direct sighting was 0.03/km, translating into approximately one individual otter in every 30 km of rivers surveyed. The sample area-wise estimates suggest higher abundance of otters across the eastern part of the Sundarbans of Bangladesh.

The Asian small-clawed otters search for the mudskippers on the exposed mudflats. If feel disturbed, the otters quickly disappear from their foraging ground of exposed mudflats into the nearby forests, but after a while they come out, and start excavating the burrows of mudskippers, *Periophthalmus* sp. with their forelimbs. During the ebb tide, when the mudskippers usually take shelter in their tunnel-like burrows, which are abundant across the river banks in the Sundarbans, the otters have developed good strategy to catch them by inserting the forearm through one end of the burrow so that the mudskippers are flushed out with the water through the other opening. Otters instantly capture the mudskippers to feed when the flood water is drained out during the ebb tides, providing extensive mudflats with lots of opportunities for feeding on the mudskippers. Although they are reported to eat a range of species from the crabs to fish, snakes, snails, frogs, small birds and octopus, the major prey species of this otter in the Sundarbans is the mudskippers. In the Sundarbans, chemical pollution of watercourses may possibly be the most critical threat to otters but the otter hunting for their skin in trade and other parts for food and traditional medicine is not common. While habitat destruction and disturbance is minimal for the otters in the Sundarbans, trapping by entanglement in fishing gear often takes place for prevalence of widespread gear fishing in the water-courses of this mangrove forest.

In the Indian landscape, the palm civet was captured only at the edges of the buffer zone of Basirhat range, adjacent to the human habited islands.

A total of 14 species of ungulates, carnivores, omnivores and galliformes were photo-captured in the Indian Sundarbans during the 2018 estimate. The wild pig and chital were apparently the most common species where it took only 5 and 10 trap nights respectively to capture one photo while the estuarine crocodile and common palm civet were the rarest photo-captured species.

It is to be kept in mind that the camera trap photo index (RAI) for various species should be viewed in the context of using lures for camera trapping in the Sundarban landscape (lures are not used at other sites). The lures enhance chances of attracting species of carnivores and wild pigs to camera traps. Therefore, these RAI can be used for future population trend comparisons within these species in Sundarbans but not for interspecies relative abundance or comparisons between other sites. However, species like chital and red jungle fowl are unlikely to be affected by lures and their RAI are comparable across sites.

Radio collaring of tiger

A radio collar is a wide band of machine-belting fitted with a small radio transmitter and battery. The transmitter emits a signal at a specific frequency that can be tracked from up to 5 kms away. GPS-Satellite and VHF radio-collars are fitted on the neck of an aberrant tiger or tigress for monitoring movement, which provide crucial information on tiger behaviour, such as their habitat use and preference (*Avicennia-Sonneratia-Phoenix-Ceriops*-open forests-water channels of different width), home ranging pattern (resident female 56.4 km^2 and resident male 110 km^2), extrapolated population density and abundance (4.6 tigers/100 km^2) as well as extent of negative human-tiger interaction around human settlements by crossing the water channels, which help in formulation of effective management plans and strategies to reduce negative human-tiger interactions in the landscape.

Field effectiveness

1. A tigress was captured in the village named Sonaga near Bidya Range Office of West range on 22nd February 2010. It was thereafter

radio-collared and released near Netidhopani camp in West range (Storekhali) on 24th February. She dropped her collar on 21st April, 2010 at Dhonar Khal near Dobanki camp in Sajnekhali Range. She was again trapped near the northwestern part of STR and soon after radio-collaring it travelled all the way south towards the BoB and finally returned to the original village Sonaga. She had a home range of 329.9 km². She travelled a distance of 120 km in 27 days from the day of her release till she dropped her collar. Earlier telemetry studies have shown that breeding tigers are territorial with an average home range size of 12.3 km² in the Bangladesh Sundarban (Barlow, 2009) as against 40 km² in the Indian Sundarban.

2. A male tiger, strayed from the Arbeshi forests, was captured from Malmelia village of Basirhat range on 20th May, 2010. It was tranquilised, radio-collared and released in the Khatuajhuri forests on 22nd May, 2010. The animal, weighing 108 kg, was blind in the right eye and bore signs of territorial fight on the head and left rear leg. There is enough prey in the Katuajhuri jungle and the forest guards, who examined the terrain, found carcasses of prey animals devoured by the tiger. The radio collar signals revealed that on the first two days, it was on the hunt and traveled only 6-7 km. But on the third day, the tiger traveled more than double that distance. The signals showed that it crossed the River Harinbhanga and left its command area and moved into a new territory at will, even if there is enough prey. So, it was surprising when the tiger suddenly started moving from south to east, towards Bangladesh. This tiger subsequently settled in Talpatti Island, which is only two kilometers from the mouth of River Harinbhanga and continued to use the entire Island till its satellite signal stopped transmitting signals on 4th August 2010. It was not possible to ground track this tiger through VHF signal since the animal was in Bangladesh.

3. Another male, weighing around 97 kg, was captured near Netidhopani camp on 20th May 2010, radio-collared and released near a fresh water pond. This tiger was recaptured on 3rd October, 2010 for replacing the one of its exhausted batteries. This tiger was tracked till November, 2010 with the help of a hand-held directional 3-element Yagi directional antennae with radio receivers (HABIT receiver model HR 2600 and Vectronics GPS Plus Handheld Terminal Unit) via boats.

4. In case of a tiger captured near Netidhopani Camp on 21st March 2010 and released near Pirkhali-7 compartment, tracking was continued till mid-April when the collar stopped transmitting probably due to exhaustion of battery.

The following data were known from the movement of the above mentioned four radio-collared tigers:

1. Core Area estimation: The average core area of tigers was estimated to be 65.33 (SE 19.31) km².

2. Distance moved per day: The average distance moved per day was 5.43 (SE 0.93) with range as 0.1- 23 km. There was no significant difference between daily distances moved by individual tigers. Tigers were observed to move more during the early hours of dawn, day and afternoon with a distinct heightened movement between 7 AM and 10 AM. For an hourly interval, the highest distance moved was 595.25 meters during 7 AM and the lowest being 31.58 meters during 7 PM per day with a peak around 4 PM in the afternoon. Tigers travelled an average of 5.06 km (SE 0.47) during the neap tide phase, when the currents are milder and substantial dry land area remains exposed allowing the tigers to explore between islands, as against 4.43 km (SE 0.40) during the spring tide phase, when the currents are stronger and less dry land area is available, per day, but there was not much significant difference observed between these two tide phases because the tigers in the Sundarbans are expert swimmers.

3. Diurnal movement pattern of tigers in Sundarban seems to be a major factor in human kills recorded during the day.

Channels crossed

Channels within 100 to 350 m in width were crossed by the tigers as per availability. Channel width more than 400 m were avoided by the tigers. The mean number of channels crossed by tigers per day was 5.19 with SE 0.99. There was no significant difference observed in mean number of channels crossed by individual tigers. There was a significant difference in the width of channels crossed by the tigers as compared to their availability.

Habitat use

The tigers preferred *Avicennia* patches whereas *Phoenix* and *Ceriops* habitat were used according to their availability, signifying their importance as critical habitat for the tigers in the mangroves. Avoidance of water was observed in the overall preferred habitat use. Its major prey species Axis deer also prefers *Sonneratia, Avicennia* and *Exocoeceria* dominated patches in the Sundarbans. Therefore, the predator spends a considerable time frequenting these places likely in pursuit of its primary prey. *Phoenix* dominated patches, growing in areas with relatively higher ground, provide the tiger a dry ground to seek refuge from very high tidal fluctuation. But in the Bangladesh Sundarbans such preferred tiger habitat is not known.

Relative abundance of prey species

The overall encounter rate for chital sign was 1.92 per kilometer. The encounter rate per kilometer for wild pig sign was 0.26. Presence of the Axis deer and Wild boar within the home ranges of the radio-collared tigers was found to be low. The encounter rate per kilometer for human sign was 0.56.

Other predators

Some experts (Curtis, 1933; Seidenstlcker, 1983; Blower, 1985) believed that leopard (*Panthera pardus*) was once found along the edge of the Sundarban (Nishat *et al.,* 2019: 23). The small carnivore community is composed of the leopard cat (*Prionailurus bengalensis*), fishing cat (*Prionailurus viverrinus*) and jungle cat (*Felis chaus*). The leopard cat is the common lesser cat in the Sundarbans, which prefers dry sandy areas to defecate. Rats and mice including the Longtailed Tree Mouse (*Vandeleuria oleracea*) are their main prey items followed by insects (grasshoppers, beetles, etc.) and small and medium birds (excluding the feathers), particularly the Red junglefowl, Agamids, crabs as well as plant materials like seeds, fruits, leaf blades, etc. The encounter rate of the fishing cat was 0.03/km^2. The pugmarks of the leopard cat were distinguished from those of the fishing cats on the basis of absence of any nail-imprint on the soil. The Jungle cats are found mainly in fringe areas. Otters were rarely sighted (encounter rate 0.009/km^2)(Mallick, 2011). But, Majrekar and Prabhu (2014) recorded higher encounter rate of the Smooth-coated otter.

In all they covered an area of 237.8 km to conduct a boat transect survey, out of which 117.3 km was within the Sajnekhali range and 120.5 km in the National Park West range. During the entire survey, a total of four sightings (three at Sajnekhali WLS and one at National Park west range) were recorded involving 11 individuals and also otter signs were marked at seven different locations (tracks at six locations and a spraint at one location). The mean group size was 2.75 (S.E. = 0.85, Range = 1-5) and in one of these smooth coated otter groups sighted in Sajnekhali range, an otter cub was also seen.

The creeks fringing the mangroves and associated mudflats harbour the diurnal smooth Indian otter (*Lutra perspicillata*) and the nocturnal small-clawed otter (*Aonyx cinereus*), which live on the fishes, crustaceans and mollusks.

The jackal (*Canis aureus*) and Indian fox (*Vulpes bengalensis*) are present in the fringe areas. Even though the Golden Jackal had been spotted by the foresters and locals earlier, we have got the first camera trapped photographic evidence of the animal, as a co-predator of the tigers, early during the last decade. It seems to be a recent coloniser restricted to the mangrove forest edges of Sajnekhali WLS and islands of SBR adjacent to the human-inhabited islands, although no such evidence is available in case of the Indian fox.

The Marsh Mongoose (*Herpestes palustris* Ghose, 1965), previously considered a subspecies of *H. javanicus*, is an endemic species confined to the wetlands. The last stronghold of this marsh mongoose in the world is the East Kolkata Wetland (EKW), a Ramsar Site, the northern part of pre-colonial Sundarbans. Moreover, during 2004-2005, a biodiversity study reported occurrence of the Bengal marsh mongooses from the Typha reed/grass beds in the semi-natural wetlands (leased out for commercial fishery purpose) on the eastern boundary of the Joka campus of Indian Institute of Management, Kolkata, South 24-Parganas district, located in between 22°26'45"N latitude and 88°18'32"E longitude (within the pre-colonial Sundarbans).

Taxonomy of *H. palustris*

Validity of the species as well as endemicity of *H. palustris* has been recognized by the Zoological Survey of India (ZSI), Kolkata. Out of forty four endemic mammals of India, *H. palustris* is the only mongoose species. The Bengal Marsh Mongoose *H. palustris* was first reported as a new natural as well as native species from the swamp of the present Salt Lake City and distinguished from the superficially similar small Indian mongoose, *H. auropunctatus*, by examining the skin and skull of the adults in the stock of ZSI.

Specimens

During a one-year survey from January to December 1965, twenty six specimens of both sexes were collected from the wetlands in EKW- Salt Lake, Bantala, Duttabad, Hederhat and Nalban. Nalban, located between 22°25'-22°40' N latitude and 88°35' E longitude, is the eastern limit of its known range. Subsequently, occurrence of the Bengal marsh mongoose was recorded from three sites in South 24-Parganas district- Diamond Harbour, Patiatala and Bajbaj. Diamond Harbour, located at 22°11'N latitude and 88°14'E longitude, is the known southern limit of its range.

Present distribution

The current survey in EKW reveals that the whole population of Bengal marsh mongoose in northern Salt Lake bheries is constituted of *H. palustris* as no other sympatric mongoose, e.g., *H. auropunctatus*, was trapped from these wetlands. Prior to reclamation, the Bengal marsh mongoose was fairly common in Salt Lake and other adjacent marshy areas in North and South 24-Parganas districts. However, at present, due to continuous habitat destruction, they are rarely seen in the converted areas, where their preferred food and shelter are not available adequately. During the current survey in EKW, the Bengal marsh mongoose was reported from the Nalban and Sahebmara bheries. But it was not found from other bheries like Choubhaga (south), Bantala (south-east), Jhagra Sisa (east) and Mahisbathan (north).

Population

Since it was a sample survey based on only trapping and observation (not direct reconnaissance) on the fossorial and semi-aquatic Bengal marsh mongoose, its population in the entire EKW could not be ascertained. However, the population density in the small sample areas could be assessed. In Nalban, twenty burrows were sealed in the evening and five were found opened in the next morning. Hence, the least count of the mongoose at that niche was taken to be five, considering that each burrow was occupied by at least an individual. Accordingly, 11 to 13 Bengal marsh mongooses were estimated in a stretch of half kilometer in Nalban during the survey. In a 50 m secluded portion of the mud-bank between Sukantanagar and Char (4) Number bheri, about 20 mongooses were reported to live closely in a number of burrows dug near to each other in 2005.

Morphological features

H. palustris is a small carnivore similar to the weasel in appearance, despite not being closely related. It has a small head, long face with a pointed muzzle and vermiform body to suit their fossorial and predatory nature as well as semiaquatic life style. Its ears, placed on the side of the head, are small, semi-rounded above and then slopes down, mostly concealed by hairs and projected slightly beyond the pelage. It has an acute sense of hearing. It has small but prominent brown eyes, glittering and snake-like, with linear erect pupils. Its powerful eye-sight helps in monitoring the environs and detecting the small preys. The Bengal marsh mongoose has light reddish hair-free nose. It often sniffs the ground in search of the prey. At close range, its fine facial vibrissae (whiskers) provide an acute sense of touch. The female has four functional mammae. Its tongue is rough as a cat's. In the Nalban, a mongoose was found to lick its lip with its tongue extended out. The well formed, relatively long, pointed and deep-rooted lower (mandibular) and upper (maxillary) canines or cuspids are fanglike and projected prominently, particularly when the mouth is wide open. These are the largest teeth of Bengal marsh mongoose. The canine teeth are used primarily for firmly holding the prey in order to tear it apart, and occasionally as formidable weapons for both offence and defense. The Bengal marsh mongoose is an occipital cruncher. The premolar teeth are thick and used for crushing the hard shell of the food-stuff. It bites into the back of the head of the prey so that the claws and teeth of the prey are kept out of the way. The tail is long and muscular, thick at the base, then tapering gradually up to the tip. The tail is, however, little smaller than the body. It has short legs with five toes on each of the four feet. The hind limbs are comparatively muscular and powerful. Its semi-plantigrade feet, exemplified when running, have short, compressed and sharp claws, which are non-retractile. These are primarily used for scratching and digging the burrow. The paws are soft and sensitive with the thumb passively enhancing grip on slippery surfaces of the muddy banks. Its fur is rough and coarsely grizzled. There is intra-specific colour variation among the Bengal marsh mongooses. It is dimorphic in colour, having both the dark and light forms. This variation perhaps depends on the moisture content of the muddy habitat the animal used to forage for longer duration. However, if seen from a distance, the dorsal skin appears to be deeper in colour. In the dark form, individual contour hair is alternately banded with black and buff-yellow, while in the light form, it is a combination of blackish-brown and straw yellow. Ventrally, it is comparatively lighter. The fur is long and coarse on the back and short about the limbs. There is little hair around the anus and on the upper lip.

During the current survey in EKW, the adult males were found to be heavier and longer than the adult females (n= 6). Whereas the range of body mass of the trapped adult females varied between 500 g and 625 g (562.5 g), their head and body length was measured between 300 mm and 320 mm (310 mm) and the tail length appeared to be between 250 mm and 270 mm (260 mm). Therefore, the total mean-max length of the adult female is 550 mm and 590 mm respectively. The adult males weighed 625–900 g (737.5 g). Their head and body length was measured 320–360 mm (340 mm) and tail was 260–280 mm (270 mm) long (total length 580–640 mm (610 mm).

Differences between *H. auropunctatus* and *H. palustris*

The stuffed skin rolls of Bengal marsh mongoose, preserved by ZSI in Kolkata, were verified for noting the major morphological differences between *H. auropunctatus* and *H. palustris*. *H. palustris* was found to be relatively longer than the sympatric Small Indian mongoose (250–320 mm). Another distinguishing feature of the Bengal marsh mongoose is that its coat is comparatively rough and the pelage is more coarsely grizzled. The skull is close to *H. auropunctatus* in respect of size, but slightly longer and stouter. The small Indian mongoose has a dark brown muzzle and its cranium gradually narrows from the orbit, whereas the Bengal marsh mongoose's muzzle is black and the cranium narrows abruptly behind the orbit so that the postorbital region appears as a constriction between the frontals and cranium.

Ecology

The intricate network of EKW now lies between the IT hub on the east and busy Eastern Metropolitan Bypass on the west (22°25′–22°40′ N latitude, 88°22′–88°55′ E longitude). The paleo-environment of this swampland was an extension of the Sundarban mangroves. Being a part of mature delta of the river Hooghly, these wetlands were originally formed as a spillover basin of the river Bidyadhari, but have now been converted into a vast derelict swamp with the cessation of tidal influx. The hydrological setup of these wetlands is completely different from any other wetland in India. There is no catchment for these water bodies and perched aquifer is found to occur below these water bodies at depth of >400 feet.

These wetlands play a significant role in the ecological security of the region. The vast low-lying area acts as a sink. EKW is a primary producer of the carbon, bio oxygen, biota like the zooplankton, phytoplankton and so on. Their role in reducing the biochemical oxygen demand (BOD) and coliform bacteria is also significant. This expanse of EKW has numerous small and large water bodies (canals and bheries), which is among the rich tropical wetland systems of the world. The variety of habitats in EKW sustains rich biological diversity. EKW offer suitable habitats to the resident and migratory birds as well as typical mammals. The surroundings of the water bodies are used by the semi-aquatic mammals and grassland, the scrub and orchards are occupied by others. The Bengal marsh mongoose is completely a wetland dependent species living in association with the fishing cat (*Prionailurus viverrinus*) and smooth-coated otter (*Lutra perspicillata*), whereas other carnivores including the rodents prefer the grassland and scrubs around the wetlands.

The scientific name of the species is often given after the typical habitat it dwells in. In Latin, *palustris*, the common species name in biological nomenclature, means 'swampy' or 'marshy', i.e., shallow low-lying areas of ground, filled with the reeds and other aquatic plants. Ecologically, *palustris* is exclusively a wetland species and prefers to live near the water sources than the dry lands because of the high productivity of the moist wetlands. It depends solely on this habitat for food (nursery of the small fishes and crustaceans), shelter (the reed beds and aquatic plants growing in the fringe of the ponds, lakes and channels) and as a breeding place too.

In fact, the Bengal marsh mongoose occupies mostly the undisturbed and long-stretched bheries of EKW. These shallow wetlands are lined by the narrow mud banks where the grasses and a few stunted trees are grown. These mud-banks are the haunts of Bengal marsh mongoose. Its ecological niche is considered distinct from other mongoose species, particularly *H. auropunctatus*, at micro level. Being an island population in the fragmented habitats of the active wetland ecosystem and segregated from the sympatric mongoose species, the Bengal marsh mongoose has developed distinct ways of living through some morphological and behavioural adaptations. The core area of EKW, i.e., the viable habitat of Bengal marsh mongoose, is composed of the fishing ponds (bheries), patches of the marshy lands or swamps and small ponds.

Habits

The Bengal marsh mongoose is semi-aquatic, foraging much time near the water sources for an easy access to the protein-rich food resources. The marshy, shallow and large water bodies or wetland bogs, swampy edges, fully or partially filled with the species of reed, the emergent hydrophytes, such as *Phragmites karka* and *Typha angustifolia*, are the ecological niche of Bengal marsh mongoose. The dense reed-beds along with the thick growth of other aquatic vegetation provide an excellent habitat for a large variety of the aquatic and marsh birds and the feeding ground of many other mammals. In the fishing areas, the fishermen do not allow any plant other than *Eichhornia crassipes* and naturally developed planktons like algae (20 species) around. Since the water bodies are exclusively used for the purpose of pisciculture, compared to the vastness of the water bodies, only a few patches of the aquatic plants have been left out for the Bengal marsh mongooses. As a result, the number of this mongoose is comparatively low in most of the bheries. The Bengal marsh mongoose lives in the self-dug burrows, wherever there is reasonable cover close to the water bodies. They are not found in relatively drier areas. Their dens are mostly found along the banks of the bheries. Sometimes it occupies the deserted burrows of other mammals. Their permanent underground burrows are short-mouthed, whereas the underground tunnel is comparatively wider with downward slope from the mouth for quick entry. The territory of Bengal marsh mongoose is not extensive, rather small, fixed, exclusive and usually spaced along the waters around where they dwell in. During the survey, it appeared that the area surrounding the occupied burrows is devoid of such clay particles unlike those of the large bandicoot rats and the fresh scratch marks were often seen near the mouths of those burrows.

Earlier, the behaviour of Bengal marsh mongoose was poorly known but it is believed to be solitary and diurnal. It exhibits a routine exit-entry movement, i.e. coming out of the burrow few hours after sunrise, following a particular route for foraging in its territorial range and returning to the burrow just before sunset. It is more active in the morning between 00:06–08:00 hrs during the summer and 07:00–10:00 hrs during the winter. After a gap of six to seven hours (probably resting period), it resumes the foraging activity in the afternoon between 16:00–17:00 hrs during the summer and 15:00–17:00 hrs during the winter. By contrast, *H. auropunctatus* is active throughout the day. The Bengal marsh mongoose is found to be very shy, generally hiding in the reedbed or long grasses and aquatic vegetation. The solitary animals were mostly sighted, but, occasionally, they were gregarious. They were found in small groups (3–4), particularly during the afternoon foraging period. Such association appears to be helpful in hunting. Pairing during the breeding season was also observed. The Bengal marsh mongoose can bark, scream and purr, but these are exceptions. It is generally silent and sometimes mews, varying with low yelps or growling in rage with bristling of the hair, particularly of the tail. If enraged, when it is trapped and cannot escape, it makes an open-mouthed fierce protest or display. It uses the tail to balance either when standing upright (laid straight on the ground), or moving fast (kept almost straight in the air). The Bengal marsh mongoose usually communicates mood and status through the tail movements, which may be variable from individual to individual. A mongoose, before entering the trap, was found to curve (angular and U-shaped) the tapering lower part of the tail outward, possibly as a gesture of threat. When it entered the trap, the tail-tip was seen to form an inward loop in defense. In another case, a mongoose fearlessly entered the trap straightway keeping its tail in the normal position when the tail-tip was touching the ground. No upward loop of the tail was, however, observed during the survey. The Bengal marsh mongoose was sometimes encountered along a path while shifting from one patch of the habitat to another. While running at a high speed, the Bengal marsh mongoose can even turn without slowing down. When photographed, it passed by the camera a few times and moved so fast that most of the pictures were blurred, or slightly out of frame. However, it paused for a moment to look around and then quickly dashed into the cover next to the path and disappeared. This type of short immobility helped focus the animal. The Bengal marsh mongoose is alert, ferocious and courageous, particularly when the pups accompany the mother. If enraged, it can make its body swell by erecting the body hairs. It can squirt out a fetid fluid from its anal glands similar to the crab-eating mongoose *H. urva*, which acts as a means of self-defense or territorial marking. Chemicals in the anal glands may constitute individual signatures and indicate its reproductive condition, sex and/or dominance rank. The food of the Bengal marsh mongoose is highly diversified. It seems to be an opportunistic feeder. Though it is generally a predator on the vertebrate and invertebrate preys, it feeds largely on the aquatic preys. The Bengal marsh mongoose preys on the crustaceans, mollusks, crabs, amphibians, small reptiles, birds including their eggs, insects and larvae. It is a good rat-hunter (*Mus* and *Bandicota* spp.) and keeps the quick-growing population of this pest under control. Occasionally, it takes poisonous or non-poisonous snakes and frogs (*Rana* and *Bufo* spp). Above all, the Bengal marsh mongoose prefers a specialized diet. It feeds mainly on the small fish and snails, particularly the common aquatic snail *Pila globosa*. Its diet also includes the bivalve mollusks like *Lamellidens marginalis* and the gastropod mollusks like *Bellamya bengalensis* and *Lymnaea* spp. These mollusks are abundant at the water's edge. The broken shells of these mollusks were also seen scattered at the mouth of the mongoose burrows.

Besides, various aquatic Hemipteran bugs (*Gerris spinolae, Sphaerodema annulatum, Ranatra elongata, R. varips, Laccotrephes griseus, Diplonychus annulatus* and *D.* [= *Sphaerodema*] *molestum*) and Coleopteran beetles (*Canthydrus laetabilis, Cybister tripunctatus, Hydrocoptus subvittatus, Hypoporus bengalensis, Eretes sticticus, Hydrophilus olivaceus* and *Berosus indicus*), the dragonfly nymphs, terrestrial grasshoppers, crickets, centipedes, crabs, etc. are taken by the Bengal marsh mongooses. They were seen taking the giant water bugs and diving beetles, which possess hard elytra and wing membranes. The juveniles feed on various terrestrial insects and small mollusks (*Lymnaea* spp). Peculiarly, after the meal, the mongooses were sometimes seen using one of the long claws of the forefeet to clean the teeth like a toothpick. Secondarily, the Bengal marsh mongoose is known to eat a variety of fruits, tubers and berries as well as those smaller mammalian species, which it may overpower. The Bengal marsh mongooses adopt different techniques for hunting different types of prey. The method of hunting is direct, open headlong attack. It often scampers off with its prey for consuming it conveniently. The most interesting part of its hunting techniques is that it frequently goes to the water body, jumps on the dense bed of water hyacinth and searches for the aquatic food, astonishingly, without being submerged into the water. Only its feet become wet in this process. It also wades along the pond banks, poking its hands into the crevices and sifting through the mud to take the frogs and crustaceans out. It also fearlessly enters water to take the underwater prey. It generally rests on the floating vegetation and reed-bed or hides behind the emergent hydrophytes during the intervening period between the morning and afternoon foraging. Before coming out of the burrow in the morning, the Bengal marsh mongoose peeps out of the burrow-mouth and observes whether there is any imminent danger. If feels safe, it comes out of the nest and moves hither and thither in the territory. In case it finds the trap with live bait, it starts encircling the trap a number of times, then approaches close to the net wall and tries to threaten the bait with a hissing sound and hit it with its paw. Although it is a ground dweller, it can climb about with skill on the cage gratings. Failing to capture the bait, it opts for entering the cage through the opening. Immediately after its entry, the string, attached to the shutter, is touched and the cage is locked. Then the trapped animal shows aggression. However, after struggling for some time, it automatically calms down. When the gate is opened after taking its measurements, it usually moves backward to come out of the trap. Creeping (justifying the meaning of its genus) is particularly observed at the time of entering or leaving the trap. The Bengal marsh mongoose, particularly the juvenile, is very inquisitive and often approaches very near to the new objects like the camera without hesitation. While coming out for hunting during the daytime, the adult occasionally sits up on its haunches or stands up on its hind legs, when the forepaws remain hanging free, for surveillance of the environs by directional movements of its head. They can also make small leaps in the air, particularly when hunting. After seizing the snail or crab, the Bengal marsh mongoose throws the prey against a hard base from standing position in order to break the shell open. For consuming the bird's eggs, it uses its fore-paws to hold the egg, cracking a little hole at the small end to suck out the yolk. It was also seen to hunt a cattle egret (*Bubulcus ibis coromandra* Boddaert), a common resident wader or marsh bird, weighing about 500 g, by seizing its throat. It drew attention of the inquisitive bird by waving its tail. When fighting with a snake, it takes care that the snake does not envenomate it. If the prey is a non-poisonous one, the task becomes easier for the predator.

In general, the Bengal marsh mongoose attains sexual maturity within a year. The females are polyoestrus. Though they usually breed twice a year, they do not exhibit a regular breeding time. The courtship and mating usually starts after winter (January) and continues up to early summer (March), when pairs could be easily seen. The female in heat finds out the partner in the vicinity. But actual mating is short and takes place repeatedly. While foraging together, the male follows or chases the female. There are very little preliminaries or ceremonial, like sniffing or licking the genital organs or vocalizations. The male suddenly jumps on the female's back and tries to penetrate her from behind. If she accepts, she lifts her tail, seen to bend to the right as much as possible and also splays the hind legs a little. After mounting, the male's forepaws rest on the female's sides or clasp her belly for support while thrusting, but griping the female's nape in jaws was not, however, seen. After a short while, they break and move around till the next copulation begins. For mating, they prefer the bushy than open areas and do not like to be disturbed in any way. Their gestation period is around 45 days and the litter size varies from two to three or rarely four. It seems that most of the birth takes place in the burrow just before the monsoon between April and June. A lactating female was collected in June. The pups were also seen during the early June. The peak birth season correlates with abundance of the small insects and mollusks in the wetlands during the monsoon, which are the normal diet of the young. Sometimes the pregnant female, before giving birth, enlarges her own burrow in order to accommodate the litter comfortably. However, during the survey, a family group of four mongooses (caretaker mother and three youngsters), peeping and coming out of a burrow, was also sighted on 29th January 2011, in the late afternoon, which indicates that the breeding may also take place in the early winter. While born, the pups are helpless and cannot protect themselves from the predators. Hence, leaving them alone in the burrow, while going out for foraging, is rather risky for the mother. Since the male does not take part in the care of the young in any way, the mother has to take the sole responsibility of rearing the neonates. Initially, she nourishes the young on milk produced by her mammary glands. When the young are able to come out of the burrows with the mother, they are acquainted with the habitat and learn the techniques of hunting small preys. The mother always protects her cubs ferociously. Even she does not hesitate to attack a large monitor lizard (*Varanus flavescens*), which often try to attack and steal her cubs. The cubs grow rapidly within 2-3 months and by August-September they are able to come out of the burrow for hunting with the mother. Just after the monsoon, in September-October, 2-3 cubs were often seen playing with each other in the evening between 16:00-17:00 hrs. They nibbled at each other's tail and ran helter-skelter near the mouth of the burrow. When they start living of their own, the mother breeds again. The Bengal marsh mongoose usually lives up to ten years. However, they often confront premature death due to anthropogenic threats or natural diseases.

Conservation

In general, the Bengal marsh mongoose acts as an indicator of the habitat it lives in and plays a crucial role in the fragile ecosystem as it directly maintains a balanced predator-prey relationship and indirectly acts as an ecosystem regulator of the plant composition, soil type and drainage too. But, since it lives in areas outside the protected area network and the habitats are usually not monitored or guarded against by the nodal infrastructure, it falls prey to the poachers and their habitat goes on being destroyed or encroached upon without much administrative interventions.

Threats

EKW is a multiple-use zone. The land-use pattern is agriculture (vegetable cultivation: 4.67 km² and paddy cultivation: 48.88 km²) and aquaculture (47.79 km²) in a series of about 300-odd fishponds, connected by the major and secondary canals. The ownership pattern of the sewage-fed fisheries is private (93.14%), cooperative (0.86%) and government (6%). The area is also infested with the wholesale markets, roads and 43 villages. EKW supports a human population of about 1.5 lakh. It includes a garbage dump, known as Dhapa Square Mile, operating since mid-1800s. Though classified as wetlands, only 23.66 km² is left out of such direct human activities. KMDA is responsible for land use and development control over the entire area. The conservation threats of Bengal marsh mongoose are catastrophic events, edaphic factors, human interferences, loss of habitat and fragmentation. During early 1930s, the halophytic vegetation of the Sundarbans largely dominated the cluster of EKW. Later on, a gradual change has taken place in water quality from polyhaline condition to almost freshwater with a change in the profile of flora and fauna of the region. EKW are vanishing at an alarming rate of 1% per year and being severely threatened by the urban encroachment and invasion of real estate dealers by unscrupulously filling up of the wetlands. The most important and massive land use change was initiated by the government in 1955 for acquisition of 173.70 acres of land for expansion of reclamation on the north of Salt Lake (Bidhan Nagar). Large-scale reclamation of this wetland started since 1960s. Out of 20,000 acres of wetlands recorded in 1945, less than 10,000 acres are left as wastewater fish ponds. By 1968, 36 km^2 of the southern Salt Lakes had been reclaimed leading to a rapid decline in the population of Bengal marsh mongoose to such an extent that its type locality has become a city, whereas earlier it was a swamp. Another satellite township (Rajarhat megacity) is growing over 30.75 km^2, destroying the same type of landscape.

Moreover, changes in the land use pattern over a period of time have led to conversion of some of the largest fish farms from pisciculture to paddy cultivation and changes in hydrological regimes are affecting the ecological balance and functions. Another problem in EKW is arsenic in the groundwater. The percentage of arsenic considered safe for consumption is 10 mg/1. But, in the surrounding area of EKW like Bhangar, Kharibari, Rajarhat, Bishnupur I and II, Gangra, Mahisbathan II, the levels of arsenic has been reported up to 15 mg/1. Additionally, large numbers of industries dump effluents without treatment into the sewers that empty into the city outfall channels flowing eastwards. This has caused a substantial amount of metal deposition in the canal sludge and rendered the waste water incapable of ensuring the edible quality of the fish and vegetables grown in the wetlands and the piscivorous Bengal marsh mongooses are also victimised. Important aquatic sports and recreational centers have recently come up along the edge of this area, causing immense disturbance and sound pollution. The Bengal marsh mongooses are very sensitive, extremely cautious and run away very fast at slightest disturbance. They usually avoid these crowded zones. The ecologically important mongooses are also illegally hunted for their bristle. The animals are trapped and then bashed, stoned or stunned to death. The hair is then plucked by hand (average 20 g/animal or 1 kg/50 animals), packed in the gunny bags and sent to the brush industry for making the paint, shaving and hair brushes. Because the mongoose hairs are pliant and the tip tapering to a fine point, these are particularly preferred for oil, tempera and acrylic painting.

Status

The Bengal marsh and other mongooses were originally included under Schedule IV of the Indian Wildlife (Protection) Act, 1972. Since they were being poached indiscriminately, particularly for their prized hairs, threatening their survival, the government of India upgraded legal status of the mongooses and placed them under Schedule II vide a notification (No. S.O.1085 (E), dated 30th September 2002) as per the Amendment Act, 2002. This has imposed complete ban on the use of any live mongoose and its parts or derivatives for any purpose. The penalty for hunting, trading and possessing this animal or any article derived from it, is imprisonment of 3 to 7 years and/or a fine up to rupees twenty five thousands. Though its earlier Red List status was Lower Risk/least concern (LR/lc), now it is 'Endangered' as per the IUCN criteria (B1+2abcd) and considered to be facing a very high risk of extinction in the wild. All mongooses are now included in Appendix III of the CITES. The Bengal marsh mongoose is now considered a rare species in EKW and identified as the endangered mammal (Rank 239) on the edge (ED' 11432, Edge 4.5997) and proposed for conservation priorities based on the threats and phylogeny. Although conservation breeding for this endemic species has been recommended on the ground that there is no captive population of the Bengal marsh mongoose in India or any other country, no follow-up action has yet been taken in this regard.

Among the civets, the small Indian civet (*Viverricula indica*) is common in the reclaimed lands along with the palm civet (*Paradoxurus hermaphroditus*), which is seen occasionally.

Prey-base

Rhesus macaque (*Macaca mulatta*), the only primate in the Sundarbans, is quite abundant. They are gregarious and found in scattered groups of 30-40 individuals. At Sudhanyakhali, this monkey is abundant seasonally and its numbers generally decline after April, when most of them migrate to the interior forests, and increase from October onwards with the start of the tourist season. Their mating season is usually between February and July. The isolated island monkeys have been changed during the course of their evolution here. They are smaller about 500 mm in length and 4 kg in weight, whereas the mainlanders are larger by at least 100 mm and double-weighed. Other islander animals also follow these adaptation rules of Sundarbans, the tidal cycle- high (2-3 hours) and low (8-9 hours), *bhara kotal* (peak period) and *mara kotal* (lean period), interlinked food crisis and alarming climatic change.

Though genetically the arboreal monkeys should have consumed only on the fruits-tender leaves-buds-flowers-grass-seeds-roots, which are available in abundance seasonally (mostly during May-October), they have also learnt to hunt proficiently and survived on the nutritious secondary aquatic diets available throughout the year. When the lowlands are submerged during the high tide or during night and storms, they are arboreal, but when the flood water recedes, the troops of rhesus come down to the banks occasionally in lure of floating ripe fruits, tender leaves, flowers, etc. but preferably the aquatic food, rich in protein-vitamin (e.g. A and D, phosphorous, magnesium, selenium, iodine, etc). For hunting the small fish in the saline water they have invented a suitable technique like the fishing cats. But, while foraging on the banks, they are always alert against the larking tigers or those in ambush. At the slightest instance of the predator's presence, the alpha male makes a shrill alarm call, thumps (red alert for other group members scattered around) and climbs the nearest tree at lightning speed followed by the females, infants and sub-adults.

Such hunting activities could often be observed in the morning (6-10 am) and afternoon (3-7 pm) at Sajnekhali (especially between the rivers Pitchkhali and Gomdi). Their typical gait to avoid the piercing pneumetaphores, the technique of fishing, standing at low water, and making them into pieces to consume, by using the limbs. By virtue of social hierarchy, the alpha male first shows the way and the other family members wait patiently on the bank for completion of his consumption, only after which they start fishing. A few years back such a fishing incident was photographed for the first time on the bank of Gomdi involving a troop of seven. But the species of small fish hunted and consumed by the alpha for ten minutes could not be identified from a distance.

Besides, they forage for the crabs because crustaceans (fiddler and mud crabs) account for the largest portion of the animal biomass. A troop was also observed to feed on the hermit crabs *Clibanarius longitarsus* (Diogenidae) exposed by the low tide on the same riverbank.

To quench thirst, the monkeys prefer the sweet water accumulated in the ponds dug near the watch towers or in-forest ditches or even drink the brackish water. They were also seen to lick the dewdrops. During the rainy season, their water requirements are met primarily by consumption of succulent plant food.

The monkeys were seen to locally migrate island to island by crossing the creeks in search of new niche rich in food sources. During the tourist season, i.e. winter, up to February, large groups of rhesus congregate around the popular watch towers for residual human food offered by the tourists. This is extremely injurious to the resident primates, detrimental to their conservation and must be discarded.

During the Phase I *khal* survey of 2018, the FD conducted boat transects for prey base density estimation. However, as the perpendicular distances were estimated ocularly, the previous cycle's effective strip width was used in tandem with 2018 cycle's encounter rates for estimating *Axis axis* and *Sus scrofa* densities; it is assumed that the detectability has remained constant (a reasonable assumption).

Axis deer was captured throughout Sundarban landscape with highest concentration of photo-captures in Sajnekhali WLS. Two of the capture hotspots in this sanctuary coincide with the location (Dobanki camp) where chital had been released in the late 2000s.

Wild pig was distributed throughout the Sundarban landscape with highest concentration of photo-captures in the BR. Prevalence of more capture hotspots in the BR concurs with higher sightings of wild pig during boat transects in this area.

Rhesus macaque was captured more or less throughout Sundarban landscape with highest concentration of photo captures in Sajnekhali beat of Sajnekhali WLS. The capture hotspot in this beat coincides with the location of the beat and range offices where there is maximum instantaneous human presence inside the tiger reserve on any given day as these offices issue tourist and fishing permits.

Monitor lizard distribution was primarily in the tiger reserve with very few photo-captures in SBR. The capture hotspots in the tiger reserve were in the Sajnekhali WLS, the buffer zone of Basirhat range and the core area of the NP viz. West range.

Camera trapped species

Prionailurus viverrinus was captured throughout the landscape. However, there were fewer photo-captures in the East range of the NP in the core area.

Prionailurus bengalensis was captured throughout the landscape and its abundance is indicated by the numerous capture hotspots both in the TR as well as SBR.

Felis chaus was captured only at the fringe areas of the mangrove forests, with its highest photo captures in the buffer zone of Basirhat range.

Canis aureus was captured for the first time in Sundarban and seems to be a recent colonizer restricted to the mangrove forest edges of Sajnekhali WLS and islands of SBR adjacent to the human habited islands.

Sundarban is home to three species of otter, viz. *Lutrogale perspicillata*, *Lutra lutra* and *Aonyx cinereus*. However, it was not possible to identify the individual species from the camera trap photographs. Hence, varying intensity of photo-captures of all three species was combined together. The capture hotspots were primarily in the core zone of NP beside wide channels.

Paradoxurus sp. is rare and was captured only at the edges of the buffer zone of Basirhat range, adjacent to the human habited islands.

Crocodylus porosus was one of the rarest species to be photo-captured, however, camera traps are not the most ideal equipment to evaluate this semi-aquatic reptile's distribution.

Rare colour aberration

A number of occurrences of colour aberration of the wildlife is recorded in the Sundarbans. These are-

1. Hypomelanism (mutations affecting melanin biosynthesis, pigment granule trafficking or membrane sorting; phenotype-Beige, brown, golden, yellowish or reddish fur; skin and eyes always normally coloured): A hypomelanistic tiger, the colour of which was uniformly brown and the stripes were not visible, was recorded from the Bangladesh Sundarban (22.222° North, 88.839° East), possibly Khulna or Bakerganj (Mahabal *et al.*, 2019). In the early 20th century (exact date unknown), this tiger was seen without visible black stripes (Praeter, 1937). It was learnt that the tiger used to hide in the open sandy area. The incident is known on October 16, 1936 from a letter published in the "Times" newspaper, which was written by W.H. Carter:

> "I was much interested in Captain Guy Dollman's letter on black tigers in The Times of October 14, having been resident in the neighborhood mentioned by him for years. In one of the official district Gazetteers of Bengal (Khulna or Backerganj) a local variety of tiger was mentioned, which had lost its stripes as camouflage in the open sandy tracts of Sundarbans. The uniform color scheme adopted was however, brown and not black, but perhaps his cousin in the hinterland found black more suited to his background. The author of the Gazetteer in question is, I believe, dead."

2. Leucism or partial albinism (total lack of both melanins in all of the hair follicles and skin due to the heritable absence of pigment cells caused by the failure of melanocytes to migrate to the skin and hair follicles; phenotype- all-white or whitish hair, pale skin; eyes and/or body extremities normally coloured) :

(i) A leucistic Asian small-clawed otter (*Aonyx cinereus*) was spotted in the Bangladesh Sundarbans in 2018. In the Indian Sundarban also a similar case was recorded when a team of five (led by Apoorva Chakraborty of Prakriti Sangsad) surveying in the 'Core' area of STR at the curved mouth of Bakultali Canal in Bagmara compartment on 29th December 2017 during the 'Asian Waterbird Census' or Bird Census of 'Wetland International' in collaboration with the Sundarban Tiger Project Authority) observed two small-clawed otter after emerging from the water and descending down the slope to enter into the forest, one of which was normal brown in colour, but the other was yellowish-white or 'leucistic'. Association of a normal and aberrant otter is really exciting.

(ii) A leucistic Collared kingfisher is also seen a number of times in STR (Adhikary and Mondal, 2019). On 20th September 2018, a colour-aberrant kingfisher was spotted near Dobanki Camp (22.01°N, 88.76°E). The bird was identified as a Collared Kingfisher (*Todirhamphus chloris*) based on its size and the colour of its upper and lower mandibles, rather than the similar Black-capped Kingfisher (*Halcyon pileata*). Subsequently, the individual was sighted several times in the area by many birders and on 29th September, it was observed at mid-day. It was calling, and its call matched that of a Collared Kingfisher- a harsh 'kee-kee-keekee-kee'. It looked exactly like a Collared Kingfisher except for its white plumage. The bird was observed for hours. It always tried to fly near a normal coloured Collared Kingfisher. It was also observed that though the feathers of the individual were all white, its eyes, feet and bill were normal in colouration; unlike an albino, which would have red eyes, and pink feet and bill. As the colour of both the mandibles, the feet, and the eyes was not affected by the pigmentation loss and remained almost true to that of a normal Collared Kingfisher, it was proved to be none other than a collared kingfisher. The bird was observed a number of times subsequently.

(iii) A leucistic saltwater crocodile was observed on 9th August, 2019 in the Indian Sundarbans. Earlier during the monsoon season, this young leucistic crocodile with some black spots and regular coloured eyes was basking on the mudflats of the river at 11.12 in the morning on 25th July, 2019. It was also photographed. In 2015, a three-year-old crocodile was released into the creek from the Bhagabatpur Crocodile Project, which grew up in the past few years. At birth, they were yellowish, and the skin was black-dotted. Thereafter, the yellow tint becomes more lighter and becomes papery white. This leucistic crocodile used to live in harmony with the rest of the crocodiles. Since its whitish colour is easily visible, its life is always at risk.

(iv) A leucistic specimen of a Long whiskers catfish *Mystus gulio* (Hamilton 1822) (Siluriformes: Bagridae) is also recorded from

Pakhiralay, Indian Sundarbans, on 21st February, 2014. The non-aberrant colouration of *M. gulio* consists of dark silvery-grey colour on the dorsum and upper portion of the flanks. It gradually turns white on the lower sides going into the ventral region. The fins have black markings on them. The present specimen collected from a local market displayed a leucistic phenotype. The body and the head possess white colour while the fins are pinkish-red in colouration. The occipital process and snout are pink. Eyes retain the original colour, black. There is an absence of melanophores in any part of the body. Leucistic animals are rare in the wild because of their easy detection by the predators due to their conspicuous colouration. Aside from that, researchers have also observed behavioural impairment with conspecifics in aberrant coloured animals. The local fisherman, who was selling the fish, said that he caught the leucistic individual in his 'Ghoni', a fish trap, from a brackish water inlet of the Bidyadhari River in Pakhiralay. It is not a reared fish. He also mentioned that differently pigmented individuals usually don't sell in the markets because buyers think that the colour variation is a result of some disease in the animal. So, fishers generally leave them out, if they find them while sorting their catch.

(v)   The Indian flapshell turtle [*Lissemys punctata* (Lacépède, 1788)]: Three yellow (leucistic) mole tortoises have been rescued from the Sundarbans in the past year. The colour of the rescued turtles was different from the normal colour of this turtles, i.e. bright yellow shell and the spots on it are not transparent. Yellow variants of the Indian flapshell turtle, which typically is brown with yellow spots and a creamy white underside, is often found in other regions.

3.   Melanism (abnormal deposition of melanin (not necessarily an increase of pigment) in the skin and/or hair follicles; phenotype-increase of black and/or reddish-brown or altered pattern): Two melanistic (black-coloured) leopard cat were also photo-captured by WWF at Bonnie camp (21.866^0N, 88.891^0E) in Ajmalmari Reserve Forest during a camera trapping exercise by WWF-India to estimate the population of tigers. It is the first photographic evidence of its presence in the Indian Sundarbans.

Symbiotic relationship

An interesting association has been developed between monkey and deer, the former hopping from tree to tree and dropping fruits, twigs and leaves which are eaten by *Axis* on the ground and the deer also get advance information about the movement of the tiger from the monkey's call. The habitat for the *Sus scrofa* is provided by the tangled mass of *Ceriops* sp. whose breaks are penetrable to man. Sweet water ponds provided at various places within forests including the watchtower points attract innumerable deer and pigs along with a few tigers. The monkey however, gets their sweet water mostly from the dewdrops from the leaves of *S. apetala* trees.

Rodents

Among the rodents, *Bandicota indica, B. bengalensis* and *Mus booduga* are quite abundant. The white-bellied tree rats (*Rattus rattus arboreus* Horsfield, 1851), nesting in bunches of leaves and twigs, were also observed. Holes and nests of the five-striped palm squirrel (*Funambulus pennantii*) in the tree trunks and branches were also discovered.

Aquatic mammals

The rivers and near-shore waters are the abode of five aquatic mammals: the dolphins and porpoises. Among them, the Gangetic dolphin (*Platanista gangetica*) and Irrawaddy dolphin (*Orcaella brevirostris*) are common along the upper part of the rivers Matla, Bidyadhari, the confluence of Raimangal and Jhilla at Bagna, Amlamati and at Sudhanyakhali. Rarely, the black finless porpoise (*Neomeris phocaenoides*) is met with. An Indo-Pacific hump-backed dolphin was sighted and photographed a few years back near Sajnekhali. There are no recent records of sighting any Pantropical spotted dolphin in the Sundarban. Dolphin sign and sighting were in general low with higher presence recorded in the Bangladesh side, especially within the declared PAs for the species.

Micro-organisms and invertebrates

Micro-organisms and invertebrates are the most numerous groups of species in the mangrove ecosystem. Invertebrates are a highly diverse group, ranging from tiny insects to giant squids and account for more than 90 percent of all animals. Micro-organisms comprise the vast and diverse range of organisms that are too small to be seen by the human eye, such as bacteria- a type, species or strain of bacteria. Both micro-organisms and invertebrates play major roles as biological control agents, and are indispensable in nutrient cycling, in the decomposition and in the recycling of organic matter in soils. Both groups are vital to food and agriculture. Aquatic invertebrates are a major source of food for different animals such as monkeys, birds, snakes, fishes and even for humans such as oysters and mussels.

Functional groups

The functional groups of invertebrates and/or micro-organisms are as follows:

(a) Pollinators, including honey bees, and (b) Biological control agents and bio-stimulants.
(b) Soil microorganisms and invertebrates, with emphasis on bioremediation and nutrient cycling organisms and microorganisms of relevance to ruminant digestion.
(c) Edible fungi and invertebrates used as dietary components of food/feed and microorganisms used in food processing and agro-industrial processes.

Trophic groups

There are four recognised groups of invertebrates in the mangroves of Sundarbans.

(1) Direct grazers such as insects and mangrove tree crabs;
(2) Filter feeders such as sessile invertebrates, which feed on phytoplankton and detritus;

(3) Deposit feeders such as mobile invertebrates that consume detritus, algae, and small organisms from the sediment surface; and

(4) Carnivores such as highly mobile invertebrates that feed upon all other groups.

Ecological role

Micro-organism and invertebrate consortium is an extremely important component of the rich biota of Sundarban. They help maintain mangrove ecosystem functions through activities such as the cycling of nutrients, filtration and biochemical processes, breaking down of pollutants and flushing toxic substances and ultimately modifying the oxidation status of the surrounding sediments. They are also an important source of food for many higher animals and also constitute a source of food for the people living there. Invertebrates are also vital to the fertilisation of a vast number of plants. In short, the life forms in the Sundarban mangroves depend on the invertebrates for their sustenance. Despite the tremendous importance of thousands of invertebrate species in running the mangrove ecosystem, few vertebrates are given priority in the conservation efforts including scientific research. They are most neglected by the public, media and policy makers.

Habitat users

Animals are not completely terrestrial or aquatic but lie along a continuum. Mangrove fauna can be divided into different types of habitat users, such as-

(1) **Aquatic animals** include crustaceans, insects, snails, mussels, fish including sharks, amphibians and reptiles. In the Sundarbans, tetrapods, for example, the small cetaceans (Gangetic river dolphin and some seasonal visitors, e.g. Irrawaddy dolphin and finless porpoise) are fully aquatic and therefore are completely tethered to a life in the water. No bird species is fully aquatic, as all must lay and incubate their amniotic eggs, as well as begin raising their young, on land. Similarly among the marine reptiles, sea turtles are almost fully aquatic, but must come ashore to lay eggs, for example, the Olive Ridleys nest in the remote and undisturbed beaches in the Indian Sundarbans during winter months. The

river terrapin *Batagur baska* also crawls out of the waters to lay eggs on land. The females are very choosy about their nesting sites. Once a year, during the full moon at the end of winter, mostly around Holi (the festival of colours), they swim hundreds of kilometers to find sandy beaches along sea-facing islands, to nest. Except for these rare occasions, it is hard to spot them on land. Unlike other freshwater species, the saltwater crocodile thrives along brackish mangrove channels (though it is also known to inhabit freshwater habitats) and can swim far out to sea. They are also semi-aquatic for the same reason. They nest on the elevated (beyond the limit of tidal inundation) banks of rivers (preferably upper reaches) and lay eggs in the mound nests usually constructed out of mud and vegetation during the wet season. They are also capable of travelling long distances between feeding and breeding grounds like the turtles mentioned above.

Lack of distributional, taxonomic and ecological information on the aquatic animals represents a major data gap for aquatic species in the Sundarbans and the database accumulated during this study will provide management inputs and build conservation initiatives.

Some of the aquatic fauna dwell in ground- or surface-water (including standing water lakes, ponds, swamps, bogs, and some wetland areas) habitats. Flowing water (e.g. rivers and streams) is also preferred by some aquatic communities. Headwater tributaries of these rivers include both perennial and intermittent streams or canals. These two divisions are, obviously, generalisations of the immense diversity of aquatic habitats that exist in the mangrove ecosystem and grade from one to another. Aquatic systems are interconnected and integrated between an aquifer to a lake or a river.

Aquatic organisms have adapted themselves to a life in the water by various means. They take in dissolved oxygen that is in the water or come up to the surface of the water to take in air. Fish and other aquatic creatures like prawns, crabs, mussel and tadpoles have gills to help them breathe underwater. Some insect nymphs have gills which enable them to live at the bottom of waterbodies. Some aquatic organisms do not stay under water all the time. They often come out of the water and move around on land for a short period of time. In addition to gills for them to breathe underwater, they also have gill chambers as in the crabs or mudskipper. In addition, some entomofauna like the young of the mosquito (the larva/wriggler and the pupa), the water stick insect and the water scorpion take in oxygen using a breathing tube (also called a siphon) found at the end of their abdomens. The great diving beetle and the water spiders get air to breathe from the surface with the help of a 'mini oxygen tank' called 'air bubbles' fitted on their body so that it can live underwater. Some aquatic worms such as tubifex worms and flatworms, amphibians such as frogs and toads use their thin and soft skin, kept wet all the time, to take in oxygen that is dissolved in water. Many aquatic animals with lungs have special adaptations in their bodies, capable of tolerating higher concentrations of carbon dioxide in their blood than most other breathing animals. This enables them to use oxygen more efficiently and stay underwater longer. Their nostrils are positioned on the tip of their snouts. When they want to breathe, they push their nostrils above the surface of water. When they are underwater, they shut their nostrils to keep water out. Some aquatic mammals like whales, dolphins and porpoises, have special nostrils called blowholes on top of their heads to breathe even when most of the body is underwater. When they need to breathe, they come up to the surface of the water, expel the carbon dioxide and take in a breath of fresh air.

They are important members of the food web. Some of them process leaves and other organic matter and they provide food for other animals, including humans. Insects also contribute tremendously to the diversity of aquatic animals. Different stages of these insects are found in all types of aquatic habitats. Some of these insects are predators. The aquatic insects often shred leaves and other organic matter, thereby serving as important food sources for many fish. Specific fish hosts are needed for the mussel to complete its larval stage and disperse. They are also useful indicators of water quality. For example, the mussels are filter feeders and highly susceptible to poor water quality. They are also major food sources for many fish, reptiles, and some terrestrial animals.

(2) **Semi-aquatic or semi-terrestrial animals**: In biology, semiaquatic are those types of animals that spend part of their life cycle or a significant fraction of their time in water as part of their normal behaviour and/or obtain a significant proportion of their food from the aquatic habitat they use. In the Indian Sundarbans, the semiaquatic animals include vertebrates, some of which have been described as aquatic animals above.

The tigers living in the islands of Sundarbans are capable of leading an almost amphibious life unlike the mainland tigers. Besides, its prey species like the Axis deer and wild boar are also semi-aquatic. The rhesus macaques were also found to migrate in search of food. Hence, the apex predator has to change their territory following its prey species by swimming from one island to another. Aquatic animals like the crabs and fishes are also eaten by Sundarbans tiger. Hence this top predator occupies the pinnacle of both terrestrial as well as aquatic food web.

The fishing cat is also semi-terrestrial. They live in wetlands and meadows near rivers, lakes, swamps and other sources of water. Unlike most wild cats, these animals are excellent swimmers, divers and fishers. They mainly eat fish instead of small mammals. The preferred hunting ground of the otters is the riverine Sundarbans, where they forage in search of fish and crabs.

Semi-aquatic birds in the Sundarbans are plovers, red-wattled lapwing, avocet, stint, curlew, sandpiper, common greenshank, gulls, terns, etc.

Other semiaquatic animals living in the Sundarbans are amphibious fish and also several types of normally fully aquatic fish that spawn in the intertidal zone, semiterrestrial echinoderms of the intertidal zone such as sea urchin, the starfish (Asteroidea), springtails, semiterrestrial malacostracan crustaceans, some amphipods and isopods. Horseshoe crabs are mostly aquatic, but spawn in the intertidal zone and the juveniles live in tidal flats.

(3) **Terrestrial animals:** The term terrestrial is typically applied for species that live primarily on the ground, in contrast to arboreal species, which live primarily in trees. Terrestrial animals are those species that live predominantly or entirely on land (e.g., cats). Some terrestrial birds are doves, kingfishers, woodpeckers, pigeons, flycatchers oriental magpie robins, red jungle fowls, owls, rose-winged parakeets, etc. Arthropods, including ants, flies, crickets, grasshoppers and spiders, are the most abundant terrestrial animals by species count. Vertebrates, arthropods, and mollusks are representatives of most successful groups of terrestrial animals.

Important free living terrestrial animals in the Sundarbans are-

(i) Rotifers (wheel animals) live in transient terrestrial water and go dormant during desiccation;

(ii) Nematodes (roundworms) and other worms;

(iii) Annelids (Clitellates) require moist habitats, highly diverse and derived from their marine relatives;

(iv) Arthropods (fully terrestrial members: Insects, Arachnids, Myriapods, Woodlice, Sandhoppers, and Terrestrial crabs, semi-terrestrial members include Water Fleas, Copepods, and Seed Shrimp);

(v) Mollusks (Gastropods: land snails and slugs); and

(vi) Chordates (Tetrapods).

Roundworms, rotifers, some smaller species of arthropods and annelids are microscopic animals that require a film of water to live in and are, therefore, considered semi-terrestrial. Some worms and annelids depend on more or less moist habitats. The arthropods, mollusks and chordates have adapted to and mostly depend on the terrestrial resources for sustenance and have no marked aquatic phase in their life cycles.

(4) Arboreal animals: Animals that live in the trees are called 'arboreal' and they have some amazing adaptations to make the most of their leafy surroundings at every level. Trees are the nesting, roosting and feeding sites for the animals. In the most diverse terrestrial plant communities the plant-animal interactions involve frugivory, nectarivory and insectivory. Few mangrove plant species tend to serve as "foraging hubs or connectors" when they are visited by many (if not most) direct consumers in a community. Tree species bearing the qualities of foraging hubs or connectors are predicted to have the greatest positive effect on biodiversity. The consumer preference is based on phenology and palatability, long fruiting season and nutritional value of fruits. For example, flowering and fruiting of most of the mangrove species is reported during May-October. The higher stands of *S. apetala* are most important from the ecological point of view because of the species' longer fruiting period. It produces more fruits per mature tree and more seeds per fruit in autumn than spring.

The fruits of *Ficus* spp. are preferred by the animals for figs are rich in minerals including potassium, calcium, magnesium, iron and copper and are a good source of antioxidant vitamins A and K. *Sonneratia caseolaris*, commonly known as mangrove apple, is the food source of moth and other insects. Insectivorous birds can prefer a tree species through the association between higher leaf nitrogen content and preferred arthropod prey.

The rhesus monkey is at home in the trees. Basically, they are reported to be frugivores and folivores. They are dependent on both mangroves and mangrove associates. Most of the mangrove leaves are suitable fodder for them. The succulent leathery leaves of Keora (*Sonneratia*), Baen (*Avicennia*), Genwa (*Excoecaria*), Garjan (*Rhizophora*), Kankra (*Bruguiera*), *Ficus* (Moraceae), etc., are taken. They were seen to pluck the tender leaves from top branches. Different types of creepers are also consumed. Feeding on the flowers and flowerbuds, petioles, and other parts of a variety of trees, shrubs and climbers was observed. They prefer good quality ripe fruits. Fruits of Sundari (*Heritiera*), Hental (*Phoenix*) and Genwa (*Excoecaria*) are also favoured by the macaques. During August–October, when large amounts of fruits are available in the Sundarbans. The monkeys were seen to move from one Keora grove to another for feeding.

When the habitat is partially submerged, the monkeys are mostly confined to their arboreal abode, rarely descending from the trees. The high trees are their nighttime sleeping sites, but during storms and rains, they were seen coming down to the dense canopy of lower branches for protection.

The civets are both terrestrial and arboreal, but mostly tree dwellers. During the day they usually rest in trees. Although the civets are omnivores, they are largely frugivores; they eat berries and pulpy fruits as a major food resource and thus help maintain forest ecosystems via seed dispersal.

Many species of birds build their nests in the mangrove trees. Herons, egrets, cormorants and darters enjoy roosting in colonies on the tall trees. Several species of birds use trunk, branches and aerial roots of mangrove as observation posts for viewing on fishes, molluscs, crustaceans and aquatic insects. Shaded by the canopy, it is darker and more humid in the understorey, providing an ideal niche for many amphibians that thrive where the air is moist. The tree dwellers must have the ability to survive against the onslaught of the winds, storms, heavy rain and exposure to scorching sun. The predatory birds (raptors) scan the canopy from a perch, looking and listening before swooping down to seize prey.

Ecological role

These animal communities utilize mangrove areas for their daily activities, such as foraging, breeding and loafing. These animals play a significant role in the management of mangrove forests and in balancing nature in and around the mangrove areas.

Unlike the plants, vertebrates actively choose preferable biotopes and do not die off upon transition between the stages of succession. The role of vertebrates in the functioning of ecosystems is determined by their contribution to matter and energy turnover and effect on the structure of ecosystems of different ranks. The transfer of matter and energy in food chains accelerates their turnover and the distribution of excrements and corpses within the ranges of animal activity and migrations modulates the pattern of matter and energy fluxes.

The effect of vertebrates on phytocoenosis of the forest zone is less apparent but no less significant. Micro-successions in forest areas (tens of cm^2 to over 1,000 m^2) dug over by wild boars (*Sus scrofa*) have shown that species diversity of herbaceous plants in freshly overgrown wild boar diggings is three times higher than in undisturbed areas. Annuals dominate, since their seeds readily germinate in bare soil. Then follow vegetatively mobile plants, which expand to diggings from undisturbed areas. Since the wild boars continue to dig different areas in search of their food, such a "rotation system of land use" by wild boars maintains the diversity of herbaceous vegetation under the forest canopy. Mammals and birds disperse the seeds of higher plants either in their digestive tract or on the body surface. The smaller the animal, the shorter its dispersal distance.

The feeding of vertebrates at each trophic level of biocenosis has an effect on the lower trophic level. This effect has two aspects: the removal and digestion of biomass as such and selectivity in its removal and digestion. A predator is not always capable of taking every potentially accessible prey. The success rate in hunting Axis deer by the tiger often decreased from 50 to 20% as the latter learned to escape attacks by the predator. On the other hand, the passage of biomass along food chains is accompanied by energy losses upon each transition to a higher trophic level, which accelerates energy turnover in ecosystems. The turnover of matter is accelerated as well. In destructive chains, food remains and excreta of predators are decomposed more rapidly than the carcasses on the mangrove floor.

In terms of the role of vertebrates in ecosystems, feeding selectivity accounts for diversification of matter and energy fluxes in addition to circulation of parasites from helminths and mites to bacteria and viruses. All wetland and water birds and, to a lesser extent, mammals, amphibians and reptiles transfer the production of aquatic ecosystems to land. The production of freshwater ecosystems is utilized on land, where it is also decomposed by the microorganisms. The released nutrients not only enrich the soils of terrestrial biogeocenosis but also are partially washed away by rainfall and return to inland water bodies for onward transfer to the estuaries flowing into BoB.

Estuarine Sundarban wetlands with their abrupt selective gradients and relatively simple biotic assemblages are highly productive. Abundant plant and animal food resources are available through both the terrestrial vegetation and the marine food chains associated with the tidal channels. On the other hand, the fauna and flora associated with salt and brackish marshes are *depauperate* because the ecosystem is, in some ways, inhospitable to the vertebrates and invertebrates, who regularly face a number of severe environmental threats here, such as fragmentation, ditching and impoundment, reduction in area, pollution, establishment of invasive species, sea-level rise, salinity and high frequency of storm surges. Hence, it is not surprising that many populations are very small and have shown rapid declines. Many of them are on various state and regional lists for threatened, vulnerable, endangered or critically endangered species. Many are extinct or on the verge of extinction.

Ecological adaptations

Adaptation is essential to the species' successful survival. Adaptations help the organisms to exist under the prevailing ecological habitat. It refers to the process of adjusting or changing in behavior, physiology, or structure to become more suited to an environment. It may also be defined as the state reached by the biological population undergoing adjustments or changes. The trait that made the species a better fit for the environment is referred to as the adaptive trait. Adaptation differs from acclimatization. Acclimatization is also related to change, but it is the physiological adjustment to the new conditions and does not involve increasing species diversity as adaptation does. For a trait to be considered as an adaptation, it has to be heritable and functional, and increase fitness.

Unlike the mainland lifeforms, the insular flora and fauna in Sundarban mangroves are able to live, grow, survive and thrive by facing the challenge in a constantly changing environment only because they have special structures and strategies to overcome those challenges including diurnal tides and flooding, too much salt in the water, too little fresh water, lack of oxygen underground, and the muddy ground being too wet and soft to support standing erect. A plant, not especially adapted to such conditions, would quickly die if its roots were in the salty, soft and soaked soil that mangroves grow in, or would topple over in the very soft ground, or would succumb due to lack of oxygen in waterlogged ground.

Floral adaptations

The mangrove flora is typically salinity resistant and has the following special root system:

There exists an extensive lateral root system for a proper anchorage against diurnal tidal inundation/scouring. Large parts of mangroves' roots can be seen above the ground, particularly during the low tide. These aerial roots (above-ground parts of the roots) have highly unusual shapes and functions, quite unlike the roots of other plants. Different mangroves have different root types. The main types are stilt roots (or prop roots). These roots grow down from branches and the trunk of a mangrove till they reach the ground and anchor into it. They can form many loops to get a stronger grip. Criss-crossed roots make an almost impassable tangle. Stilt roots are common in *Rhizophora* spp. These stilt roots are meant for support not for respiration. Another type of root system is found in the Sundarbans, which are called stilted or spike roots (like *Avicennia* spp.) because while sticking out of the ground, they look like upward-pointing spikes. They are inundated during the high tide, but they can breathe air during low tide to make up for the deficiency during the high tide. However, some roots that emerge above the ground surface, bend like a knee (hence called knee roots like and *Bruguiera* spp.) and grow down into the ground again. They twist and connect with each other to cover the ground around a tree. The fourth type, i.e. buttress roots (as in *Heritiera* spp.) looks like straight or bent planks that grow out of the trunk and provide additional support to the tree. These buttresses have many small roots that grow into the soil.

The lateral roots get submerged and breathe oxygen. Pneumatophores have developed such breathing facilities that are also adapted by throwing stilt roots from branches and stem, studded with lenticels in case of *Rhizophora mucronata*. *Excoecaria agollocha* shows perforated burr formation on its stem in more inundated areas, for breathing facility. Vertical knee roots from horizontal lateral roots are thrown by *Lumnitzera, Bruguiera gymnorhiza, Kandelia candel,* etc.

The plant cells exert very high osmotic pressure in order to draw water from outside salt solution. This cell sap is rich in organic electrolyte in case of *Rhizophora* etc. and inorganic electrolyte in case of *Suaeda* spp.

Existence of water storage tissues in the leaves of mangrove species make them mostly succulent and glossy. Stomata occur on both the leaf surface and the salt resistant plants have longer and wider stomata. Higher length and breadth of stomata induce salt resistant character and with higher value of epidermal cells, resistance to salinity gets increased. The leaves are normally thick and often contain salt excretory channels to deposit crystals of various compositions on the leaves and also waxes. *Lumnitzera racemosa* and *Sonneratia* spp. accumulate extra salt in their salt glands on both sides of the leaves as a result of which their leaves are succulent or thick. Salt hairs on leaves of *Potracea coarctata* burst to excrete salt. Star-shaped lepidote structures are also present on the abaxial leaf surface of *Heritiera* forms. The arms of these structures cover the stomatal pits and render lower transpiration. Through salt glands on leaves *Aegiceras corniculatum, Avicennia alba, A. marina, A. officinalis, Acanthus ilicifolius* and *Aegialitis rotundifolia* actively secrete excess toxic salts efficiently. *Rhizophora* spp. also excrete accumulated salt from the leaves. The deciduous species like *Xylocarpus* spp. and *Excoecaria* sp. exclude salt at the time of senescence of the old leaves before new growth, flowering or fruiting and foliage formation. The tumor-like outgrowth on the trunk-base of *Excoecaria* sp. performs the same function.

The mangroves and other halophytic species suffer from low productivity and retarded growth in the high salinity zone, but in fresh water and low saline habitats they flourish. To the contrary, *Avicennia* spp. can grow better in higher saline soil and regularly inundated tidal areas compared to the less saline area because it can accumulate Na^+ ion on its leaf tissue ten times higher than K^+ ion, whereas *Heritiera fomes* grows well in low saline soils because they can accumulate more of K^+ ion than Na^+ ion.

Mangrove reproduction starts with flowers, which vary in size (tiny to large) and colour (white, red, yellow, etc.). They are adapted to attract animals (insects, birds, bats) that pollinate them and help make flowers into fruits. The fruits of *Rhizophora, Bruguiera* etc. germinate right on the tree and fall like a dart on the mud flats to get anchored against tidal inundation (Vivipary).

A single mangrove produces hundreds of seeds each year. Not all of those seeds survive to become trees, but many do if the conditions are right. A single mangrove growing in a suitable place for many years will become surrounded by smaller trees, gradually creating a mangrove forest. For the purpose of successful reproduction, the fruits, seeds and seedlings are all adapted to the intertidal habitat. Mangrove seeds come in a variety of shapes and sizes and can float in water and have a fleshy or woody layer that protects the seeds inside. Some fruits burst apart when they dry up and release the seeds inside. Some sprout while still hidden inside the fruit, making them ready to quickly grow as soon as the fruit opens up. Seeds of some species sprout while still on the parent tree. The seedlings eventually drop into the water below and float away. When they reach soil, roots grow quickly to keep the seedling from drifting away.

The central section of Mangrove patch of Sundarbans delta between rivers Thakuran and Harinbhanga is typified by the accelerated geomorphic action of ingressing back waters, which does not get any upstream resistance of sweet water. This has resulted in movement of plant association within outer, inner and mid estuaries.

Some other plants inhabit the mangroves, but they are not considered mangroves. They are smaller in size and also salt-tolerant plants, including mosses, ferns, vines, grasses and orchids. They grow on the ground and cover or hang from the branches and trunks of mangrove trees.

Based on the habitats and the corresponding adaptations of plants, they are classified into (1) Hydrophytes, (2) Xerophytes, (3) Mesophytes, (4) Epiphytes and (5) Halophytes.

Abnormal floral adaptations

The role of geomorphology has become a very important factor in Sundarban owing to accelerated erosion and deposition created by undeterred tidal activities. The resultant action has been a movement of plant association which has given rise to overlapping of inner estuarine zone and mid-estuarine zone by outer estuarine species association. Thus, such detrital mangals also have started manifesting quaint adaptations (being out of normal domain), as in case of *Avicennia* species, which never throws stilt roots and is an outer estuarine species, when found in the mid-estuarine creeks at the foreshore, gives rise to stilt pneumatophores in order to combat the higher velocity and undermining effect of water. Both stilt roots and normal pneumatophores of *Avicennia* are histologically alike and contain chlorophyll unlike other stilt rooted mangroves. The species like *Excoecaria agallocha* (an inner estuarine species), which normally does not have pneumatophore or stilt root, gives rise to perforated 'Burr' formations on the lower stem in order to ensure gaseous exchanges at places, where tidal amplitude is more severe i.e. like the mid-estuarine environment.

Thus the mangals of Indian Sundarbans exhibit an unique pattern of species movement setting the whole consociation in a dynamic state of phyto-plasma, overlapping the normal estuarine modes and developing resultant adaptational abnormalities.

The islands and reclaimed areas have been colonized by terrestrial insects, spiders, oligochaetes and vertebrates. The freshwater forms include molluscs, insects, crustaceans and amphibians. Seventeen animal phyla have their representatives in the estuary. Among the mammals, the dolphins, porpoises and whales are truly aquatic. There are also some semi-aquatic mammals in the estuary. The abundant invertebrate communities of the estuary are the favourite food of the ages. It was observed that aquatic fauna of the estuary constitute about three-fourth of the total faunal species known. Moreover, the fauna is not uniformly distributed throughout the estuaries.

Faunal adaptations

The habitat pattern in SBR is uniformly monotonous, not broken by such landscape features that is known to provide specific cover for the mainland tiger in other areas. Mangrove forests provide a characteristic type of habitat suitable for animals inhabiting vast tidal swamp areas. Because of their intimate association with the estuarine environment, a sizable portion of aquatic and semi-aquatic animal communities are inter-related with the animals inhabiting the land areas.

Terrestrial, brackish water and marine fauna constitute the major faunal components in this ecosystem. Mangrove fauna occupies, in general, three different biomes, namely littoral or supra-littoral forests, inter-tidal mudflat and estuary. The littoral or supralittoral forest biome is typically a terrestrial environment, which includes both aerial and arboreal forms and the soil inhabitants. Arboreal animals and soil dwellers may be categorised as epifauna and infauna. The intertidal mudflats are essentially semi terrestrial or semi aquatic habitat supporting mainly the soil forms and the benthos, while the estuary is inhabited by aquatic forms, consisting of planktons and nektons.

The pattern of distribution of animals in the mangrove ecosystem is influenced by the substratum, salinity, tidal amplitude, vegetation, light, temperature, etc.

The supra-littoral forest habitat includes an area where water may or may not reach at all and is essentially densely covered with halophytes. It offers forest floor, roots, stems, branches and leaves of trees as the abode. The intertidal mudflat proper is recognised as areas having periodical inundation and exposure i.e. either exposed or submerged with tidal flush for a certain period for the day and is sparsely vegetated or with no vegetation at all in the lower littoral zone. The third biotope is the estuary or the creek, which has a characteristically low-saline brackish water system.

The aerial and arboreal forms mostly belong to different groups of vertebrates. While the other faunal components in the mudflat and estuary can broadly be divided into zooplankton, nekton and benthos. Several species of crustaceans and larvae of fishes form the main component of the zooplankton in this region.

Migration and residency: Function of ecological conditions

The fauna in the region may be divided into two categories on the basis of their duration of stay in a particular location- resident or non-resident (seasonal migrant). Residents are mostly invertebrates, e.g. molluscs, polychaetes, crustaceans, etc. who are detritivores and help in detritus-formation, like the meiofauna, peanut worms, crabs etc. They are also abundant, where the conditions are favourable. There are also a few vertebrate residents. Among the non-residents, particularly during the breeding season, the species are mostly planktonic or nektonic. There are certain euryhaline species, which migrate into the estuary. The diversity of avian migrants, more during the winter and few during the summer, to this estuary is also about one-third of the total number of species using the rich resources in the estuary. Among the reptiles, the aquatic species are turtles, some snakes and crocodiles. *Batagur baska* and *Crocodylus porosus* are two important species of the estuary. The amphibians are not estuarine here. Among the fish, freshwater, estuarine and marine species are found and the latter species migrate into the estuary. There are a few species (gobiids, blennids and eels) which are typical to the estuary. Eels live as adults in the estuary and migrate to sea for breeding. Some of the gobiids have almost adapted to terrestrial life, e.g. *Periophthalmodon schlosseri*. Many species of ray sharks are found here. Thread fins and catfishes seasonally ascend the river for spawning in the freshwater zone of the estuary. The confluence happens to be the breeding ground for "Hilsa", which moves upstream from the seas to the river seasonally. Though Hilsa occupies a prominent place in the estuary, their occurrence is markedly reduced due to salinity and pollution.

The distribution of molluscs is influenced to some extent by the nature of sediment and the organic content. The regions with fine sediments are rich in organic carbon and the fine texture of soil reduces evaporation and provides a habitat for burrowers. An unstable substratum, strong currents and long exposure to air are not conducive for burrowing bivalves. There is complete depletion of benthic fauna in regions which experience a high rate of erosion. In the study area dominant gastropod molluscs are able to withstand exposure to air. Some of these gastropods, especially the tree-dwelling prosobranchia and air-breathing ellobiidae, have developed adaptations for terrestrial existence and converted their mantle cavities into lungs. Here, *Cerithidea cingulata*, a common species to these estuaries, prefers sand mixed with clayey substratum. Sediment grain size is important for the settlement of detritus and microorganisms, which form the food of detritus feeding molluscs like *C. cingulata* and other potamidids. In general, it is seen that detritus feeders (potamidids, assiminids) and amphibious species are the dominant fauna in the estuary. Amphibious molluscs and the burrowing forms are able to penetrate a greater distance up the estuary than those which are affected by salinity fluctuations in the water. Majority of the species, especially bivalves occur in the vicinity of mangals (mangrove swamp). Molluscs dominate the macrobenthos of Prentice Island, having large concentrations in mangrove biotope. About one-third of the species occur either within or near the mangal area. Mudflats seem to support more molluscan species than any other biotopes. Molluscs occur in the mangrove fringed muddy area. The species composition of mangrove bivalves constitute less than half of the total number of bivalves recorded from the estuary. Bivalves being filter feeders take water through inhalant siphon. The environment, which has shifting sediments and heavy silt load, is detrimental to their existence.

The classification of the zooplankton species according to ecological categories has revealed that a large number of species belong to marine-brackish category followed by marine and brackish. In fact, estuaries, in general, contain mainly marine elements, which adapt themselves to certain levels of environmental fluctuations. They remain in the brackish water as long as they are able to tolerate the fluctuations and retreat back to their original zone during adverse conditions. The number of such 'opportunist' species ought to be more as they explore wider areas for their fortunes as compared to those who have settled in the brackish water by enhancing their adjustment capabilities. While species of Pseudodiaptomus, Acartidae, Temoridae and Oithonidae exhibited a wide range of salinity tolerances, the members of family Diaptomidae were quite susceptible to higher salinity levels. Similarly, many marine species also possessed very little flexibility in their salinity thresholds. The population maxima of copepods always followed the abundance of diatoms. Besides, phytoplankton, detritus supply was also sufficient throughout the year. The oxygen supply, pH and temperature were also moderate. Hence, it may be concluded that salinity was the single most important factor, which governed the dynamics of the copepods in *the estuarine system.*

Morphological and behavioural adaptations

Biological and physiological adaptations of many species in the estuary are very interesting. The most favoured fodder species is Keora (*Sonneretia* spp.) whose fruits and leaves are liked by Cheetal and Monkey. Fresh shoots of Hental are browsed by Cheetals. Pangas fish have been found to eat Keora fruits. Baen and Genwa are the next favourite among forage flora. *Oryza* grass, although serves as good grazing grounds for deer, but except for very succulent tips, remains are untouched.

The Sundarban tiger exhibits certain distinctive morphological adaptations that make it particularly suited to the mangrove habitat. In STR, two adult female and three adult male tigers were captured, radio-collared and released back into the forests in 2010. Of them, a six-year old male (Netidhopani tiger) was captured, while straying into inhabited areas, and weighed only 97 kg. Its neck was measured 53 cm against the normal size of 65-75 cm. Another adult tiger (Khatuajhuri male), which was caught at Malmelia village of Basirhat range, was also found to be frail and underweight (108 kg) with a neck-size of 58 cm.

During 2010-2011, three tigers were radio-collared in STR. On 19 May, a male tiger, which strayed out to Kalidaspur village on the same day from Jhilla-2 was immobilized, radio collared and released at Harinbhanga-1. On 22 May, another male was trapped at Netidhopani camp (Netidhopani-1), immobilised, radio-collared and released at Netidhopani-1 on the same day. On 2 October, again the same tiger was trapped, immobilised, the radio collar was changed and released at the same location.

Ear tag and micro-chipping for tracking

The microchips are being introduced to track tigers that repeatedly stray into inhabited areas. Since July 2009, some of the straying tigers in STR were ear-tagged with microchips. In the first case (July, 2009), a pregnant tigress, which entered Adivasipara on Kumirmari Island and, opportunistically, killed some livestock, was caught, examined and tagged with a microchip placed at the base of her tail before release.

While male occupants of a territory change more, females tend to hold a territory for longer periods. The territory of a male also overlaps the territories of more than one female. The cubs are reared exclusively by the mother and remain attached to her for up to two and half years. The female cubs often occupy territories or home ranges adjacent to that of their mother.

Unlike the mainlanders, these islanders are much smaller (suggesting insular dwarfism), more muscular with leaner frame and lesser body mass- in case of the adult male it was ±100 kg, while the female weighed <70 kg, compared to the minimum weight of mainland tigers, i.e. 140 kg. The total length of the four captured tigers in the Indian Sundarbans varied between 239 and 261 cm (generally 270 to 310 cm for the male and 240 to 265 cm for the female). Another unique feature of the Sunderban tiger is its smaller girth of neck, chest and shoulder height. The neck girth of the captured tigers ranges from 48 to 57 cm against the normal size of 65–75 cm, that of chest from 87 to 92 cm (generally about 150 cm) and shoulder height from 67 to 94 cm (generally about 109 cm).

This morphological adaptation is advantageous for the Sundarban tiger as a smaller animal needs less food. The Sundarban tiger has adapted to changed food habits and in addition to its natural terrestrial prey species like cheetal, wild boar, rhesus macaque, otter, lesser cats, birds and others about 17.5 per cent of its food comes from the aquatic sources, which include the fish, crab, water monitor lizards, turtles, etc. It has been reported that the tiger also preys upon the kill of the crocodile or even consume the carcass. It is also known to consume the insects like grasshoppers in crisis. The tiger has been observed to take honey (not a regular feature) and, before breaking the combs, smeared themselves with mud to avoid the stings. Their scats sometimes glitter with silvery scales. It often eats the shellfish and even the rats and frogs, when pressed for immediate food. Because catching the small, scattered prey (rats, frogs or fish) involves energy loss higher than the energy, which can be squeezed out of such prey, the tiger first tries to meet most of its needs from the large packets of energy- the ungulates. However, the ungulates are scarcer than the rats or frogs and can only be killed infrequently through spending lots of energy.

Thus, the Sundarban tiger has adapted to a combination of terrestrial, arboreal and aquatic prey base. During 1980s, 60% of tiger excreta showed wild boar hairs, because the wild boar is less alert than the deer. Whereas a decade later, the prey percentage in the tiger scats was reported as cheetal (32.35%), wild boar (38.23%), rhesus macaque (2.94%), fish (2.94%), crab (4.41%), water monitor lizard *Varanus salvator* (7.35%), turtles (1.47%), fishing cat (1.47%), birds (2.94%), leaves and fruits of *Phoenix paludosa*, a study, conducted during early 21st century, shows that the tiger's diet consisted of chital (53.79%; the consumption figure remarkably increased), wild boar (19.31%; now decreased compared to the higher figure during 1990s), rhesus monkey (6.89%; increased consumption), monitor lizard (9.65%; increased consumption), fishing cat (0.68%, reduced consumption), fish (4.13 per cent; increased consumption), crabs (4.82 per cent; almost same) and turtle (0.68 per cent; slightly reduced figure). In the Indian Sundarbans, the wild rhesus macaque is not a normal diet of the tiger, but the seized monkeys released in the wild become the easy prey for the tiger. It was also reported that a diseased tiger eats grass. Some of the tigers stray into the villages for killing the cattle, but more number of cattle are killed than required for consumption, but the human is not usually killed. The *modus operandi* is peculiar because in its own territory in the mangrove forests, it does not spare a human prey approaching nearby.

Normally the tiger is not a snake-eater. However, once a tiger completely ate up a large python *Python molurus* (length up to 8 m and weight 2 quintals), which earlier had swallowed a cheetal stag. On 18 July 2009, after the onslaught of 'Aila', autopsy of the carcass of an adult female tiger with no external injury revealed that it ate two poisonous snakes. The undigested parts of three uncommon prey species- a king cobra *Ophiophagus hannah* (sex unknown, probable length 2 m+), a cobra (*Naja keutia*) (probable length 1 m+) and a half-digested crab were found in the stomach content of an adult dead tigress at Netidhopani in STR (Mallick, 2011). The post-mortem revealed that it could have died due to an infection, as its necrotic focci in the heart, lungs and other body parts had enlarged. Difficulties faced in hunting may be the cause of such alternative food habits. As this tigress was old (12-14 years) and the habitat became hostile and inundated that it could not hunt the ungulates by chasing. Some opined that when the king cobra consumed the cobra, the tiger came on hunting. But if that was so, normally there should have a biting sign of the king cobra on the head of cobra, but such a sign was, in fact, could not be seen. However, hunting of the live venomous snakes successfully in the wild proves the exceptional skill of the Sundarban tiger.

The intelligence of a straying tiger was also observed, while it frequently managed to hunt the bait from the trap cage. First, it started moving all around the cage, but did not enter it nor make any attempt to take the bait away. The field staffs think that the female is comparatively more intelligent. However, after a few days, it had taken away the bait without stepping on the lever, which leads to closure of the cage. In another area, the animal avoided the main cage door, but hit the cage repeatedly from the back till one of the nut bolts became loose and it was able to secure a small opening. It then somehow squeezed its paw inside the cage through that small opening to hit and pull out the bait in such a manner that half part of the goat was left inside the cage.

Being lighter in the Sundarbans also made the tiger easier to move around in the muddy terrain or negotiate the rivers and creeks of the Sundarbans. Moreover, the ratio of their skull circumference to neck girth is almost equal or could be lesser than one, as compared to the mainland tigers, which is greater than one. This may be the reason that, in some cases, the standard radio collars brought to track the Sundarban tigers did not actually fit them and often loosened and dropped.

The stress factor associated with changes in their natural habitat and the availability of the smaller prey species is often related to such a phenomenon. Crossing the channels and maneuvering between the islands effectively reduces the body fat by exhausting considerable energy reserves leading to lean, muscular and agile tigers. Over a period of time, natural selection seems to have favoured survival of the smaller and lighter individuals with increased fitness to inhabit these islands.

Apart from the long-stretch swimming, the tiger has adapted to changed food habits. An earlier study on the scats of the tiger at fifty three locations in STR has revealed that in addition to its natural terrestrial prey species like the cheetal (32.35%), wild boar (38.23%), rhesus macaque (2.94%), its food included the fish (2.94), crab (4.41%), water monitor lizard, *Varanus salvator* (7.35%), turtles (1.47%), fishing cat (1.47%), birds (2.94%) and *Phoenix* leaves (Sanyal, 1993 & 1999c). The tigers also eat honey and, before breaking the combs, smear themselves with mud to avoid the stings. Remains of the venomous snakes (*Ophiophagus hannah* and *Naja keutia*) were also detected in the stomach of a dead tiger at Netidhopani, after the onslaught of the Aila. In case of the man-eating within the forest areas, the victims were mostly fishermen. Only 5% of the tigers were reported to be man-eaters. 'Man-eater' may not be the right word to describe a tiger after all and in its latest standard operating procedure released last year National Tiger Conservation Authority in India (NTCA) has discarded the use of the word man-eater for 'conflict tiger'!) because its frequent straying into the reclaimed villages in search of the domestic animals or shelter often causes retaliatory killing by mobs including poisoning and electrocution. This negative attitude is intensified for their resentment against the strict enforcement of the laws regarding entry into the forests. In the late 20th century, at least 12 tigers were killed by the villagers in the Indian Sundarbans. Sometimes thousands of people from the neighbouring villages gather to drive the straying tiger back to the forest by using drums, crackers, fire, etc. Sometimes the tiger goes back to the forest on its own. The forest staff and local Panchayet members always try to protect the animal and persuade the villagers not to harm the animal. But, when the mob becomes arrogant, the situation often goes beyond control and they kill the strayed tiger. Sometimes all the evidences of killing were quickly removed or destroyed by the villagers to avoid punishment. Three incidences on 29 July 2001 at Pakhiralaya, 2 October 2001 at Kishorimohanpur and 15 December 2001 at Kumirmari in the transition zone of STR were the last cases of such retaliatory killings of tigers by the villagers. Previously, in isolated cases, the straying tigers were killed by the villagers in self-defence, although all the straying animals are not man-eaters. However, the negative attitude of the villagers towards tiger conservation has now been changed with the strengthening of joint forest management (formation of 14 Ecodevelopment in STR and 51 Forest Protection Committees in STR and South 24-Parganas Forest Division protecting 641.16 km² forest areas). As soon as a straying tiger is detected, in most cases, the villagers inform the forest staff. However, if the location is distant from the nearest forest office, it often takes hours to reach the spot. Under the circumstances, the villagers jointly try to drive the straying tiger away. The majority of people living in the fringe villages of the Sundarbans now perceive that tiger protection efforts serve their own interest. But there may be possibility of unrecorded cases of poaching because the chronicles of seizure of body parts of the tiger and prey species from the Sunderbans has established the notion that poaching is a silent killer.

In a recent report, the West Bengal Crime Investigation Department (CID) had suspected the role of Bangladeshi money behind the illegal tiger hunting trade in the forests of the Indian Sunderbans. According to the study, Bangladeshi money lenders are financing poachers, equipping them with local boats and firearms to hunt the tigers in Sunderbans, under the guise of Indian fishermen. Inadequate protection camps at strategic locations, coupled with old weapons and slow moving boats are some of the major challenges Sunderbans needs to overcome.

Microbes

A microbe or microorganism is a microscopic organism that comprises either a single cell (unicellular); cell clusters; or multicellular, relatively complex organisms. They live in water, soil, and in the air. The mangrove forests, occurring at the interface of terrestrial and marine ecosystems, provide a unique environment, harbouring diverse groups of microorganisms such as bacteria, fungi (yeasts and molds), algae and protozoa; microscopic plants (green algae) and animals such as rotifers and planarians. Some microbiologists also include viruses, but others consider these as non-living.

Microorganisms live in different habitats where there is liquid water, including soil. Most importantly, these organisms are vital to humans and the environment, as they participate in the Earth's element cycles, such as the carbon cycle and the nitrogen cycle. These microorganisms also play vital roles in the functions and integrity of ecosystems with respect to maintenance and restoration. Microbes make an essential contribution to the productivity of the mangrove ecosystem. They are involved in recycling of nutrients, other organisms' dead remains and waste products through decomposition as well as conservation of natural habitats. They produce and consume gases that affect the climate, destroy pollutants and treat wastes.

Microbes have an important place in most higher-order multi-cellular organisms as symbiote (an organism in a partnership with another, such that each profits from the other). They are also exploited by people in biotechnology, both in traditional food and beverage preparation and in modern technologies based on genetic engineering. Many microbes are either pathogens or parasitic organisms. They are harmful since they invade and grow within other organisms, causing diseases that kill plants, animals and humans. Moreover, microorganisms from mangrove environments are a major source of antimicrobial agents and also produce a wide range of important medicinal compounds, including enzymes, antitumor agents, insecticides, vitamins, immunosuppressants and immunomodulators.

Although the mangrove ecosystem of SBR is very rich in microbial diversity, most of the species have not been enumerated before; in many cases neither their ecological role nor their application potential is known. Recently, developed technologies in molecular biology and genetics offer great promise to explore the potential of microbial diversity in the mangroves. Hence, the present study makes an attempt to review the microbial diversity and their ecological role in the mangrove ecosystems of Sundarbans.

In the east-west estuaries in Sundarbans, availability and distribution of the microbes depends on the seasonally variable temperature and water salinity along with total dissolved solid and total suspended solid deposited as well as the ecological gradients, i.e. gradual change of temperature, salinity, elevation, pH, CO_2, etc. vis-à-vis corresponding change in species composition and distribution. Whereas the temperature showed positive correlation with microbial groups in the Hooghly and Saptamukhi estuaries, negative correlation was reported in case of the Matla estuary. The salinity also shows inverse relationships with the pathogenic microbes, so also dumping of effluents especially sewage from the coastal and upstream human habitations and tourist complexes enables the microbes to flourish in the Hooghly estuary followed by the Saptamukhi and Matla estuaries.

Bacteria

Bacteria are unique organisms. They are single-celled, prokaryotic organisms in comparison to animals and plants which are multicellular, eukaryotic organisms. Because bacteria are prokaryotic, they do not have a nucleus and no membrane-bound organelles. In contrast, plants and animals, distributed in most of the habitats, are made up of eukaryotic cells, i.e. they have a nucleus and membrane-bound organelles like mitochondria or golgi apparatus.

Typically a few micrometres in length, bacteria have a wide range of shapes, ranging from spheres to rods and spirals. Bacteria were among the first life forms to appear on Earth, approximately 3 billion-4 billion years ago, and are present in all ecosystems. Free exchange of genes through conjugation, transformation and transduction, even between widely-divergent species, coupled with a high mutation rate and many other means of genetic variation, allows microorganisms to swiftly evolve (via natural selection) to survive in new environments and respond to environmental stresses.

The microbial community of marine estuary water from Sundarbans was dominated with bacteria occupying more than 96% of total community and archaea represented only 4% (Dhal *et al.* 2020).

Domain Bacteria Woese, Kandler & Wheelis, 1990 (prokaryotic microorganisms)

Phylum Proteobacteria Stackebrandt *et al.*, 1988, Garrity *et al.* 2005 [Gram-negative bacteria including a wide variety of pathogenic genera and free-living (nonparasitic) as well as many of the bacteria responsible for nitrogen fixation)]

Class: Gammaproteobacteria Garrity *et al.* 2005 (changeable little stick bacterium)

The diverse microbial community (for example, sulfate-reducing bacteria, methanogenic bacteria and subsurface bacteria) is particularly important in controlling the chemical environment of the Sundarbans Mangal. The community is also responsible for nutrient cycling and plays a vital role in productivity, conservation and rehabilitation of mangrove ecosystems leading to maintaining the health of the mangrove ecosystem.

Diverse bacterial communities thrive in the unique ecological environment of the mangroves. After reviewing the records available, it appears that the virus species in SBR has reached 67 after deducting the multiple strains. These species are under 48 genera, 33 families and 17 orders. The bacterial strains were isolated from the decomposed litters, detritus and from different animals, e.g. guts of *Mystus gulio, Uca* sp, *Boleophthalmus* sp. and haemocoel fluid of *Anelassorhynchus branchiorhynchus.*

The bacterioplankton communities found in the mangrove ecosystem indicate presence of numerous phyla such as *Acidobacteria, Actinobacteria,*

Chloroflexi, Cytophaga-Flavobacterium-Bacteroides, Firmicutes, Flexibacteria, Gammatimonadates, Planctomycetes in addition to overwhelming abundance of *Proteobacteria* (alpha, beta, gamma, and delta). Surface sediments of this area are dominated by *Deltaproteobacteria* followed by *Gammaproteobacteria, Alphaproteobacteria, Betaproteobacteria,* and *Epsilonproteobacteria* under phylum *Proteobacteria*. Abundant bacterial orders are *Desulfobacterales, Desulfuromonadales, Myxococcales,* and *Bdellovibrionales.*

Among the families Hyphomicrobiaceae, Rhodobacteraceae, Pseudo-monadaceae, Erythrobacteraceae, Kordiimonadaceae, Hyphomonadaceae, and Ruminococcaceae are dominant. Bacterial strains are known to be involved in a variety of biodegradation/biotransformation processes including hydrocarbon degradation, and heavy metal resistance.

Multiple environmental/anthropogenic stressors (like salinity, pollution, eutrophication, land-use) affect the water bodies of Sundarban estuary and the microbial communities too. Bacterial counts are generally higher on attached mangrove vegetation than they are on fresh leaf litter.

Fungi

Mangroves are inhabited by a large number of fungal communities including lower fungi (oomycetes and thraustochytrids) and higher fungi (ascomycetes and basidiomycetes), known as manglicolous fungi. They are mostly marine fungi; however, a small group of terrestrial fungi also occurs in mangrove environments and can be categorised into saprophytic, parasitic, and symbiotic fungi.

Mangroves being detritus-based ecosystems, substantial fungal populations play an important role in nutrient cycling and decomposition of organic matter by production of variety of extracellular degradative enzymes such as cellulase, xylanase, pectinase, amylase, and so on, which are isolated from the mangrove fungi and harnessed for several biotechnological applications.

In mangrove habitats, fresh-water and terrestrial fungi are involved in litter decomposition after at least up to six months under low saline conditions. Twigs (live or dead) of mangrove canopy harbour terrestrial fungi. When such substrata gets into mangrove waters during monsoon, these fungi dominate at least for a few months. Endophytic fungi of mangrove leaves, stems and roots consist of more terrestrial than marine fungi. Even after one year of immersion of wood in the mangroves, an amorphic fungi dominates during the subsequent monsoon.

The fungal composition depends on the seasonality in the mangrove habitats. Under low saline conditions, fresh-water and terrestrial fungi are involved in litter decomposition. The anamorphic fungi dominate during the monsoon period (June-September), whereas the increased salinity during summer (February-May) help the marine fungi flourish on detritus. Most species (endophytic) growing on mangrove leaves, stems and roots are more terrestrial than marine fungi and obtain their nutrients from dead organic matter. Some fungi are symbionts or parasites on other organisms.

Like higher plants, most fungi are attached to the substrate they grow (stumps, logs, twigs, straws, compost, residues, debris and other organic materials like seeds, shells and leaves) on because the substrate is the source of their nutrients, moisture and energy so that they can produce mushrooms.

Fungi recorded in SBR may be classified into five divisions along with the extant orders against each which are described below.

1. **Zygomycota** (conjugated fungi): Example (one order): Mucorales. These are mainly terrestrial and most species are saprobes feeding on plant detritus or decaying animal material; a few are parasites, particularly insects.

2. **Ascomycota** (sac fungi): Examples (fifteen orders): Xylariales, Eurotiales, Pezizales, Sordariales, Chaetosphaeriales, Capnodiales, Mytilinidiales, Helotiales, Hypocreales, Hysteriales, Pleosporales, Trichosphaeriales, Microthyriales, Meliolales. These fungi are found in every type of habitat, including both freshwater and saltwater environments. In aquatic habitats, ascomycetes typically live as a parasite on algae or other living organisms, whereas some species live on decaying matter. They serve an important function as decomposers, breaking down organic material, so it doesn't build up. This is a function for some ascomycetes in non-aquatic environments as well.

3. Basidiomycota (club fungi): Examples (ten orders): Agaricales, Auriculariales, Russulales, Dacrymycetes, Polyporales, Geastrales, Gloeophyllales, Hymenochaetales, Boletales, Phallales. Found in virtually all terrestrial ecosystems, as well as freshwater and marine habitats. Some species of Basidiomycota are pathogens for both plants and animals. Some form symbiotic relationships with the roots of vascular plants. Ecologically, they are vital for decaying dead organic matter, including wood and leaf litter.

4. Myxomycota (slime fungi): Example (one order): Physarales. These are common but inconspicuous inhabitants of moist dead wood, rotting logs, damp soil, leaf mold, moist sawdust, bark of trees, decaying fleshy fungi or other organic matter, spending their entire life within the substrate and only come out to produce sporangia. Following monsoon they may be seen on the blades of grasses or leaves of other plants.

5. Deuteromycota (imperfect fungi): Example (one order): Moniliales. Most members live on land, with a few aquatic exceptions.

As an important group of saprophytes, fungi play a great role in the nutritive cycle to sustain the mangrove ecosystem of SBR. They are adapted to adverse environmental conditions such as high salinity, frequently inundated soft-bottomed anaerobic mud and they contribute a nutrient cycle by producing a plethora of extracellular enzymes such as xylanase, pectinase, cellulose, laccase etc which have great ecological and economical importance.

They were found in the decayed woods or sediments three different zones, namely- (1) dense forest region, (2) mid littoral zone (rooted), where pneumatophores were present and (3) lower littoral zone, where pneumatophores were absent (unrooted). It was observed that the fungal population was smaller in the deep forest region with respect to the other two regions, probably due to less availability of water in the deep forest region than in those of other two regions. The former seldom gets inundated by the tidal waters, which does not lead to formation of organic carbon-rich sediment like those of the low-lying areas, which are regularly flooded and plastered with thick sediments. In all the sediments deposited, organic carbon content was found maximum during post monsoon followed by monsoon and pre-monsoon.

Most of the recorded fungal forms are thermophilic and others are mesophilic. Both these forms were available more in the northern part and less in the southern tip of SBR. The majority of species were of *Deuteromycotina, Ascomycotina* including ascosporic *Aspergilli* and *Penicillia, Zygomycotina* and *Mycelia sterilia. Aspergillus orchraceous* and *Penicillium frequentus* were the most abundant fungi species found in the sediments of Sundarban mangroves, when compared to other fungal genera from the same environment. Other genera include *Chaetomium, Curvularia, Fusarium, Trichoderma, Mucor* and *Pestalotiopsis*. Some genera were rarely found like *Acremonium, Neosartorya* and *Cladosporium*. Fungal population was found to be maximum during pre-monsoon and minimum during monsoon.

Twenty-eight orders, fifty-nine families, hundred and seven genus and one hundred eighty species are recorded in SBR. The checklist shows below family-wise species dominance along with habitat ecology of each species. Three dominant families with more than ten species are Trichocomaceae (Eurotiales), Polyporaceae (Basidiomycota) and Agaricaceae (Basidiomycota). Trichocomaceae holds the prime position because the saprobes taxa are aggressive in colonising a wide spectrum of habitats, common on stored grains and adaptable to extreme environmental conditions now prevailing in the mangrove habitats. Moreover, they are ubiquitous in soil and common associates of decaying plant and food material. The most familiar and, phylogenetically, closely related fungi like *Penicillium* and *Aspergillus* are members of this family.

Algae

Characteristics

The term algae, meaning "seaweed" (Latin), are organisms of the kingdom Protista that belong to Domain Eucarya and distinct from animals by being photosynthetic. However, they differ from the vascular plants by lacking true roots, stems, and leaves. Algae are abundant and diversely distributed in different brackish water and estuarine habitats of the Indian Sundarbans.

Types

There are three recognised types of algae based on their habitat in SBR-

1. Aerial and terrestrial algae- Epiphytic on trees (for support only, not food) and subterranean in moist soil;

2. Aquatic algae- Most of the algae are aquatic (a) freshwater (still or running water forms) or (b) marine (sea water forms);

3. Algae of unusual habitats- (a) Halophytic (in water containing high concentrations of salt), (b) Lithophytic (attached to stones and rocky areas), (c) Epiphytic (attached to other algae or higher plants), (d) Aerophytic (growing on leaves, bark or land animals termed respectively as epiphyllophytes, epiphoeophytes and epizoophytes), (e) Akinets (characteristic of *Cladophora* and *Pithophora* where the entire cell gives rise to a new plant and grows in size with a thick wall), (f) Autospores (resting spores develop structures exactly like parent cell except in size and give rise to adult forms); (g) Auxospore (specialized structure in Diatoms); (h) Endospores and (i) Cysts (common in Vaucheria which are formed during unfavourable conditions and they are many layered).

Ecological divisions

1. **Planktonic** (microscopic single celled organisms floating or freely swimming in water), composed of green algae (Chlorophyta represented by *Rhizoclonium, Cladophora, Ulothrix*), blue-green algae or cyanobacteria like Nostoc, Oscillatoria, Microcystis, Anabaena, diatoms (Chrysophyta such as *Pinnularia, Surirella, Cyclotella* etc), and euglenas;

2. **Benthic** (mainly bottom-dweller macrophytic angiosperms like mangrove trees, marshgrasses, and seagrasses attached to one or the other substratum);

3. **Halophytic** (found in water containing high concentrations);

4. **Lithophytic** (growing in or on rocks);

5. **Epiphytic** (living on other plants);

6. **Soil** (growing on or in soil).

Ecological importance

Algae turn carbon dioxide and water through photosynthesis into biomass (organic food molecules) and as a by-product of photosynthesis release oxygen. Ecologically, they are at the base of the food chain. The biomass they produce is consumed by the organisms in the higher trophic levels. Of them, single-celled phytoplanktons are largely responsible for the primary production. Phytoplanktons are mostly eaten by the small zooplanktons (mostly small crustaceans like copepods) that drift near the water surface. The smaller zooplanktons are in turn fed upon by the larger zooplanktons, small fish and filterfeeding krills. Energy is then transferred when the larger fish eat the smaller ones. At the top of the open-water food web are fish-eating birds, dolphins, very large fish such as sharks and humans. As the algae die, their bodies are consumed by the decomposers (mostly fungi and bacteria). The decomposers consume decaying plants and high-energy molecules essentially remineralizing the biomass into lower-energy molecules that are used by other organisms in the food web.

Diversity

More than nine hundred algae species are recorded from SBR. The dominant algae classes in SBR are Cyanophyceae in freshwater, estuarine and marine habitats followed by Chlorophyceae. Bacillariophyceae are rare in fresh water but are represented by good numbers in the marine zone. Blue green algae dominate in the entire SBR.

The predominant species observed in SBR included *Coscinodiscus radiatus, C. eccentricus, C. lineatus, Ceratium tripos, Chaetoceros affinis, Pleurosigma elongatum, Rhizosolenia alata, R. styliformes, Skeletonema costatum*, etc. In general, *the Cyanophycean* community with species of *Oscillatoria, Lyngbya, Phormidium* and *Microcoleus* occurs on bare mud flats and muddy soil between phanerogams. The pneumatophores of mangrove plants are covered with a number of blue green algae viz., species of *Calothrix, Anabaena, Lyngbya, Hydrocoleum*, along with some red algae viz., *Caloglossa, Catenella* and *Bostrychia*. Several epiphytic blue-green algae like *Dermocarpa, Xenococcus, Chaemosiphon* are also recorded. The Planktonic blue green forms are dominated by species of *Trichodesmium, Synechococcus, Aphanothece, Gloeocapsa, Gloeothece, Merismopedia, Oscillatoria, Fohannesbaptistia* and *Microcystis*. These Planktonic species presumably contribute very much to primary production of the estuary. In partly reclaimed areas, the water-logged rice fields or brackish water fish tanks are also colonised by a number of cyanophycean algal forms viz., species of *Aulosira, Spirulina, Arthrospira, Gloeotrichia, Calothrix, Nostoc, Anabaena, Oscillatoria, Aphanocapsa, Myxosarcina, Crinalium, Polyclamydum, Lyngbya, Rhaphidiopsis* and *Microchaete* etc., along with some salt tolerant green algae like *Enteromorpha* and *Ulva*. These forms provide a significant contribution to the soil fertility and nutrient balance in the wetland ecosystem.

Dense algal association is recorded from the forest floor, tree bark, pneumatophores, knee roots and stilt roots of the mangrove plants like *Avicinnia alba, Bruguiera gymnorhiza* and *Heritiera fomes* etc. A number of epiphytic and epixylic algae are also recorded from these plants. An endozoic unicellular alga *Chlorococcum infusionum* is recorded inside the body of Slugs-the shell-less molluscs. Among the filamentous microalgal flora, *Rhizoclonium* and associated genus *Lola* appeared as the most dominant genus with 9 species all together viz. *L. capillaris* and *L. tortuosa*. *Cladophora* with three species (*C. glomerata, C. crystallina, C. nitellopsis*) and *Chaetomorpha gracilis* are the other dominant green algal genera of Sundarbans flora along with two species of *Pithophora* (*P. cleveana* and *P. polymorpha*). Four species of *Spirogyra* are recorded from the study area among which only *S. pseudoreticulata* is recorded from brackish water zone, whereas *S. jaoense, S. hymerae* and *S. punctulata* are collected from the fresh water pond of these islands. Tree barks and the pneumatophores of mangrove forest remain covered with *Rhizoclonium* at the marine region of the estuary exposed to the daily tidal cycle. But the aerial portion of the mangrove plants remains covered by orange thallus of *Trentepohlia*.

Habitat ecology

Most algae are found to show substratum preferences. The blue green algae are noted to prefer a soft, hydrophilic, biologically active mud, rich in organic matter; whereas, the green algae prefer a more consolidated type of soil rich in nutrients. Similarly, the red and brown algal groups seem to prefer hard consolidated soil in supra littoral zones or peripheral zones which are regularly inundated. At the nutrition level, individuals prefer specific ecological levels. For example, *Polysiphonia* spp. and *Gelidiella* sp. grow under high phosphate levels, but *Chaetomorpha* spp. is found in low saline areas, whereas *Rhizoclonium* spp. and *Polysiphonia* spp. are found in moderate saline areas, although *Gelidiella* sp. are restricted to high saline areas.

Their habitats may be classified on the basis of abiotic factors as follows:

- pH spectrum –
  Acidobiontic: Organisms occurring below 5.5 pH.;
  Acidophilus: Organisms occurring between 5.5 and 6.5 pH.;
  Indifferent: Organisms occurring between 6.5 and 7.5 pH.;
  Alkaliphilus: Organisms occurring between 7.5 and 9.0 pH.; and
  Alkalibiontic: Organisms occurring in alkaline water with pH above 9.0.
- Temperature spectrum-
  Euthermal: Organisms occurring at temperatures above 30°C.
  Mesothermal: Organisms occurring in temperatures between 15 - 30°C.
  Oligothermal: Organisms occurring below 15°C.
- Saprobien spectrum-
  Saprobiontic: Organisms occurring in heavily polluted water.
  Saprophilic: Organisms usually in polluted water but also in clean waters.
  Saproxenous: Organisms usually in clean water but also in polluted water.
  Saprophobic: Organisms occurring in clean water habitats only.
- Current spectrum-
  Limnobiontic: Organisms of lentic environment only.
  Limnophilus: Found in lentic habitats but also present in lotic habitats.
  Indifferent: Common in both lotic and lentic water.
  Rheophilous : Found in lotic habitats but also present in lentic habitats.
  Rheobiontic : Found in a lotic environment only.

Protozoa (Biswas and Bandyopadhaya 2016; Rakshit *et al.* 2016)

The protists, mostly unicellulars, are eukaryotic organisms living in water, damp terrestrial environments like soil or even as parasites and traditionally considered as the first eukaryotic forms of life and a predecessor to plant, animals and fungi. They are autotrophic or heterotrophic (derive nutrition from plants or animals) in nature. They are broadly classified into five subdivisions:

1. **Protozoa- The most well-known** examples of protozoans are amoeba, paramecium, euglena. They have four groups-
   (i) Amoeboid protozoans – Mostly found in water bodies, either fresh or saline. They have pseudopodia (false feet) which help change their shape and in capturing and engulfing food, e.g. amoeba.
   (ii) Flagellated protozoans – They can be free-living as well as parasitic, e.g. euglena.
   (iii) Ciliated protozoans – They are always aquatic, e.g. Paramecium.
   (iv) Sporozoans – They are exclusively parasitic in habitat.
2. **Slime molds-** Generally these saprophytic organisms live in damp forest floor and under rotting logs and damp leaves or water such as fish tanks, etc.
3. **Chrysophytes-** These photosynthetic organisms including diatoms (Class Bacillariophyceae) and golden algae are found mostly in freshwater sources or marine waters.
4. **Dinoflagellates** (Class Dinophyceae)- These photosynthetic organisms occur in all aquatic environments: marine, estuaries, mangroves and freshwater. They are also common in a benthic environment. Dinoflagellates commonly encountered in the Sundarbans are *Ceratium, Gonialax, Protoperidinium, Noctiluca*.
5. **Euglenoids-** These organisms like *Euglena, Trachelomonas*, etc. live in fresh, estuarine and marine waters and can be found in moist soil or mud in hard or soft water habitats of varied pH and light levels - mainly marshes, swamps, bogs, mires and other wetlands with abundant decaying organic matter.

Ecological role

Protozoa: Protozoan ecology is the interaction in space and time of protozoa (flagellates, amoeboid organisms and ciliates) with other living organisms and the physical environment (temperature, salinity, nutrients and levels of pollutants). As components of the micro- and meio-fauna, protozoa play an important ecological role by-

(i)   providing food source to micro-invertebrates;
(ii) transferring bacterial and algal production to successive trophic levels;
(iii) preying upon unicellular or filamentous algae, bacteria and micro-fungi;
(iv) being both herbivores and consumers in the decomposer link of the food chain;
(v)   controlling bacteria populations and biomass to some extent.

Slime mold: Their role is categorized below.

(i)   They contribute to the decomposition of dead vegetation by feeding on bacteria, yeasts and fungi.
(ii) They also help in recycling minerals and important compounds such as carbon and nitrogen.
(iii) They provide favourable substrates and shelters for various species of fungi and insects, mainly Coleoptera (beetles), Latridiidae and Diptera (flies). Whereas some beetle species use the spores and plasmodia of slime molds as a nutritional source, the Minute Brown Scavenger Beetles lay their eggs in sporophores, aethalia and plasmodia. Some Diptera larvae live on the slime mold plasmodium and feed on them, with some remaining there as pupae. In fact, some flies can be reared on slime molds as their only food.

Chrysophytes: They are vital to an aquatic ecosystem food chain as the primary source for vital organisms and their presence contributes to the abundance of higher food sources in the chain, e.g. for zooplankton. The zooplankton is then consumed by larger organisms as a vital link in the ecosystem's food chain.

Dinoflagellates: They lead up the food chain in aquatic ecosystems. They are predators on a wide array of prey items, including phytoplankton, copepod eggs and early naupliar stages. They are, in turn, important prey for some metazoa. Some are predators of and simultaneously prey for other dinoflagellates. This prey-predator relationship is fundamental to control populations within the ecosystem. Its presence is crucial to create a balanced aquatic ecosystem via carbon fixation from photosynthesis. The blooming of these microorganisms is considered harmful under certain consequences.

Euglenoids: They are able to photosynthesize, take in carbon dioxide and release oxygen into the atmosphere so that other organisms can survive. They are indicators of aquatic superiority.

Diversity: A total of 104 protozoan species from Sundarban mangrove ecosystem was recorded during the late 20th century (Nandi *et al.* 1993). These protozoans belong to four major phyla viz., Sarcomastigophora (45 spp.), Apicomplexa (24 spp.), Myxozoa (4 spp.) and Ciliophora (31 spp.). The phylum Sarcomastigophora, includes 19 species of the subphylum Mastigophora and 26 species of Sarcodina. Only one species of gregarine, Cephaloidophora metaplaxi, has been reported so far from Sundarban. The phylum Myxozoa whose members are well known fish parasites is represented by four species only. The phylum Ciliophora was represented earlier mostly by entocommensal ciliates of shellfish while several species of free living ciliates have been recorded from water and soil samples of mangrove region. Out of 104 protozoan species, 41 species represent free living forms, 68 species parasitic forms and 5 species as symbionts.

Rotifer

The largest group of Rotifers, the microscopic aquatic animals, is called Monogononta. They are found mostly in freshwater but also in soil and marine environments. Littoral rotifers are more abundant in the tidal creeks than that of the mangrove swamps. On the contrary, the diversity is more in the mangrove swamps in comparison to that of the tidal creeks. Their community structure depends on the salinity and temperature. They are found in freshwater environments, estuarine wetlands and moist soil. The average salinity is little more in the mangrove swamps than that of the tidal creeks, whereas the temperature is little higher at tidal creeks than that of the mangrove swamps. The average total phosphorus is a little higher at mangrove swamps than that of the tidal creeks, whereas nitrogen is a little higher at tidal creeks than that of the mangrove swamps.

Here, they inhabit the thin films of water that are formed around soil particles. They also live in still and flowing water environments, such as lake bottoms and rivers or streams respectively. They commonly settle on mosses and lichens growing on tree trunks and rocks, in rain gutters and puddles, soil or leaf litter, on mushrooms growing near dead trees, in tanks of sewage treatment plants and even on freshwater crustaceans and aquatic insect larvae.

The rotifers are also ecologically important in the mangrove ecosystem. Rotifers colonize habitats quickly and convert primary production (algae and cyanobacteria) in a usable form for secondary consumers, making energy available for the next trophic levels. In interstitial water (aqueous solutions that occupy the pore spaces between particles in rocks and sediments) from swampy soils, they contribute to nutrient recycling.

Sponge

The poriferans, commonly known as sponges, are represented mostly by marine and some freshwater species, where they are found in the peripheral shallow margins of the water-bodies. Sponges are aquatic, sessile filter-feeding invertebrates. These are known to provide safe sanctuary to a variety of animals ranging from minute protozoans to lower chordates. It is a source of food for many animals. It has a symbiotic relationship with the protists and bacteria. It has a role in recycling calcium in the sea. It is used for decorative purposes and grown by aquaculture. BoB and Sundarbans will change if sponges are taken out of the ecosystem because many other animals depend on them for shelter and food.

Plankton

Plankton is one of the major assemblages in the mangrove estuaries of Sundarbans and thus contributes substantially to the gross productivity of the region. Planktons provide a crucial source of food to many small and large aquatic organisms, such as bivalves, jellyfish, etc. Diverse planktons consist of unicellular and multicellular organisms with different sizes, shapes, feeding strategies, ecological functions, life cycle characteristics and environmental sensitivities. The plankton species composition and abundance in SBR depend on a variety of water quality parameters- viz, temperature, pH, salinity, nutrients, dissolved oxygen and total suspended solid during different seasons- pre-monsoon, monsoon and post-monsoon.

The water temperature varies between 20.5 and 32^{0}C, where the lowest temperature was recorded during the monsoon and post-monsoon periods owing to additional discharge of headwaters and higher temperature value recorded in the pre-monsoon because of decreased discharge from the headwaters and heat raising temperature of surface water.

The pH of water ranged between 7.8 and 8.7 with marked rise during the pre-monsoon period. The riverine network of the central part of the estuary showed higher pH i.e, toward alkalinity than eastern or western regions.

The alkalinity of water varies between 80 and 192 mg/lit with higher values during pre- and post-monsoon period.

Similarly, the dissolved oxygen (DO) ranges between 4.5 and 8.7 mg/lit. The highest DO is recorded during the post-monsoon and lowest during the pre-monsoon period.

All the three nutrients viz, NO_3^- (nitrate ion), PO_4^- (phosphate) and SiO_4^- (silicate) showed conspicuous seasonal and temporal variations. The concentration of nitrate nitrogen varies from 0.2 to 2.26 mg/lit with very lower values in the central part of the estuary almost round the year. The silicate content also varies from 0.5 to 5.0 mg/lit with a similar trend like that of nitrate nitrogen. The phosphate content also varies between 0.02 and 0.21 mg/lit with comparatively higher values towards western part of the estuary.

The total suspended solid also ranges between 10 to 814 mg/lit with higher values on eastern and western part of the estuary due to major riverain discharge.

There are marked salinity variations ranging between 1 ppt to 29 ppt. The higher salinity value prevailed during the pre-monsoon followed by the post-monsoon season. The central part of the estuary showed higher salinity value than eastern or western part of the estuary in almost all seasons. Among them, salinity and nutrients are the most important biological factors governing their composition and abundance.

It transpired that whereas there is positive correlation of the planktons with pH, alkalinity and salinity, the nutrients like NO_3^-, PO_4^- and SiO_4^- are less significant.

Planktons usually inhabit water. They may primarily be divided into marine and freshwater planktons. Marine planktons include bacteria, archaea, algae, protozoa and drifting or floating animals that inhabit the saltwater of oceans and the brackish waters of estuaries. Freshwater planktons are found in the freshwaters of lakes and rivers. There are also airborne versions, known as aeroplanktons, that live part of their lives drifting in the atmosphere including plant spores, pollen and wind-scattered seeds, as well as microorganisms swept into the air from terrestrial dust storms and oceanic plankton swept into the air by sea spray.

Planktons are also functionally (trophic level wise) divided into five groups as follows:

(a) Phytoplanktons are autotrophic prokaryotic or eukaryotic algae living near the water surface, where there is sufficient light to support photosynthesis, such as the diatoms, cyanobacteria, dinoflagellates and coccolithophores.
(b) Zooplanktons are small protozoans or metazoans (e.g. crustaceans and other animals) that feed on other planktons. Some of the eggs and larvae of larger nektonic animals, such as fish, crustaceans, and annelids, are included in this group.
(c) Microplankton include fungi and fungus-like organisms, which, like bacterioplankton, are also significant in remineralisation and nutrient cycling
(d) Bacterioplankton refers to bacteria and archaea, which play an important role in remineralising organic materials down the water column (prokaryotic phytoplankton are also bacterioplankton).
(e) Virioplanktons or viruses are more abundant in the plankton than bacteria and archaea, although they are much smaller.

The freshwater stretch of the estuary in SBR supports a relatively higher plankton population followed by high and low saline stretches. For example, the plankton population of Hooghly main channel fluctuated within the range of 198-514 u/l. River Matla and Ichhamati harboured higher density of plankton in the range of 390-468 u/l and 308-437 u/l, respectively.

The Jharkhali stretch of river Matla seems to have a relatively better status in terms of the proliferation of plankton, both phyto and zooplankton as an impact of the mangrove vegetation in the surrounding river basin. The species belonging to Chlorophyceae are more pronounced in the freshwater stretch, whereas those of Dinophyceae are more restricted, mainly to the high saline stretches and mangrove areas. *Cyclops* sp and *Nauplii* constituted the bulk of the zooplankton population followed by rotifers, protozoans and cladocerans. *Nauplii* are often extremely abundant in plankton samples and are generally from copepods, but can occasionally be from barnacles.

Bottom biota

During the early 21st century, the average annual population of macro-benthic fauna in the Hooghly estuary fluctuated between the range of 3,581 and 5,089 number/m^2. The seasonal variation of benthic population in the Hooghly estuarine system including the distributaries like Matla and Ichhamati indicates much larger community size during monsoon followed by summer. Gastropods emerged as the most predominant macro-benthic invertebrates followed by bivalves. South-western Harwood-point (north of Sagar Island) followed by northern Dhamakhali appears to be more congenial for the greater colonization of macro-benthic forms. However, the distribution of bivalves is more common in coastal Bakkhali followed by western Jharkhali and north-eastern Hasnabad. Matla and Ichhamati rivers may be linked to the tendency of harbouring a relatively higher population of Chironomids and Oligochaetes and indicative of increasing organic load in these systems ascribable to various anthropogenic activities.

Lichen

Lichens are plantlike organisms consisting of a symbiotic association of algae (usually green) or cyanobacteria and fungi (mostly ascomycetes and basidiomycetes), occurring in a variety of environmental conditions including mangroves called manglicolous lichens. Lichens use the trees as structural perches not for extracting nutrients or water from tissues of the trees because they are photosynthesizers themselves but for getting better access to sunlight as well as moisture from the lofty trees.

In some cases, the lichens are colourful (white, green, grey, orange, yellow, red, brown and black), cover a large area and therefore are easily noticed but in other cases, they are cryptic and microscopic. They are found on the exposed wood, branches, bark or leaves of trees as well as rocks or soil, which are considered their ecological niche. Bark of the mangrove trees are preferred by the lichens because of their smoothness. The terrestrial lichens occupy the upper part of the trees and marine species occupy the lower part.

Morphologically, lichens are generally categorised into four types or growth forms, namely crustose (crusty), foliose (leafy), fruticose (shrubby) and squamulose (scaly). There are some other forms like dimorphic forms between foliose and fruticose. Since high levels of salinity affects the lichen population in the mangroves, the crustose lichens are abundant on the bark of mangroves with fewer foliose lichens in such habitat. Among the foliose lichens, the genera *Dirinaria* and *Pyxine* are most common encounters in mangrove forest.

Lichenised fungi differ from all other fungi in the formation of complex, persistent vegetative thalli. They are found in the terrestrial as well as aquatic habitats. Their growth depends on substrata like rock surfaces, woody plant bark and wood, soil and dead organic matter, light and moisture availability. Lichen-forming fungi form stable symbiotic association with photosynthetic partners, such as algae and/or cyanobacteria. Lichenicolous fungi live exclusively on lichens, most commonly as host-specific parasites, but also as broad-spectrum pathogens, saprotrophs or commensals. Allied fungi share space and colonise together in mutualism.

A large number of invertebrate or vertebrate species use lichens in various ways. The non-human vertebrates eat terrestrial or arboreal lichens and crabs also eat fruticose species. Lichens are eaten by many small invertebrates, including species of bristletails (Thysanura), springtails (Collembola), termites (Isoptera), psocids or barklice (Psocoptera), grasshoppers (Orthoptera), snails and slugs (Mollusca), web-spinners (Embioptera), butterflies and moths (Lepidoptera) and mites (Acari). But lichens are low in protein and also generally poor in various minerals. However, they have appreciable digestible carbohydrate content and produce bourgeanic acid, a fatty acid. Many bird species are known to use lichen fragments in their nests, most likely for camouflage. For example, when a small bird builds its twig nest in the dappled light found within many shrubs, such a setting of a nest built just of twigs would be discernible as an eye-catching dark mass in an otherwise dappled setting and could arouse a predator's curiosity. Fragments of lighter-coloured lichen thalli, affixed to the outside of the nest, would give it a dappled look, better able to blend with the dappled light, which would not usually attract predators. A variety of invertebrates shelters within lichen colonies that grow on rocks, bark or on the ground. Many other invertebrates are so coloured that they are inconspicuous when resting on lichen colonies. Various invertebrates also carry lichen fragments as camouflage. The mites often lay eggs in lichens and during egg laying a softening substance is excreted by the female and this causes the lichen to swell, grow over the eggs and so protect them. After hatching, the larvae feed on the lichen thallus.

Twelve orders of lichens have been recorded from SBR- Arthoniales (lichen-forming or parasites on other lichens), Gyalectales (lichenised fungi), Lecanorales (mostly lichen-forming fungi), Opegraphales (lichenised and lichenicolous fungi), Ostropales (lichenised ascomycota), Peltigerales (lichen-forming fungi), Pertusariales (pertusarialean lichenised fungi), Pyrenulales (saprotrophic and lichenised fungi), Rhizocarpales (crustose, lecideoid, lichenised fungi), Teloschistales (mostly lichen-forming fungi), Trichotheliales (lichenised fungi associated with trentepohlioid algae) and Verrucariales (mainly lichenised fungi). Thirty-eight families were found under the above orders and one family could not be ascertained. Of them Graphidaceae is most rich in terms of genus and species (17:52), followed by Pyrenulaceae (4:40), Arthoniaceae (4:32), Opegraphaceae (4:26), Trypetheliaceae (7:24), Roccellaceae (5:18), Ramalinaceae (5:15), Caliciaceae (4:14), Physciaceae (3:10) and so on.

Maximum lichen species is recorded from the latex bearing mangrove species like *Excoecaria agallocha* compared to the hosts like *Avicennia* spp., *Rhizophora* spp. or *Ceriops* spp. Dominance of crustose lichens in the forest was observed. Among the *Vicennia* spp. *A. alba* hosts the lichen species and *A. marina* and *A. officinalis* are not preferred hosts of these lichens. Fruticosa indicates wetter habitat and closed canopy. In the outer estuarine mangrove vegetation in SBR the trunks and branches of the trees like *Avicennia* spp, associated with *Aegialitis rotundifolia, Bruguiera cylindrica, B. parviflora, Ceriops tagal* and *Sormeratia griffithii* are mostly dominated by the crustose form, e.g. *Anisomeridium consobrinum, Arthothelium distendens, Bacidia convexula, Lecanora leprosa, Stirtonia obvallata, Trypethelium eluteriae, T. tropicum*, etc., followed by the foliose forms viz. *Dirinaria* spp., *Parmotrema dilatatum, P. saccatilobum, Relicinopsis dahlii, R. malaccensis* and fruticose forms viz. *Ramalina leiodea* and *R. pacifica*.

On the other hand, concentration of the lichen species in the inner estuarine mangrove vegetation of this region may be categorised under the following distinct zones in the RFs as well as the human habitations in the fringe areas:

True mangroves

Here the dominating crustose forms are *Anisomeridium consobrinum, A. leptospermum, Arthonia cinnabarina, Arthothelium distendens, Bacidia convexula, Bactrospora metabola, Buellia curatellae, B. lauricassiae, Cryptothecia culbersonae, Lecanora leprosa, Myriotrema compunction, Ochrolechia subpallescens, Diorygma hieroglyphicum, D. megasporum, D. pruinosum, Pertusaria velata, Phaeographis subtricosa, Polymeridium proponens, Pyrenula concatervans, P. mamillata, P. ochraceoflava, Sarcographa labyrinthica, Stirtonia obvallata, Trypethelium nigrorufum, I nitidiusculum, T. tropicum,* etc. The foliose forms are represented by *Dirinaria confluens, D. leopoldii, Leptogium denticulation, Parmotrema dilatation, P. saccatilobum, Physcia aipolia, Pyxine retirugella, Rimelia reticulata,* whereas the fruticose species are *Ramalma leiodea* and *R. pacifica.*

Semi mangroves

Here the dominant lichen species are *Anisomeridium ubianum, Arthothelium atro-olivaceum, A. distendens, Bacidia convexula, Buellia curatellae, B. lauricassiae, Catinaria bengalensis, Diorygma hieroglyphicum, D. megasporum, D. pruirtosum, Dirinaria confluens, Fissurina dumastii, Graphis apertella, Lecanora leprosa, Myriotrema comptmctum, Ochrolechia subpallescens, Parmotrema dilatatum, P. saccatilobum, P. tinctorum, Pertusaria leucosorodes, Phaeographis subtricosa, Physcia aipolia, Pyrenula introducta, P. ochraceoflava, P. parvinuclea, Pseudopyremila subnudata, Pyxhte retirugella, Ramalina leiodea, Sarcographa tricosa, Stirtonia obvallata, Trypethelium eluteriae, T. tropicum,* etc.

Hinterland mangrove zone

This zone is represented by *Anisomeridium tamarmdi, Arthonia antillarum, A. cirmabarina, A. ravida, A. subvelata, Arthothelium abnorme, Bacidia convexula, Bactrospora myriadea, Buellia lauricassiae, Enterographa pallidella,*

Graphis apertella, Lecanora leprosa, Opegrapha dimidiata, Phaeographis subtricosa, Pyrenula nitida, P. ochraceoflava, P. parvinuclea, Sarcographa tricosa, Stirtonia obvallata, Trypethelium eluteriae, T. tropicum, etc. The foliose forms are few and represented by the species like *Dirinaria applanata, D. confluens* and *Pyxine cocoes.*

Coastal mangrove zone

Here the dominant species are *Amandinea insperata, Anisomeridium tamarindi, Bacidia medialis, Bactrospora myriadea, Dirinaria applanata, D. confluens, Graphis apertella, Lecanora leprosa, Pyremda nitida, P. ochraceoflava, P. panhnuclea, Sarcographa tricosa,* etc.

Non-mangrove zone

This zone is represented by *Arihonia citmabarina, A. subvelata, Arthopyrenia majuscula, Bacidia convexula, Buellia lauricassiae, Dirinaria applanata, D. confluens, D. picta, Graphis apertella, Lecanora leprosa, Phaeographis meduciformis, Pyrenula anomala, P. nitida, Sarcographa tricosa, Trypethelium eluteriae, T. tropicum,* etc.

Vascular plants

Vascular plants (Tracheophytes) form a large group of plants with vascular tissues, particularly xylem and phloem for conducting water and integrating food, respectively. Examples are pteridophytes (lycophytes, horsetails and ferns), gymnosperms (including conifers), ferns and angiosperms (flowering plants). All the angiosperms are seed plants (dicots or monocots).

The pteridophytes are called 'botanical snakes' as they evolved after bryophytes (the amphibians of plant kingdom).

The plants included in phylum angiosperms are living in terrestrial, freshwater and marine habitats.

Habitats

The vascular plants are homoiohydric, i.e. they can regulate water concentration within the cells and tissues. Thus, the vascular plants are found in different habitats. Nevertheless, they cannot bear desiccation and would die due to water deficiency.

Classification of plants

Most plants can be classified into three categories: herbs, shrubs and trees based on their simple morphological characters or growth habits- height, nature of stem and branches.

Very tall plants with thick, woody and hard stems called trunks as well as many branches in the upper part, much above the ground are called trees.

Medium sized (6-10 m) plants, taller than herbs and shorter than a tree with developed branches near the base of stem, which is bushy, hard and woody, but not very thick, are called shrubs. Although stems are hard, they are flexible but not fragile. The life-span of these plants usually depends on the species.

Plants with soft, green and tender stems without the woody tissues are called herbs. They are usually short and complete their life cycle within one or two seasons. Generally, they have few branches or are branchless. These can be easily uprooted from the soil. Herbs contain enough nutritional components, including vitamins and minerals. Their stems are made up of cellulose, but not as rigid as lignin. Hence, they decompose relatively easily. Herbaceous plants are again categorised into annual, biennial or perennial on the basis of lifespan. An annual herb completes a whole life cycle in one growing season. Annuals have to be seeded every year. Annuals might be summer or winter growing. Summer annuals grow well during warmer periods of the year. While winter annuals germinate in the cold seasons and survive through the winter.

In addition to these three categories of plants, there are two more types which need some support to grow. They are specifically called climbers and creepers.

Ecological roles of tracheophytes

The diverse wildlife greatly depends on the habitat richness with food, nesting and breeding environment. Herbs, grasses, shrubs and different palatable parts of trees, i.e., leaves, flowers, fruits, and seeds having nutritive values, i.e., protein, carbohydrate, fiber, etc., constitute fodder of wild herbivores. Hence, the wild fruit and fodder-producing plant species under Poaceae, Cyperaceae, Fabaceae, Moraceae, Myrtaceae, and Zingiberaceae families play a great role in maintaining ecosystem food supply. Production of these resources vary considerably during different seasons. Insects, birds, chordates, and reptiles have different nesting and breeding natures which vary widely from each other.

Angiosperms (vascular seed plants)

Ancestral mangrove plants are believed to have reinvaded marine environments in multiple episodes from diverse angiosperm lineages culminating in today's mangrove flora. While some angiosperm genera found in SBR are exclusively mangroves, the majority of the genera include non-mangrove species. The angiosperms are primary producers. The basis of terrestrial food chains starts from the angiosperms because they convert solar energy into chemical energy and glucose through photosynthesis. Angiosperms are the sources of food both for the animals and humans. Fruits, vegetables, nuts, and cereal grains are types of food that angiosperms produce. They are a good source of medicine, fiber, timber and non-timber forest products and are economically beneficial for the people. Angiosperms impact the ecology by keeping the environment clean, increase rainfall in forest areas, protect the soil from erosion and help decrease global warming.

Pteridophytes (seedless vascular plants)

Seedless vascular plants are also beneficial in the ecosystem in providing food resources to animals and humans. Ecologically, the ferns provide several benefits:

    (i)   Weathering of rocks.
    (ii) Reduction in erosion (due to rhizomes spreading in the soil).
    (iii) Preparation of the topsoil.

(iv) Housing nitrogen-fixing cyanobacteria to provide the nutrient for aquatic life (water ferns).

(v) Absorption of toxins from the soil.

Lycophytes are not well represented in the Sundarbans.

Dilemmas of classifying mangrove species

Mangrove species are broadly classified as 'true mangroves' (Barik and Chowdhury, 2014) and 'mangrove associates', which is not universally accepted by the scientists, particularly in case of some fringe species found mainly on the landward transitional zones of mangroves. Hence, they classify the mangrove species as (i) true mangroves, (ii) mangrove associates and (iii) controversial species. Whereas the exclusive mangrove species are confined only to the mangrove environment (referred to as strict mangrove, obligate mangrove or true mangrove), the non-exclusive mangrove species inhabit the terrestrial or aquatic habitat, but also occur in the mangroves, for which these species are called semi-mangroves, back mangroves or mangrove associates. True mangroves are woody plants, facultative or obligate halophytes. The criteria of true mangroves are-

(a) Grow only in the mangrove forests and do not extend into terrestrial communities;

(b) Play a major role in the structure of the mangrove community, sometimes forming pure stands;

(c) Have typical morphological features like aerial roots to counteract the anaerobic sediments, support structures such as buttresses and above-ground roots, low water potentials and high intracellular salt concentrations and viviparous propagules;

(d) Physiological mechanism for salt exclusion and/or salt excretion through leaves and buoyant; and

(e) Taxonomic isolation from the terrestrial versions.

The major components of true mangroves cover all species of genera *Avicennia, Lumnitzera, Bruguiera, Ceriops, Kandelia, Rhizophora* and *Sonneratia*, including *Nypa fruticans*. In addition, there are some minor components of mangrove forests such as *Xylocarpus, Aglaia, Aegiceras, Excoecaria, Aegialitis, Heritiera, Scyphiphora, Acanthus, Brownlowia, Cynometra, Salacia, Clerodendrum* and *Phoenix*. The mangrove associates includes the genera- *Hibiscus, Thespesia, Caesalpinia, Dalbergia, Derris, Sarcolobus, Pentatropis, Suaeda, Salicornia, Tamarix, Heliotropium, Porteresia, Myriostachya, Hydrophylax, Ipomoea, Cerbera, Dodonaea,* etc. Mangrove habitat ferns are *Acrostichum, Pyrrosia, Drynaria* and *Stenoclaena*. There are epiphytes and parasites like *Dendrophthoe, Hoya* and *Viscum*. Many back mangroves (glycophytic plants and brackish water with low salinity and higher vegetation density) are also there- *Erythrina, Premna, Terminalia, Calophyllum, Manilkara, Saccharum, Vilex, Lannea, Hewittia, Crotolari, Canavalia, Tinospora* and *Tylophora*.

However, the transitional species are classified as controversial species (Wang *et al.* 2011). Some of the controversial species are-

Acanthus spp. grows better in areas with more freshwater input. The Specific leaf area (SLA) of *Acanthus* is higher than that of the true mangroves, but lower than that of the mangrove associates. In terms of all other parameters it is closer to the true mangroves.

Acrostichum spp. also grows in the landward edge of mangrove environments. *A. aureum* also survives in non-saline conditions and its growth is better in freshwater. In comparison to the true mangroves, it is closer to the mangrove associates in terms of all the parameters and is classified as mangrove associates.

Excoecaria agallocha inhabits the landward edge of mangrove forests and low-salinity areas.

Clerodendrum spp. is found both on the landward edge as well as deep inside the mangrove environment. It behaves like both true mangroves and mangrove associates. But it's SLA is similar to the mangrove associates. Hence, it is considered as mangrove associates.

Excoecaria spp. inhabits both on the landward side and the mangrove environment. Except for leaf osmolality, which is higher than the average value of mangrove associates and similar to true mangroves, all other parameters are close to the mangrove associates. Hence, it is classified as mangrove associates.

Heritiera littoralis inhabits back mangals and occasionally landward fringe of the mangroves, flooded by spring or equinoctial high tides. But it reportedly occurs in terrestrial habitats or inlands. It grows better in the freshwater. Except for leaf SLA and Nitrogen mass, all other parameters match with the features of mangrove associates. Therefore, it is classified as mangrove associates.

Xylocarpus granatum occurs on the landward edge of mangrove forests. It is closer to mangrove associates than any other true mangrove. Its leaf succulence and Cl content show the features of true mangroves, but its K/Na ratio (1.73) fell into the range of mangrove associates (0.56-2.01); it's SLA and Na content ranked between the mean levels of true mangroves and mangrove associates. So it is classified as a true mangrove.

Non-vascular plants

Bryophyte (tree-moss, oyster-green), taxonomically placed between algae and pteridophytes, is an informal group consisting of three divisions of non-vascular land plants (embryophytes): the liverworts, hornworts and mosses. Bryophytes produce enclosed reproductive structures (gametangia and sporangia), but they do not produce flowers or seeds. They reproduce via spores. They are characteristically limited in size and prefer moist habitats although they can survive in drier environments. Hence, they are one of the important components of biodiversity in moist environments and wetland ecosystems.

Habitat ecology and functional role

Among the plant kingdom, Bryophytes, representing a heterogeneous assemblage of plants, are designated as the 'Amphibians of the plant kingdom' because they were the pioneers to colonise a terrestrial habitat from an aquatic one, in the process of their adaptation to a terrestrial mode of life, but partially water is indispensable in one stage or another in their lifecycle. They are also called 'resurrection' because of their remarkable capacity of absorbing water to become fresh in no time.

Usually they prefer to inhabit microclimatic niches. Their colonies are generally found on the bare rock surface (rupicolous), on the stones or gravels (saxicolous), soil (terricolous), as epiphytes (epiphyllous) on the bark of vascular plants (corticolous), decaying logs, stumps, leaf litter and forest floors (lignicolous) and upper surface of leaves (folicolous) with cyanobacteria, fungi and lichens. Interestingly, liverworts are the predominant constituents of the epiphyllous communities. They provide a spongy bed or carpet in every possible habitat. They usually inhabit narrow ecological niches with preference for damp, shady and humid conditions near water bodies. They grow abundantly during the rainy season. Streams, lakes and bogs are home to many species of bryophytes. Some of the bryophyte species found on damp soil near water can also tolerate drier areas, while others cannot survive away from a moist environment. Many bryophytes are found in association with freshwater, but there are no marine bryophytes. A few species are found in brackish water of river estuaries. Stream and lake bryophytes are typically attached to some substrate, usually stationary (such as the beds or sides of the streams or lakes) but some mosses may be found growing on the shells of living freshwater molluscs and there are also a few floating species. The bryophytes in such sites are frequently splashed with water and, from time to time, may even become submerged for relatively short periods. But there are no obligatorily aquatic bryophytes. Those found in water are essentially terrestrial species, albeit with varying degrees of adaptations to a watery environment. All the species found growing in water including those found submerged can also be found growing on land. However, some physical or physiological differences between the land and water bryophytes may be seen.

Bryophytes are used in pharmaceutical products, in horticulture, for household purposes and are also ecologically important. Liverworts and mosses have antifungal, antibacterial, antiviral, anti-inflammatory and antioxidative potential. They provide vital ecosystem services like soil formation, habitat modification and nutrient cycling and are useful in air-borne and water pollution detection and monitoring. Some bryophytes grow only in a narrow and specific pH range and, therefore, their presence can be used as an indicator of soil pH. Because of their high water holding capacity and permeability to air, when they envelope the forest floor and tree trunks, they aid in moisture conservation and prevent growth of weeds. Some mosses play the role as inhibitors of soil erosion due to their trample-resistant structure and their regenerative ability. Moreover, they accumulate minerals (K, Ca and Mg) from rainfall. Some help fixation of nitrogen in suitable substrates.

Since most of the bryophytes seem to absorb water and mineral nutrients directly into leaves and stems, they are extremely vulnerable to air-borne pollutants in solution.

In spite of their enormous ecological role, the bryophytes are one of the most neglected (least studied) groups of plants in the Sundarbans.

Zooplankton (Rakshit *et al.* 2016; Nandy *et al.* 2018; Nandy and Mandal 2020; Purushothaman *et al.* 2020)

Ecological profile

Zooplankton plays an important role as secondary producers in coastal and marine ecosystems, representing an important link between phytoplankton, microzooplankton and higher trophic levels such as fish. Spatial and temporal distribution and population of zooplankton communities in SBR like Copepod, mysid, lucifer, gammarid amphipod, cladoceran, ostracod, cumacea, Cnidaria, Rotifera, hydromedusae and Chaetognatha among holoplankters and larval forms of Decapods and Cyclopods in the Hooghly and Matla estuaries showed significant ecological fluctuations associated with monsoonal and hydrological parameters. Along with phytoplankton, they also form a major part of the trophic states. The enrichment of inshore waters from rivers might have generated enough food sources for the ctenophores (echinoderms, sponges and cnidarians) which lead them to migrate into river channels of Sundarbans. The highest microzooplankton abundance in surface water has been recorded from Sagar Island, compared to other stations in the northern BoB. Similarly, Ciliates and dinoflagellates had the highest abundance in the Sagar Islands. Hence, abundance of different groups also varies in the estuaries of Sundarbans. For example, whereas ctenophore forms 42% of the mesozooplankton community here, copepods held the most abundant status, particularly during June-July and January-February, with an average of 48.40% of the total mesozooplankton composition, followed by chaetognaths (3.35%) and others (Chaetognatha, Decapoda, Ostracoda and Appendicularia). *Acartiella tortaniformis, Pseudodiaptomus serricaudatus, Paracalanus parvus, Acrocalanus gracilis, Bestiolina similis* and *Oithona similis* were reported to be the most dominant copepod species at intermediate and low salinity zones because they have the capacity to develop even in eutrophicated and polluted areas. It was reported that *Euchaeta marina, Euterpina acutifrons* and *Clytemnestra scutellata* were exclusively recorded at Jambu Island, where the salinity level varied from 21.0 to 34.5 ppt. Efficient ecological adaptations have made the copepod abundant in the coastal Sundarbans.

Here, ctenophore abundance was high in the stations of low sea surface temperature, low dissolved oxygen and less biological productivity, i.e., low nitrate concentration and low Chlorophyll.

Crustaceans are an extremely diverse group including animals such as crabs, lobsters, isopods, shrimp and barnacles and have adapted to nearly all environments. Most crustaceans are aquatic, living in marine or freshwater habitats, but some groups are completely terrestrial.

Copepods are a group of small crustaceans found in both freshwater and saltwater habitats as well as limnoterrestrial and other wet terrestrial places, such as swamps, under fallen leaf in wet forests, bogs, ephemeral ponds and puddles, damp moss or phytotelmata. They form the bulk (nearly two third) of zooplankton diversity and biomass with seasonal fluctuation, i.e. highest in pre-monsoon (e.g. April), declining during monsoon and again rising in the late winter. They form the food of both larvae and adults of fishes and crustaceans whose abundance in a particular area has been directly related to the availability of either a particular species or assemblage of few copepod species. The parasitic copepods are also attached to bony fish, sharks, aquatic mammals and many kinds of invertebrates such as molluscs or tunicates. They live as endo- or ecto-parasites on fish or invertebrates.

Species diversity within the copepods based on ecological parameters was also observed. The abundance of most of the species of Copepods is related to salinity. Pseudodiaptomidae and Acartiidae can tolerate a wide range of salinity whereas Diaptomids are generally intolerant to higher levels. Herbivorous copepods were found to be dominant over the omnivores and carnivores throughout the year. Calanoid copepods are most dominant followed by cyclopoids, whereas monogeneric harpacticoid copepods represent very few species. However, during monsoon, oligohaline species such as *Pseudodiaptomus binghami, Halicyclops tenuispina; Mesocyclops* sp. and *Cyclops* sp. assemble along with the greater flow of freshwater in the estuary from the upper stretch. Adequate population of the euryhaline species was more or less marked throughout the year (e.g. *Acartia* sp., *Acartia spinicauda, Acrocalanus* sp., *Eucalanus elongatus, Eucalanus subcrassus, Labidocera euchaeta, Laophonte* sp., *Microsetella rosea, Oithona* sp., *Paracalanus* sp., *Pontella andersoni, Pseudodiaptomus annandalei. Pseudodiaptomus hickmani,* "*Saphirella*" cf. *indica*). Existence of another distinct group (*Acartia sewelli, Centropages dorsispinatus, Cladorostrata brevipoda, Clytemnestra scutellata, Corycaeus crassiusculus, Euchaeta marina, Euchaeta wolfendeni, Euchaeta* sp., *Euterpina acutifrons, Harpacticus* sp., *Macrosetella gracilis, Pseudodiaptomus aurioilli*) fluctuated depending on the salinity profile in the estuary. There are also a casual group of migrants here, such as *Acartiella keralensis, Candacia bradyi, Canthocalanus pauper, Centropages furcatus, Corycaeus agilis, Corycaeus catus, Cosmocalanus darwinii, Cyclops* sp., *Eucalanus* sp., *Euchaeta concinna, Euchaeta tenuis, Halicyclops tenuispina, Labidocera* sp., *Labidocera minuta, Mesocyclops* sp., *Neodiaptomus schmackeri, Oncaea venusta, Pontellopsis herdmani, Pseudodiaptomus binghami, Pseudodiaptomus masoni, Pseudodiaptomus tollinqeri, Pseudodiaptomus dauqhlishi, Pseudodiaptomus* spp., *Tachidius discipes, Temora turbinata, Temora discaudata, Tortanus gracilis* and *Tortanus forcipatus.*

Earlier, a total of 77 species under 35 genera and 26 families of Copepods have been reported from SBR with largest number such as *Acartia (Acartia) danae, Acrocalanus longicornis, Labidocera acuta* and *Oithona similis.*

Molluscs (Haque and Choudhury 2015)

Molluscs are one of the dominant invertebrate groups in the mangrove community having high density and biomass and play an important ecological role in the ecosystem together with decapod crustaceans. They are terrestrial, semi-aquatic or aquatic and diurnal or nocturnal or both. Depending on their dependence on the mangrove resources, they may be divided into native, facultative and migrant.

Types

Four classes of molluscs are found in the Indian Sundarbans and hinterland. They are-

1. Gastropods ("stomach foot") comprising snails, slugs, conchs etc. living in salt water, fresh water and land, from the bottom of aquatic habitat to the canopies of mangrove forests.
2. Bivalves ("bivalve foot"), comprising clams, oysters, cockles, mussels, scallops and others living in saltwater or freshwater. Most bivalves bury themselves in sediment, while others lie on the floor or attach themselves to hard surfaces.

3. Cephalopods ("head foot") composed of squid, octopus, cuttlefish and nautilus that are marine animals, although exceptionally can tolerate brackish water but never venture into fully freshwater habitat; and

4. Scaphopods ("boat foot"), usually called tusk or tooth shells, are marine animals living in soft substrates offshore (usually not intertidally).

Habitat ecology

The gastropod and bivalve mollusks occur more abundantly in the river mouths or confluences and backwaters of the Sundarbans than the Cephalopods and Scaphopods. They are found to live mostly in aquatic habitats like marine, estuarine and freshwater. Mangroves are also zoned both horizontally along land-water interface and vertically at different heights from the ground level. The macro benthic molluscs of the estuary, can be broadly grouped under three categories-

(i) those living attached to stems, pneumatophores and leaves of living plants (arboreal),

(ii) those living attached to or in the crevices of dykes, bricks, wooden pillars, jetties etc. and

(iii) those living on the muddy substratum, either moving freely on it (epifauna) or burrowing into it (infauna). They also occupy animal burrows, but not deep into the sand. A few of the gastropod species may have overlapping habitats. But, in general, it has been observed that the species which are arboreal do not usually occur on the ground except for a short duration. Those living in the crevices of dykes, jetties etc., do not usually forsake the crevice-dwelling habit. But there are certain exceptions; *Nerita* spp. and *Potamacmaea fluviatilis* are usually found attached to mangrove trees but in the absence of those trees the snails cling to the crevices in jetties, etc. *Columbella duclosiana* is usually found attached to pneumatophores but often found crawling.

Importance

The molluscs play a great ecological role in the detritus-based mangrove ecosystem composed of coastal areas, estuaries and riverbeds. They prefer to use the mangroves during different stages of their life for abundant food, shelter and breeding. They hydrodynamic mangroves retain their immigrating larvae and juveniles. They suitably adapt to the various macro-habitats of this ecosystem. Some species are found in the bottom substrate of the water bodies. Some also morphologically adapt to the mangroves by eliminating gills and converting mantle cavity into lungs. The tree dwellers (about one-fifth of the diversity), on the other hand, have resorted to reabsorbing calcium carbonate from internal shell structures so as to survive in this tough habitat. Most of them are ground dwellers to live in mangrove roots, trunks and leaves fallen on the forest floor. They shift upland and lowland following the rhythm of high-low, spring-neap and day-night tidal fluctuations, even climb trees during flooding to save themselves from drowning.

They are important ecosystem engineers, helping to structure aquatic bottom environments and providing complex and heterogeneous habitat, protection and food to a wide array of other organisms. They are a potential storehouse of carbon in their body. Many of the species are edible and consumed by the birds, fish and human beings. They also help in bioremediation of water and significant carbon sinks. Some molluscs cause damage to plants and woods. Two terrestrial species- *Achatina fulica* and *Macrochlamys indica* are common crop pests. Some species are animal parasites like *Lymnaea (Pseudosuccinea) acuminata*, *Lymnaea (Pseudosuccinea) luteola*, *Indoplanorbis exustus*, *Fasciola gigentea*, *F. hepatica*, *Schistosoma indicum*, *S. nasalis*, *S. spindalis* and others. Substantial amounts of dead wood in the intertidal zone of mature mangrove forests are tunneled by teredinid bivalves. They create many such tunnels in large woody debris. When the tunnels are exposed, a wide range of cryptic animals are able to use those tunnels, significantly cooler than air temperature, as suitable refuges, as per availability. Some common bivalve species causing damages to the mangrove plants as well as wooden structures and vessels are *Bactronophorus thoracites*, *Bankia companellata*, *Bankia nordi*, *Bankia rochi*, *Dicyathifer mann* and *Nausitora dunlopei*. The calcium rich species are traditionally exploited for preparation of lime, medicines and poultry feed and thousand tons of oyster shells are collected from the river beds and canals of different parts of Sundarbans and brought to the factory at Canning for crushing them into powder. The mostly exploited commercial and medicinal species are -

Freshwater: *Filopaludina bengalensis*, *Bellamya dissimilis*, *Pila globosa*, *Lamellidens corrianus*, *Lamellidens marginalis*, *Parreysia favidens;*

Estuarine and Marine: *Telescopium telescopium, Pirenella alata, Pirenella cingulata, Volegalea cochlidium, Tegillarca granosa, Crassostrea cuttackensis, Crassostrea gryphoides, Geloina bengalensis, Meretrix meretrix, Pelecyora trigona.*

Gastropods and bivalves are considered as the main molluscs of mangrove forests. The gastropods have high diversity because they are more mobile, whereas the bivalves are settled in narrow zones for feeding, larval growth and rich sediment characteristics of low salinity and high organic concentrations.

Crustaceans (Gokul 2017; Sau *et al.* 2017)

Crustaceans include all the invertebrate animals of the Crustacea (*crusta=* shell), such as the crabs, lobsters, shrimp, krill, copepods, amphipods and more sessile creatures like barnacles. The ecology of the crustacean differs from one type to another. Generally, they live in aquatic and terrestrial environments. Most of them are marine, estuarian or freshwater species, but a few groups have adapted to life on land, such as terrestrial crabs and terrestrial hermit crabs burrowing in the sand of beaches near access to water. Some freshwater crustaceans are crawfish and fairy shrimp. Crawfish live in lakes and rivers hidden under rocks and sand. Fairy shrimp are found in vernal ponds which are temporary puddles during the rainy season.

Benthic forms occur in all the aquatic environments from the abyss to the shoreline. They also occur in estuaries, lagoons, freshwater lakes, ponds and streams, salt lakes and damp vegetation. Different species are found in rocky, sandy, and muddy areas. Some species living in the spaces between sand grains are so small that they are inconspicuous. Others tunnel in the fronds of seaweeds or into wooden structures.

Sau *et al.* (2017) has proposed eight habitat types of the macro-benthic crustaceans in the Sundarbans:

1. Muddy substratum of upper, middle and lower littoral zone inhabited by the genera *Uca, Dotillopsis, Metaplax, Scylla* and *Squilla.*
2. Sandy substratum inhabited by *Uca lacteal.*
3. Muddy and sandy mid and lower littoral zone inhabited by *Dotilla* spp.
4. Mid littoral zone is represented by *Thalasinna* and *Parasesarma.*

5. Mid and upper littoral zone is inhabited by *Sesarma, Sesarmos, Perisesarma* and *Episesarma.*
6. Hard substratum with mangrove palm is inhabited by *Sesarmoids longipes.*
7. Dead trees, stumps and mud flats are used by *Balanus* sp.
8. Coastal area is inhabited by *Clibanarius padavensis.*

Clibanarius padavensis, Scylla serrata and *Scylla tranquebarica* were found to be the most common species along the coastal areas as well as far inside in the brackish water areas and abundantly encountered in Sundarban coast. They are nocturnal crabs. The adults usually hide during the day in burrows or remain concealed under logs or other objects. They make ellipsoidal burrows usually with a single external opening, either straight or slightly slanted with a depth of 0.7 to 1.45 m.

The fluctuation of salinity in the Sundarban estuaries displayed both significant positive and negative correlation with the species diversity and population dynamics of the crustaceans. For example, the species like *Austruca annulipes* or *Tubuca dussumieri* were encountered mostly from the moderately saline niches than the high saline ones, which indicates their inability to tolerate high salinity and the species are typically euryhaline. Hence, salinity seems to play a decisive factor in seasonal occurrence of the benthic species in this vibrating environment.

The importance of temperature as a factor like salinity for distribution of these fauna in a particular environment or niche in the brackish water, i.e. exclusion of species in areas with unsuitable temperature condition, is also emphasised. For example, during low to moderate temperature, the fiddler crabs are generally active and feed during the time of low tide in the midday, but during the warmer months of the year, they are exposed to temperature on the open mudflats which may sometime reach to even lethal level because of direct solar radiation. Moreover, the burrow temperature remains almost uniform throughout the seasons. During the pre-monsoon months, when the temperature is high, the fiddler crabs were seen to pay frequent visit to their burrows to replenish the water loss in the body due to their surface activity and evaporation, whereas during the post-monsoon months (December-January), when the temperature is comparatively low, the surface activity of the fiddler crabs were very much reduced mainly because of the cold environment persisting on the substratum. From the above, it transpires that both positive and negative correlations to air and soil temperatures respectively were prevalent among these crabs.

In SBR, the average concentration of dissolved oxygen was found highest during monsoon, followed by post-monsoon months and lowest during the pre-monsoon months. The increase in the concentration of oxygen is associated with the increased freshwater inflow in the estuary during the monsoon months. In the fiddler crab community, for example, *Austruca triangularis* showed significant positive correlation with dissolved oxygen.

Substrates are important in determining the species composition of the various habitats. Deposit feeding crabs inhabit substratum types ranging from pure sand to pure mud. It was observed that *Austruca annulipes* restricted their distribution only to those areas where the percentage of sand is higher, but *Tubuca acuta* were encountered in large numbers from all the habitats and *Austruca triangularis* recorded in moderate density in most of the habitats. *Austruca annulipes* is much better adapted and dispersed among sand grains than are its congeners e.g. *Tubuca acuta* and *Austruca triangularis*. This may be one of the reasons for restriction of *Tubuca acuta* and *Austruca triangularis* to predominantly muddy habitats. Therefore, sediment with high sand contents plays the key role in favouring the distribution of *Austruca annulipes* and restricting the distribution of *Tubuca dussumieri* and *Austruca triangularis*. Widespread distribution of *Tubuca acuta* in different ecotones suggests that they are both euryhaline and eurycious.

The population of brachyura presented a trend of seasonality. Among the fiddler crabs in SBR, *Tubuca acuta* was encountered more during pre-monsoon, while *Austruca annulipes* during post-monsoon and *Austruca triangularis* during the monsoon only. *Tubuca dussumieri* did not show any seasonal preference.

The impact of seasonality was also marked among different sex and age classes of the crabs. Juvenile crabs are usually ready to settle on an intertidal flat within about one month i.e. spring tides, when the tide current would ensure up estuary transport towards suitable adult habitats. The ovigerous females appeared at the onset of monsoon, continued to increase, attained to its maximum during August-September and then declined steadily in numbers during monsoon and early phase of post-monsoon. Therefore, pre-monsoon always experienced higher abundance of large-sized crabs or higher biomass output followed by post-monsoon and monsoon. But high population density of fiddler crabs in some monsoon months was attributed to colonisation of large numbers of juveniles and their consequent development into small-sized crabs. Surface- activity of the fiddler crabs was higher in pre-monsoon months, but significantly reduced during the post-monsoon months. There is a possibility that some female crabs might have escaped from the predation (e.g. by birds) as they used to take shelter deep into the burrows during the last phase of monsoon and post-monsoon months.

In the Sundarban mangroves, a large proportion of the leaf litter is directly consumed by the crabs, particularly those in the family Sesarmidae, which increases the quantity of mangrove biomass into the food chain in three main ways:

1. Shredding: As the crabs feed on leaf litter, they shred it into fine particles, increasing the surface area for leaching and microbial colonisation. A part of the material processed by crabs is dropped without being ingested, but even this is shredded into fine particles.

2. Accelerated leaching: Their faeces contain about 85% of the consumed leaf litter. Processing of this material in the crab gut reduces about one-fourth the content of unpalatable tannins in faeces than that of the freshly fallen leaves. As a result, in the next stage, the decomposing microbes colonise the processed leaves within hours or days, but such benefits of quick transformation would not have been available without prior contribution of the crabs.

3. Assimilation: Estimates for the amount of litter consumed by crabs vary. A part of the leaf litter processed by crabs is assimilated as crab biomass because some large predators then feed on these crabs, including a number of important fish species. Additionally, crabs will invest a proportion of the energy assimilated in reproduction, producing large numbers of crab larvae which are an important food source to the smaller predators. This "short circuit" of the mangrove food chain allows production from mangrove trees to reach the important fish species without passing through the detrital pathway.

Insects (Mitra, 2017; Mitra, *et al.* 2016a, b, 2018, 2019a, b; Roy, *et al.* 2016, 2017, 2019)

Insects are adapted to every terrestrial and aquatic habitat, where adequate provisions of food are available. Many live in brackish water up to 1/10 the salinity of seawater, a few live on the surface of water bodies and some fly larvae can live in pools, where they eat other insects that fall in. The intertidal mangroves provide a viable habitat for the insects, where some species prefer a certain niche than the others and consequently occur more abundantly and feed voraciously. The entomofauna use all the plant parts in the mangroves. Functionally, they may be grouped into four functional groups- pollinators, predators, burrowers and herbivores. Mangroves and insects have strong ecological relationships in which mangroves provide a suitable habitat (roots, stems, bark and leaves) for the insects. For example, some insects are stem borers, the mosquitoes inhabit in the holes of mangrove trees, such as *Avicennia* spp. or some species like the herbivores live on the canopies, particularly those which are not inundated during the high tides. The termites burrow inside the trunks and branches of mangrove trees. Some species occupy others' burrows. Some beetles and moths excavate tunnels through mangrove stems and these tunnels are used by other insect species, such as ants, mites, cockroaches, termites, spiders and scorpions.

The insects through their feeding activities on leaves, flowers, seeds or mangrove propagules contribute to the development and sustainability of the mangrove ecosystem. Their impact was marked on the leaf chemistry, nutrient level, defoliation, nitrogen level, plant growth, etc. By eating dead wood or decaying leaves and breaking down dead plant tissues, the detritivores help increase the fertility of soil or sediment. Beetles are scavengers that feed on dead animals and fallen trees and thereby recycle biological materials into forms found useful by other organisms. Some are known pollinators of the angiosperms. While some species like mosquitoes are vectors of diseases, some predators and parasites play key ecological roles in the ecosystem. Some are also important in human and veterinary medicines.

Mitra *et al.* (2016) have recorded 591 species in SBR under 404 genera of 100 families belonging to 13 orders. In terms of dominance, the order Lepidoptera (butterflies and moths) was the most speciose, i.e. 163 species, followed by Diptera (flies): 133 species, Hymenoptera (sawflies, wasps, bees, and ants): 96 species, Coleoptera (Beetles): 81 species, Orthoptera (grasshoppers, locusts and crickets): 19 species, Dermaptera (Earwig): 8 species, Isoptera (termites): 7 species, Neuroptera (lacewings, mantidflies, antlions, and their relatives): 6 species, Thysanoptera (thrips): 2 species, Blattodea (cockroaches): 2 species and Phthiraptera (lice): 2 species. Formerly, the termites were considered a separate order, Isoptera, but genetic and molecular evidence suggests an intimate relationship with the cockroaches, both cockroaches and termites having evolved from a common ancestor.

Lepidoptera

The order Lepidoptera is divided into three groups- (1) butterflies, (2) skippers, and (3) micro (smaller)- and macro (larger)-moths. Their significance in the mangrove ecosystem is great as they act not only as herbivores and pollinators, but also provide food for insectivores including birds, small mammals, reptiles, etc. Their larvae cause damage to the plants and agricultural crops.

Butterflies (Biswas *et al.* 2013, 2016a; Chowdhury 2014; Ghosh and Saha, 2016)

Ecology

Farmland near forest is the preferred habitat of the species. There it occurs along field margins, paths and irrigation ditches.

Butterflies play a vital role in maintaining the mangrove ecosystem by pollination along with the bees and birds. For example, butterflies are known to visit and pollinate the flowers of *Bruguiera gymnorhiza*. They were also seen to lick salt from its branch, which can serve as an alternative nutrient source for them. Butterflies are also diurnal pollinators of *Acanthus ilicifolius*, *Algeciras corniculatum*, *Aristolochia* sp., *Avicennia officinalis*, *Calotropis procera*, *Ceriops decandra*, *Excoecaria agallocha*, *Ipomoea illustris*, *Mikania scandens*, *Nypa* sp., *Sarcolobus globosus*, *Sonneratia apetala*, *Tylophora indica*, *Wedelia chinensis*, *Wedelia biflora* and *Xylocarpus mekongensis*. Additionally, they are also a food source to predators like birds, spiders, lizards and other animals. A few butterflies are agricultural pests.

Diversity

Butterfly species in SBR are distributed among six families. These are Hesperiidae, Lycaenidae, Nymphalidae, Papilionidae, Pieridae and Riodinidae. The dominant families are Nymphalidae followed by Lycaenidae, Pieridae and Hesperiidae within the superfamily Papilionoidea. Other families represent less than ten species. Most species rich genera are *Papilio, Graphium, Prachliopta, Appias, Catopsilia, Eurema, Rapala, Mycalesis, Ariadne, Euploea, Junonia, Ypthima, Danaus,* etc. Majority of the species are rarer than common. Large numbers of butterflies were observed mostly during summer (April to June) and post-monsoon season. Many species are periodic visitors to the mangroves from the nearby places. This is because the area is covered with thick mangrove vegetation, which is very moist or wet due to proximity to the level of tidal water and favour butterfly activities and larval growth.

Moths (Biswas *et al.,* 2016b, 2016c, 2017a, b, 2018, 2019; Mitra and Panja 2017)

Ecological role

Moths perform both positive and negative roles in the mangrove ecosystem.

Adult moths and their caterpillars are food for a wide variety of wildlife, including other insects, spiders, frogs, toads, lizards, shrews, skunks, bats and birds.

Moth caterpillars have a great impact on plants by eating their leaves. This leads to many types of plants evolving special chemicals to make them less appealing to caterpillars to limit their damage.

The nocturnal adult moths also benefit plants by pollinating flowers while seeking out nectar and help in seed production. Nocturnal flowers with pale or white flowers heavy with fragrance and copious dilute nectar, attract these pollinating insects; for example, ample nectar producers, with nectar deeply hidden, such as morning glory, gardenia, etc. The giant moths fly upwind, tracking the airborne fragrance trail to a clump of flowers.

Moths play a vigorous and dynamic role in ecosystems as nutrient recyclers.

Moths can easily be used to illustrate the flow of energy released by their relentless consumption of living, senescent and dead coarse organic material in a natural system.

Moth larvae, commonly known as caterpillars, are especially good examples of primary consumers, specifically herbivores, as many species feed directly on the leaves, stems, flower heads, seeds, and roots of plants (primary producers).

Caterpillars are "shredders" because they break apart coarse organic matter, releasing carbon into the atmosphere (carbon cycle) and enriching soil as they reduce plant parts into humus where smaller "detritivores" play their role.

Moths are meteorologists of the environment in many ways, such as the effects of new farming practices, pesticides, air pollution and climate change.

Many new moths, considered serious pests to the agricultural crops and agroforestry (e.g. *Acherontia lachesis, Actias selene, Altha nivea, Amata passalis, Antheraea paphia, Anua coronata, Argina astrea, Ceryx godartii, Cnaphalocrocis medinalis, Creatonotos gangis, Cryptographis indica, Eupterote hibisci, Hippotion celerio, Hymenoptychis sordida, Hypsipyla robusta, Parapoynx diminutalis, Scirpophaga incertulas, Spilosoma obliqua, Spirama retorta, Spodoptera litura, Syngamia abruptalis, Theretra silhetensis, Thosea cana, Utetheisa lotrix* and *Utetheisa pulchella*), were seen migrated to the Sundarbans. *Hymenoptychis sordida* and *Hypsipyla robusta* are known as fruit borers in the Sundarbans. *Hyblaea puera*, a defoliator, was reported from the mangrove forests, especially *Avicennia* spp. in Kultali, LWLS, Bhagabatpur, Prentice Island, Ajmalmari and Jharkhali of SBR.

Flies (Maity *et al.* 2016; Roy *et al.* 2018)

Forms

Mangroves with their high salinity represent a harsh environment for most insects including the dipterans. There are three groups of Diptera (flies, mosquitoes, and related insects)- Nematocera (e.g., crane flies, midges, gnats, mosquitoes), Brachycera (e.g., horse flies, robber flies, bee flies) and Cyclorrhapha (e.g., flies that breed in vegetable or animal material, both living and dead).

Ecology

Diptera are recorded from different terrestrial habitats, such as soil, mud, decaying organic matter, dung, plant or animal tissues, carcasses of different animals, sweet and meat shops, fish markets, debris and flowering plants from different parts of SBR in different seasons. Only a small proportion of species is truly aquatic at the larval stage.

Dipterans play an important role at various trophic levels both as consumers and as prey. Whereas they are primary consumers in mangrove ecosystems, they feed on soil invertebrates and larvae of several other insect groups in premature stages. Negatively, they are involved in mechanical as well as biological transmission of a number of deadly viral, protozoan and bacterial diseases in livestock and wild lives.

The dominant families dipterans in SBR are Calliphoridae (Brachycera), Ceratopogonidae (Brachycera), Muscidae (Brachycera), Syrphidae and Tabanidae. Among the insect genera and species of SBR, the dipterans hold the second rank after the order Lepidoptera, but the number of diptera families is highest amongst all other insect orders.

Wasps

These social wasps are associated with the insular ecosystem of mangrove swamp, where a variety of ecological and climatic conditions allows them to have different nesting habits. Besides the lower vegetation diversity, the mangrove areas are under strong pressure of the harsh ecological factors that makes settling of most species difficult and favours dominance of the opportunistic species. Populations with wider ecological tolerances and under stronger pressure by harsh ecological factors could quickly colonise the area.

Honey bees

Rock bees (*Apis dorsata*) migrate from the Himalayas to the Sundarbans forest every year and make bee-hives in the mangrove plants. The flowering mangrove trees attract them during summer months which is the main flowering season. Flowering starts with the bloom of *Aegiceras corniculatum* during the end of March and is followed by the flowering of *Acanthus ilicifolius, Avicennia* spp., *Sonneratia apetala, Rhizophora* spp. This continues for two months- April and May. The density of honey depends on the number of salt-excretory glands available on the tree. Khalsi (*Aegiceras corniculatum*) having 19 glands per mm² produces the best honey. This honey has reddish color and sweet in taste and locally it is called Padma modhu (Lotus honey). The Garan produces the maximum quantity and the minimum is obtained from Gnewa. It has been found that Gnewa harbours about 39% of honey comb and Baen 16%, Garan 11%, Garjan 10% and others 24%. The bees prefer the Hental-Gnewa association for construction of hives. Beehives are usually seen in those branches of trees that have an angle between 45Â° and 90Â° and the height of the honeycombs from the ground is from 5' to 30'. Sometimes beehives are seen in those branches of trees that lean into the waters of rivers, canals and some other water bodies. The beehives in those branches of trees get destroyed and the beetles die during the high tidal water. Generally the bees make their hives in the branches of *Jana* and *Kankra* tree because those trees contain more leaning branches. The huge number of hives is also seen in the branches of *Passur* and *Dalchaka* trees, even in the branches of old trees. But the beehives are not normally seen in the branches of *Sundari* and *Chanda* trees. Even the existence of beehives is rarely seen in the bushes of *Gol* tree and *Bonlebu* trees because these trees have less wide spaces for making hives. Besides, the branches of *Banlebu* are small and branches of *Gol* tree are slippery. The number of pollen grain in the flower of *Sundari* tree is comparatively huge and the bees, though do not collect honey from the flower of this tree, they find important elements in the flower of *Sundari* for protecting their hives. After accumulating honey in the hives, the bees cover the hives with such pollen grain so that it does not fall down or get sticky. This pollinating spore is called *gutli* by the forest-dependent people. The colour of the *Sundari* and *Keora* tree pollen spore is yellowish, while the colour of the *Baen* and *Garan* flower pollen is brown.

The bees mainly play an important role in the natural pollination process of plants and trees as the movement of bees is found between the middle and top layer of the Sundarbans, where the flowers of all kinds of trees are also available. The bees live on the pollen and honey. The birds also live on the bees.

A single honeycomb can give from 2.5 kg to 12 kg of honey, but most commonly the yield varies from 10-12 kg per comb. Around 25,000 kg of honey is collected each year from the Indian Sundarbans. Nearly 20,000 tonnes of honey are produced every year by bee-keepers in the Sunderbans, of which about 20% caters to domestic markets. The largest chunk of honey produced in the Sunderbans heads to markets in the US, Germany, UK, Saudi Arabia, Belgium, Morocco, Canada and UAE.

Ants

Ecology

Ants, having astonishing risk-prone colonising ability, are important ecological community in the less favourable and challenging mangrove ecosystem from subterranean, soil nesting species to arboreal species nesting inside the dead vegetation, in the canopy of different trees down to the base, hollow (excavated or natural) twigs or logs, small branches, air roots, prop roots, leaf litters, grasses and periphery areas, adjacent to shallow brackish water. But their status is uncertain because they are least studied.

Nest not only helps in ant reproduction, but also provides safe refuge during fluctuating external environmental conditions in the Sundarbans. Some species have astonishing ability to adapt to extreme environmental conditions in the estuarine micro habitats like high temperature, regular flooding twice a day, occasional unpredictable flooding by salt waters and prolonged monsoon season. When the advancing tide reaches the nest, loose soil collapses into the entrances, blocking them and preventing water from penetrating the passages beneath. The mud dwellers survive during hours of flooding due to their low metabolic activity and by staying in the nest dug deep in the mud soil, where small air bubbles are trapped in the cavity. Some also drift to the dry environment during flooding or move up in the trees. Therefore, the ants nesting in the tidal part of the mangrove need to select an efficient escape tactic or nesting behaviour which prevents them from drowning and have to restrict their activities periodically. Majority of the terrestrial species generally live above the tidal level or remain arboreal by making nests above the highest tide. For example, *Oecophylla smaragdina* is strictly arboreal and is able to avoid the impact of tidal flooding. Not only that, being aggressive predators of other insects, in ant-plant interaction or mutualism, they serve as biological control agents by protecting the host and surrounding plants potentially against pest insects like leaf beetles or sesarmid crabs. Further, while foraging in the tidal zone, the workers of *Paratrechina* or *Solenopsis* escape the approaching high tide in a typical way. When touched by an incoming wave, they contract the legs and bend the abdomen to the ventral part of the thorax to form a nymphal posture so that they are able to float over the foam of the wave and surf the rolling wave. But many of the mud living ants die due to submergence in brackish water, particularly during the fortnightly spring tides or flooding during cyclones.

Most species are carnivorous, omnivorous, predators and some species are pests on the crop plants. Their feeding habits are oriented according to their preferred habitats. For example, the arboreal ants feed on honey dew, plant seed, nectar, fruits, etc. Abundance of the ants varies seasonally depending on the availability of their food. Ants are mostly prevalent during monsoon, followed by post-monsoon and pre-monsoon. However, a few species are found throughout the year. It was observed that the species are more abundant in the undisturbed habitats than the disturbed habitats in the forests. Restricted forest zone is the preferred habitat of most of the species. It was also revealed that the arboreal species are less abundant compared to the ground dwellers including those living under leaf litter.

Diversity

The number of ant species in the Sundarbans is relatively low. Low sampling efforts may have limited detection of the ants in the mangrove forests. Frequent flooding may also reduce the species diversity here. Species under the following genus were found in all the microhabitats- in the soil, under litter and on tree or leaf folders: *Dolichoderus, Oecophylla, Crematogaster* and *Tetraponera*. Some species under the following genus were not seen to use arboreal habitat, but conveniently remain a ground dweller living either in the soil or under leaf litter- *Camponotus, Leisiota,*

Polyrachis, Carebara, Myrmicaria, Pheidole, Brachyponera, Diacamma and *Pseudoneoponera*.

Beetles

Beetles may live beneath the ground, in water or as commensals in the nests of social insects, such as ants and termites. Many small Carabidae (ground beetles) live in moss on tree trunks. They play an important ecological role in the mangrove ecosystem. For example, the carabids are predators and help control the populations of many insects by feeding on caterpillars and larvae, many soft-bodied adult insects and their eggs. Similarly, both larvae and adults of the ladybird beetles (Coccinellidae) feed on plant-sucking insects (Homoptera), such as aphids and scale insects. Only a few coccinellids (e.g., Epilachna) feed on plants. Scarabaeidae (scarabs or scarab beetles), Silphidae (carrion and burying beetles), and Dermestidae (dermestid or hide beetles) are called scavengers because they break down materials such as dead logs, dead plant and animal matter, excrement and other waste products. Some are carrion feeding species. But the beetles are harmful also. For example, the leaf-beetle (Chrysomelidae) larvae feed on leaves, stems or roots of plants, while most adults chew leaves. On the other hand, the bark beetles cause damage to the living trees. The long-horned beetles (Cerambycidae) bore into stems, trunks and roots of living and dead trees and large semi woody herbs; tender new bark is often fed by the adults. The scarab beetles (Scarabaeidae) are crop pests. Knowledge concerning the role of beetles in transmitting plant diseases is lacking.

There are few predators of the well-armoured beetles. Some beetle predators feed particularly on beetle larvae, although many beetle larvae that feed on plants and in the ground probably are distasteful to birds and other predators.

Termites

Although the termites are often called "white ants", they are not closely related to ants. The termites are broadly classified into two groups-

Wood-dwellers- (i) Damp wood termites and (ii) Dry wood termites;

Soil-dwellers- (i) Subterranean termites, (ii) Carton-nest building termites and (iii) Mound-building termites.

Termites are detritivores, consuming dead plants at any level of decomposition. They also play a vital role in the ecosystem by recycling waste material such as dead wood, faeces and plants. Many species eat cellulose, having a specialised midgut that breaks down the fibre and are considered to be a major source of atmospheric methane, one of the prime greenhouse gases, produced from the breakdown of cellulose. They rely primarily upon symbiotic protozoa and other microbes such as flagellate protists in their guts to digest the cellulose for them, allowing them to absorb the end products for their own use. Most higher termites, especially in the family Termitidae, can produce their own cellulase enzymes, but they rely primarily upon the bacteria. Termites are consumed by a wide variety of predators, such as birds, mammals like pangolin, arthropods like ants, centipedes, cockroaches, crickets, dragonflies, scorpions and spiders, lizards, frogs and toads. Termites are less likely to be attacked by parasites because they are usually well protected in their mounds.

Spiders (Basu, 2019)

Spiders are one of the integral parts of the biodiversity amidst different mangrove and non-mangrove ecosystems in the study area. Generally, top-down (predatory) and bottom-up (productivity) factors are key determinants of consumer abundance. Hence, from an ecological point of view, spiders play a pivotal role as predators in different trophic levels. They are part of various insular food chains and food webs amidst different ecosystems. Prey availability and predation act in concert to set insular spider abundance. Species-specific predatory activities of the spiders are influenced by ecological characteristics, such as foraging and prey availability in micro habitats. Majority of the species are obligate carnivores hunting the insects, arthropods and other spiders, but few are herbivores and nectarivores too. Moreover, the spiders are also fed by the predators like the birds, reptiles, amphibians, arthropods like scorpions and mammals. Such predation also regulates spider numbers. Though most spiders are terrestrial- ground dwellers or arboreal, many species are also ecologically important within the semi-terrestrial and wetland ecosystems of the transition zone from freshwater to salt marshes, dunes or beaches and even in agro-ecosystem.

The typical biotopes of SBR have a spider fauna that is rich in terms of species diversity and individuals including many new discoveries (sp. nov.) and new records, as revealed by the present review of the recent data up to 2020. Altogether 267 species were recorded at different ecosystems in and around SBR. The predominant families (Genus: Species) are Araneidae followed by Lycosidae, Salticidae, Thomisidae, Oxyopidae, Clubionidae, Theridiidae, Sparassidae and so on. If explorations in the comparatively undisturbed core area are done with the same enthusiasm like those of the buffer and transitional areas, it is expected that the known diversity would have been much more in SBR.

Whereas most terrestrial species are ubiquitous, distributed in both natural and man-made habitats, a few species live near water bodies and brackish reeds in the marshy wetland. A few species live along shores or on the surface of shallow fresh or salt water and hunt in slow-moving water. Moreover, the biodiversity of spiders is also associated with agro-ecosystem in the fringe areas where the spiders are considered a significant agent in reducing the insect crop pests. These insectivores are mostly prevalent during post-monsoon followed by pre-monsoon and monsoon which is in conformity with the incidence of the insect species. The flooding regime is the major factor controlling the zonation of these invertebrates. Habitat reduction becomes highest for species of the lower salt marsh zone and at the slopes of the dikes.

Mites and Ticks (Ghosal 2017; Gupta *et al.* 2017; Kar & Karmakar 2021)

The Acarina or Acari are a taxon of arachnids that contains mites and ticks.

Ecology

Oribatid or Cryptostigmatial mites, commonly called 'beetle' or 'moss' mites, appear to be predominant among the common soil inhabiting arthropods. These mites occur in soil, leaf litter, nest and burrows of other animals, debris, humus and compost heaps. They are also found to inhabit other habitats like tree trunks, moss, lichens, etc. Population of the phytophagous and predatory mites fluctuates seasonally in mangrove vegetation and agro-horticultural crops in SBR. Maximum abundance, positively correlated with the temperature, RH, and rainfall, was observed during post-monsoon.

Role

The mites, especially the soil-dwelling species, are ecologically important because they take active part in predation and decomposition of organic matter promoting soil fertility. They also disseminate plant diseases mainly root rots and fungal diseases and also act as intermediate hosts of tapeworms of domestic ruminants.

Nematode

Nematodes or roundworms occur as parasites in animals and plants or as free-living forms in soil, fresh water, marine habitats and water-filled cracks deep within the soil. Most of the nematodes reside in the top 15 cm of soil. They are part of the benthic meiofauna and mainly serve as food for the higher trophic levels in addition to other ecosystem processes. They do not decompose organic matter, but they can effectively regulate bacterial population and community composition, even up to 5,000 bacteria per minute. Also, they can play an important role in the nitrogen cycle by way of nitrogen mineralization. Moreover, this group of metazoan fauna is considered to be reliable bioindicators as they respond to natural and anthropogenic disturbances within their environment.

Marine worms

Species of the marine worms live in burrowed substrata and found in the temporarily exposed intertidal limits to the very deep seas as well as some estuaries. Though they are pelagic, they adapt themselves to the estuarine habitat.

Annelids (Bhowmik *et al.*, 2021)

Annelids, also known as ringed worms or segmented worms, including ragworms, earthworms, and leeches. They exist in and have adapted to various ecologies. Annelids live in a diversity of freshwater, marine and terrestrial habitats. Some species live in marine environments as distinct as tidal zones and hydrothermal vents, some in fresh water and others in moist terrestrial environments. Some burrow while others live entirely on the surface, generally in moist leaf litter. They are important in aeration and enriching of soil because when burrowing they loosen the soil so that oxygen and water can penetrate inside. Both surface and burrowing worms help produce fertile soil by mixing organic and mineral matter and accelerating the decomposition of organic matter and thereby making it more quickly available to other organisms and by adding mineral inputs and converting them to the much needed materials that plants can use more easily for their growth. The burrowers also encourage growth of populations of aerobic bacteria and small animals alongside their burrows.

Oligochaeta lives mostly in the freshwater or as terrestrial species. They are usually deposit feeders that eat soil and sediment while burrowing the ground. Some species are predatory too. In general, predatory oligochaetes are smaller than the soil-consuming relatives. Some species of oligochaetes are also vulnerable to infection by internal parasites, including by species of other worm phyla, such as flatworms and nematodes.

Earthworms are oligochaetes that support terrestrial food chains as prey of the birds like storks, smaller mammals and invertebrates. They also serve humans as food and bait.

Earthworm feces, called worm casts, are very rich in plant nutrients like nitrogen and phosphates. They are deposit feeders. They burrow through the ground, eat soil and extract organic matter from it. They are capable of forming organic compounds by digesting dead plant matter and microorganisms that make soil suitable for vegetation. They do this by digesting dead plant matter and microorganisms, many of which feed on dead animal carcasses and convert them into simpler molecules that can be absorbed and used by living plants.

Polychaetes are primarily marine organisms that either live a sedentary, burrowed existence or are active predators that move by crawling on the ocean floor or swimming. Polychaetes are sedentary filter feeders. The diet of polychaete species shows large variation. Some species are active predators, while others are scavengers. Among the burrowing species, there are species of deposit feeders and filter feeders. A number of annelid species, including many polychaetes, engage in commensal symbiotic relationships (commensalism). For example, a polychaete species may live on the body of an echinoderm. The echinoderm is unharmed, but the polychaete benefits from the stray food particles that can be found around the mouth region of echinoderm.

It may be noted that the burrowing of oligochaetes is much different from that of sedentary polychaetes because oligochaetes are not sedentary. Oligochaetes form long tunnel-like burrows through the soil, while the sedentary polychaetes do not crawl through the sand or soil, but simply dig a hole to live in, where the worm often forms a small tube slightly larger than the worm's body length. For this reason, most sedentary polychaetes are called tube-dwellers. Some species will remain in this tube throughout their lives. The tube can be formed from a variety of materials such as substances secreted by the worm, sand and debris, or a mix of these. In order to eat, they either extend feeding tentacles out of the tube to trap food particles or just eat the sediment they are burrowed in. However, not all tube-dwellers form their tubes burrowed in a hole. Some attach themselves to a rock or some other solid support within the ocean and then secrete a tube in which to live.

The leeches of the subclass Hirudinea are all carnivores. They are either predators or parasites. As predators, they capture and eat whole organisms, often other invertebrates. The remaining species are external parasites that feed off the blood of their vertebrate hosts. However, some transmit flagellates that can be very dangerous to their hosts, such as some small tube-dwelling oligochaetes that transmit myxosporean parasites causing whirling disease in fish.

Mantid

Mantids are entirely terrestrial and inhabit primary and secondary forests and grasslands. Praying mantids have a tritrophic niche, i.e., they simultaneously occupy two consumer trophic levels in natural ecosystems by virtue of feeding on herbivores, other carnivores and pollen. Most mantids are cryptically coloured to blend with their environment. Many species are diurnal. However, other species still fly at night as there are fewer predators.

Ecologically, most of the mantids are considered top arthropod predators in the food chain. They are generalist feeders and can catch and consume arthropods primarily of equal or smaller size including butterflies, moths, crickets, grasshoppers, flies and other insects. Rarely, larger mantids have been known to ensnare small mice, lizards, frogs, fishes and birds. Hatchling mantids typically feed on aphids and other insects of similar size. Since mantids are cannibalistic and also feed on other beneficial insects, their value as biocontrol agents is probably rather limited.

Bugs (Hassan and Biswas, 2014; Mitra, 2017; Mitra *et al.*, 2016; Mitra *et al.*, 2019)

The order Hemiptera (bugs, cicada, leafhoppers, scale insects) constitutes a large part of the insect fauna, both in terrestrial and aquatic ecosystems of SBR. Hemiptera occur in all forest strata (e.g., soil, leaf-litter, understorey, overstorey). There are a series of aquatic and semi-aquatic families that show a gradual transition of habitats from damp shores to subsurface waters. The majority of the aquatic forms belong to Nepomorpha and Gerromorpha (water bugs) under the suborder Heteroptera (true bugs or typical bugs). They colonise a broad range of habitats, including deadwood structures of living trees as well as standing and downed logs of different diameters and decay stages. Some remain confined to the shore, some venture further and abound the floating algal mats and other floating objects and some skate rapidly over the surface of water. The true aquatic species are normally found below the surface. Although they are mostly herbivores, the group exhibits a wide spectrum of feeding habits, including predators, fungivores, and parasites. The herbivores have variable host specificity levels-some species are highly monophagous (feeding only from one species of plant), whereas others are highly polyphagous (feeding from different families of plants), preferring the plants with high nutrients. For this reason, Hemiptera are sensitive to changes not only in habitat structure, but also floristic and plant chemistry. Similarly, seasonal changes can also influence the structure of hemipteran assemblages at spatial scales ranging from individual plants up to forest plots.

Study on the occurrence of this insect fauna is not studied in depth in the mangrove forests, rather mostly focuses on the non-mangrove habitats.

Aphid or plant lice

Aphids are soft-bodied insects that use their piercing sucking mouthparts to feed on plant sap. Aphids colonise herbaceous plants (undersides of tender terminal growth), mosses and ferns. Diversity is highest on trees, where more feeding niches are available. They feed on soft stems, branches, buds and fruit, preferring tender new growth over tougher established foliage. They pierce the stems and suck the nutrient-rich sap from the plant, leaving behind curled or yellowed leaves, deformed flowers, or damaged fruit. Most aphids feed on a wide variety of plants, although some species are specific to certain types of plants. Aphids secrete a sticky substance called honeydew that attracts ants, so following a trail of ants into a plant can often lead to a discovery of an aphid infestation. Ants are known to protect aphids from natural predators and even herd them into tight colonies so they can harvest the honeydew easier. Honeydew also creates a favorable environment for sooty mold to grow and spread. Some aphids are very important vectors of plant viruses.

Echinodermata ("spiny skinned")

Echinoderms are exclusively marine animals. They occur in various habitats from the intertidal zone down to the bottom of the deep sea trenches with little and limited tolerance to narrow ranges of salinity variations. Hence, their occurrence in truly estuarine, brackish water or freshwater stretches is very rare. Echinoderms serve as hosts to a large variety of symbiotic organisms including shrimps, crabs, worms, snails and even fishes.

Encyrtid wasps

Encyrtid wasps are biologically diverse and widespread in nearly all habitats and are extremely important as biological control agents. Their hosts include spiders, ticks and other insects. Some species are primary parasitoids while several species are also hyperparasitoids on a wide range of other parasitoids.

Ostracods

The ostracods occur within the benthos of nearly every conceivable aquatic system from temporary ponds to large rivers to groundwater habitats, but they rarely enter the plankton even though many species are good swimmers. Almost all are free living and most are herbivores on attached algae or detritivores.

Springtails

The soil springtails occur in all terrestrial habitats. They are epedaphic living above the ground on mangrove vegetation, the canopy or interstitial species living in sand in supralittoral zone. Sometimes they are hemiedaphic in forest living in the litter and the humus, or may be euedaphic living deep in the soil. The species living in the intertidal zone are adapted to higher salinity and the osmotic pressure. The epineustonic species live on the surface of fresh water. Rarely some live in commensals, with termites in the closed environment of termites nest. Springtails are eurythermal with rather broad thermal tolerance. They can also tolerate drying of the substrate to some extent. Some species lay eggs in water. Majority of them are polyphagous. They feed on organic remains, foliar parenchyma, decomposed wood, animal excreta and especially, pollen, algae, mycelium and spores of mushrooms-fungi and bacteria. Such detritivorous feeding habits enable them to play an important role in biological breakdown. They are also predated by the centipedes, spiders, opiliones, pseudoscorpions, mites, insects, such as flies (Diptera), beetles (Coleoptera) and ants (Formicidae) and vertebrates including lizards, frogs and birds.

Water fleas

Most forms are found in freshwater habitats, such as vernal pools, lakes, ponds, wetlands and even roadside ditches but a few occur in marine environments. Apart from a few predatory forms, water fleas feed on microscopic particles of organic matter, which they filter from the water. In addition to detritus, they collect Algae, Bacteria and Protozoa. Water fleas are eaten by the fishes, larvae of damselflies and aquatic beetles.

Comb jellies

Ctenophores are exclusively marine; found in coastal and brackish waters, mostly planktonic; play a significant role in the pelagic food web as they feed on fish eggs and larvae, and compete with juvenile fish for food by preying on smaller zooplankton such as copepods.

Fish (Chatterjee *et al.* 2013; Mandal *et al.* 2013; Gupta *et al.* 2016; Roy *et al.* 2016; Chakraborty *et al.* 2018, 2020, 2021; Mandal and Deb 2018; Mitra 2019; Sen and Mandal 2019)

Fishes are typically divided into three groups: superclass Agnatha (jawless fishes), class Chondrichthyes (cartilaginous fishes) and superclass Osteichthyes (bony fishes). The latter two groups are included within the infraphylum Gnathostomata, a category containing all jawed vertebrates.

Fishes are aquatic and found in all types of waters in the Sundarban mangroves- freshwater (*Labeo*), marine (*Stromateus)* and brackish waters (*Chanos*). Marine and estuarine fishes are the dominant vertebrate necton that live in saltwater, brackish water for rich food resources or even migrate in the fresh water system for breeding purpose and use as nursery sites, especially during early juvenile stages of growth. Fishes lay their eggs in extensive roots of mangrove trees and after hatching they feed on detritus and other food resources which are easily available in mangrove areas. The water in mangrove areas is turbid and rich in detritus which provide instant food items for juvenile fishes and also reduce predator's attack. Presence of mangroves and other associated floral species also regulate distribution of these fish in their habitats. A variety of fish species use mangrove areas for foraging, i.e. feed on amphipods, isopods, crabs, snails, insects, spiders, copepods, shrimp, and organic matter. In addition, fishes are sources of food for a variety of wildlife species such as birds, reptiles and amphibians, mammals, carnivore fishes and invertebrates. On the other hand, fish is highly nutritious and a major source of human diet, i.e. proteins, vitamins and micronutrients, particularly for the fringe people living below the poverty line.

Fishes in the Chondrichthyes or cartilaginous fish (e.g. whale shark, dog shark, sting rays, guitar fishes, etc) and Osteichthyes or bony fish (e.g.

Tenualosa ilisha, Scatophagus argus etc) are considered to be more advanced fish found in the Sundarbans.

They may also be classified into two types on the basis of their mobility-resident, such as *Arius jella, Coilia ramcarati, Harpodon nehereus, Hilsa toli, Ilisha elongata, Lates calcarifer, Mugil parsia, Mugil tade, Otolithoides biauritus, Pama pama Polydactylus indicus, Polynemus paradiseus, Setipinna phasa, Setipinna taty* and *Sillaginopsis panijus* and migrants (upstream and downstream), which are again divided into three sub-groups- (i) Marine forms that migrate upstream and spawn in freshwater areas of the estuary like *Tenualosa ilisha, Polynemus paradiseus, Sillaginopsis panijus* and *Pama pama,* (2) Freshwater species, which spawn in saline area of the estuary like *Pangasius pangasius,* and (3) Marine species, that spawn in less saline water of the estuary like *Arius jella, Osteogeneious militaris* and *Polydactylus indicus* (Saha *et al.* 2018).

The fish diversity and abundance in the Sundarban mangroves depends upon a suite of factors-

  (1) the salinity regime along the shoreline and inner estuaries;
  (2) the water depths within the mangrove forest interior;
  (3) the dominant mangrove species and the structural complexity of the root system;
  (4) whether the channels are blocked or choked due to different biotic and abiotic reasons;
  (5) the type of substratum (ranging from soft mud to hard sand and rock); and
  (6) proximity to open water habitat.

The reasons for such dependency are-

(i)   reduced predation linked to depth, structure, and turbidity;
(ii) increased food supply for post-larvae and juveniles; and
(iii) shelter in quiet (non-turbulent) waters for post-larvae and juveniles.

Therefore, composition of the mangrove fish is determined by interplay of a host of factors including structural diversity of the habitat, hydrological features of current speed, tidal range, turbidity and salinity and the nature of adjacent waters. The importance of the forests varies according to the age of forests, i.e. the invertebrate epifauna including fish diversity is highest in the mature forests, medium in about 15 year-old forests and low in the recently cleared forests. Hence, the quality of forests is more important in the invertebrate diversity and abundance considering the variation of productivity.

The mangrove waters, usually rich in detritus, are highly suitable for fishing. Mangroves enhance fish production by providing rich food and shelter. Mangroves forests are highly productive considering that the mean levels of primary productivity here is close to the average for tropical terrestrial forests. Their leaves and woody matter (detritus) form a key part of the marine food chains that support fisheries. Decomposers of this detritus include microorganisms such as bacteria and oomycetes, as well as some crab species. They process the leaves and woody matters into more palatable fragments for other consumers. Contribution to this productivity is also made by the periphyton and phytoplankton, occurring on mangrove trees, in soils and water columns, which typically have lower rates of productivity than the trees themselves, but are nutritionally more beneficial to the consumers. Moreover, the incoming nutrients from rivers and other adjacent habitats in the form of dissolved and particulate organic carbon and living biomass, such as planktonic larvae and maturing fish and invertebrates enrich this productivity.

In SBR, the dominant fish species include, among others, *Amblypharyngodon mola, Bregmaceros mcclellandi, Coilia ramcarati, Escualosa thoracata, Gonialosa manmina, Ilisha kampeni, Planiliza tade, Periophthalmus novemradiatus, Puntius sophore* and *Salmostoma bacaila*. The dominant orders are Perciformes, Clupeiformes, Gobiiformes, Pleuronectiformes and Scombriformes. There are also other orders which are not well represented. These are Siluriformes, Mugiliformes, Cypriniformes, Scorpaeniformes, Anguilliformes, Tetraodontiformes, Beloniformes, Myliobatiformes, Aulopiformes, Cichliformes and Gadiformes. There are more than one and a half dozen new records from the region.

Amphibia (Purkayastha *et al.* 2019)

Amphibians live in both terrestrial and aquatic environments. Mangrove frogs are predators that may eat almost every small living thing such as insects (e.g., beetles, bees, ants, termites, crickets and bugs), snails, smaller toads, prawns, and fishes. Hence, they play a very important role in the food chain of both terrestrial and aquatic ecosystems and are widely considered to be useful as bio-indicators. They perform this function as secondary consumers in many food chains. They are herbivorous to omnivorous. Tadpoles have a significant impact in nutritional cycling. They are valuable for determining the overall health of both these ecosystems. Since they are sensitive to climatic change, particularly temperature and moisture conditions, they may be used as a bio-indicator of environmental degradation, fragmentation, ecosystem stress, impact of pesticides and various anthropogenic activities. Changes in morphology or ecology of these species may indicate high levels of pollution and other threats to their survival.

Amphibians, particularly the anurans, control the crops pests like the snails. Invertebrates and vertebrates also predate them. But during the 20th century, the frogs were exploited randomly for food, model organisms in ecological, embryological, physiological and genetic research and their population had decreased alarmingly. This has resulted in an increased insect pest population. Because of their importance in the ecosystem, decline or extinction of their population has a significant impact on other organisms along with them.

Reptiles

Reptiles, traditionally considered one of the major classes of the phylum Vertebrata, include three major groups: turtles (Order Testudines), crocodilians (Order Crocodylia), snakes and lizards (Order Squamata). Mangroves provide ideal habitats for a rich reptile fauna. Reptiles in the study area are mostly inhabitants of the inland waters- permanently aquatic or semi aquatic. Inland waters can be apportioned into habitats that are categorised as fresh, brackish and even saline (rare). Only a few specialized species are permanent residents of brackish waters. Inland fresh waters can be classified as lentic or lotic or intermediate habitats. Brackish water systems can also be estuaries or coastal marshes that have both lentic and lotic components.

More than 99% of lizards are wholly terrestrial, many being primarily arboreal, and only a few species have aquatic tendencies. Most turtles are partially or totally aquatic, living in fresh waters or brackish waters, while all crocodilians can be found in inland waters although some venture into marine habitats. Most snakes are terrestrial, including many species that are arboreal or fossorial. However, many species have affiliations with freshwater habitats. Overall knowledge of the ecology of snakes lags behind that of many other groups of vertebrates.

Turtles

All turtles are placed within the order Testudines. When chelonians are referred to, it indicates turtles, tortoises and terrapins as a group. Turtles inhabit a wide range of habitats from oceans to freshwater ponds and rivers, swamps, brackish water estuaries and marshes. Most turtles require both an aquatic zone for swimming and feeding, and a land area for basking. They utilise mangrove areas, estuaries and creeks for foraging and breeding purposes due to the richness and diversity of plankton and benthic food resources. They use sandy beaches for breeding purposes. Tortoises live on land in forests and grasslands in the Indian Sundarbans.

Snakes

Mangrove snakes are aptly named for the areas they inhabit: mangrove forests, riverine areas and coastal lowland forests. The secretive nature of snakes in the wild makes them seem rare in the dense mangrove swamps. The snakes live usually near water. The carnivorous snakes feed on reptiles, birds, amphibians and small mammals and play a very important ecological role in their environment, both as predator and prey species in Sundarban. They help control populations of small mammals, birds and reptiles they prey on. Apart from this, the poisonous snakes have immense medicinal values for their venom. The snakes are also eaten by fishes, crocodiles and eagles.

The snake populations in Sundarbans are declining due to habitat loss caused by tremendous population pressure, netting by fishermen and retaliatory killing to avoid snake bite because thousands of people are bitten and die from snakebite every year. The common poisonous snakes are Common krait (*Bungarus caeruleus*) and Monocled cobra (*Naja kaouthia*). Nearly 66% of the snake bite deaths were due to Common krait bite and about 34% by Monocled cobra, most of them occurring on the floor bed in the months of June to September. Only one case of death was due to sea snake bite, but there is no reported death due to Russell's viper (*Daboia russelli*) bite. The most affected areas are Basanti Block followed by Gosaba, Kultali, Patharpratima, Jaynagar, Mathurapur, Namkhana and

Sagar Blocks. Hospital cases are better treated. Still the villagers go for the local quacks due to ignorance and traditional belief system (superstition). Awareness campaigns are often arranged by the NGOs at the grass root level. Whenever rescued, the species are handed over to the FD for release in the wild.

Crocodiles

Sundarbans was the home to three species of crocodiles as detailed below. These crocodiles principally live in the aquatic habitats ranging from forest streams, rivers, marshes and swamps to salty waters of mangroves or estuaries, where salinity changes with different seasons and distance upstream.

1. The saltwater crocodile (*Crocodylus porosus*) is abundant in the Sundarbans, but was heavily exploited for their skins until the 1970s and, consequently, greatly depleted. The movement of saltwater crocodile towards the coastal shore, between river systems or within tidal and non-tidal habitats, including homing associated with released captive-bred individuals with tags appears to be related sex and ontogenetic changes in social status.

They often come to the creeks in the fringe villages in search of food. Even the ferocious saltwater crocodiles in the rivers or creeks of the Sundarbans do not spare even the king of the forest, the tiger. The tigers often swim across the Dobanki Canal to travel between the forests of Pirkhali-5 and Pirkhali-6 of the Sajnekhali WLS. A case of tiger-crocodile conflict was reported from STR.

On a full monsoon day of August 2011, a crocodile, about 14 feet long, was being observed in the canal-month, where it took shelter a few weeks ago, when two deer were also seen swimming across the canal. Instantly, the crocodile attacked them. A deer escaped with injuries. The crocodile dragged the other into the deep water. Later, the victim's half-eaten body was seen floating in the canal. But when the victim is a strong full-grown tiger, about seven and a half feet long, it is not so easily hunted by the crocodile. But, the tiger could not resist the crocodile for long in the muddy canal because the crocodile attacked the tiger from behind while it was swimming across the canal; as a result, the backside of the tiger bore the signs of crocodile's biting. As reported by Pandit and Guha (2015), on

9th August, 2011 at about 8.45, carcass of a tiger was recovered by the staff at Pirkhali- 5 compartment near Dobanki camp. After post-mortem, it revealed that the tiger is a female although the vital reproductive part of the animal was eaten by a crocodile, but only four slightly large teats were observed. Age of the tigress was estimated to be 5-6 years. All four canines were found intact. Its length from nose tip to lumber vertebra was 108 cm and height was 84 cm. Tail and hind portion had been eaten by the crocodile so full body length could not be measured. In all 22 deep wounds were marked on the body- nine on the right upper abdomen, two on the left upper abdomen, seven on the right dorsal foreleg, three on the dorsal thoracic region and one on the dorsal posterior neck region. Jaws of the crocodile were so powerful that it could penetrate through the rib of the dead tigress. The portion beyond the second lumbar vertebrae was totally chewed off. Only a flap of skin held the dislocated femur up to paw of the left hind leg. Lungs and both the kidneys were found to be intact and a large torn vent was observed in the diaphragm. All visceral organs like heart, trachea and esophagus were in normal condition; pleural cavity was completely damaged through the big diaphragmatic vent. Blood tinged fluid was observed in the pleural cavity. The morphology of the liver was normal with only a few patchy whitish marks were observed. The gall bladder, ducts and lymph grand could not be identified due to massive loss of texture of the visceral organ. The pericardial sac was full of blood tinged fluid. The heart-muscle was not decomposed, numerous haemorrhages were marked in the pericardium of both auricles. All the three chambers of heart were empty except the left ventricle, which was full with clotted blood. In the esophagus, a piece of tail-skin of the water monitor lizard was observed. The small and large intestine were not found during the post-mortem because these were eaten up by the crocodile. Similarly the reproductive organ and the ordinary bladder were also not traced. From the second vertebra to the posterior the skeleton disappeared except the left hind leg, which was detached from the tabular cavity and hinged to the skin. Some of the wounds are thought to have occurred due to the crocodile's tail-lashings because after biting the victim. Once overpowered, the victim was dragged under the water. Then the injured victim died of suffocation. As per the post-mortem report, the cause of death of the tigress was crocodile attack.

On July 5, 2021, while returning from the 'Vaccine on Boat' program from the Kumirmari Island, the staff of Gosaba Block Health Department on board saw a full-grown crocodile jumping into the Raimangal River near Bagna Forest with a deer held in its mouth.

It has been observed that out of 60-70 eggs laid by this species, 40-50 or less number of crocs actually survive. After hatching, the young crocodiles used to jump into the water by running or riding on the mother's back. Up to 5-7 weeks the mother crocodile takes care of her babies. The crocodile's eggs are a favorite food for the pigs, snakes, and other wild animals. Even when the flood water enters the nests, most of the eggs are destroyed. The continuous increase in salinity in the rivers of Sundarbans is also adversely affecting reproduction of the crocodiles.

2. The Marsh crocodile (*Crocodylus palustris*) was extirpated from the Bangladesh Sundarbans during 1980s for hunting and loss of habitat. The last open habitat of this freshwater crocodile, which lived for nearly six hundred years, in a waterbody locally called Thakur (Khanjeli) Dighi measuring two hundred acres (about 25 km north of the Sundarbans) adjacent to the shrine of Hazrat Khan Jahan Ali at Bagerhat (then named 'Khalifabad').

3. The Gharial (*Gavialis gangeticus*) was almost extirpated from the Sundarbans during the 20th century due to habitat destruction, lack of natural sites to lay eggs, spoilage of eggs, death of young being entangled in the nets used by fishermen, reduced water flow in the river during the dry season and degradation of the aquatic environment. However, on December 27, 2018, a juvenile male of this rare species was seen basking at about 8 am on the bank of a narrow creek in STR. Presence of a baby gharial means existence of the adult gharials in the wild.

Saltwater or estuarine crocodile *Crocodylus porosus* appears to have a substantial estuarine trophic role. Adult estuarine crocodiles are common in the tidal estuaries, coastal lagoons and mangrove swamps. Often, the crocodiles are seen in the mudflats while basking in the sun. Although they are the top predator in the aquatic ecosystem of Sundarbans, they are relatively less efficient foragers on fish or other immersed prey and snap at schools rather than attacking individuals. On the other hand, they are highly successful water's edge specialists that are capable of hunting terrestrial birds and mammals. Adult crocodiles will also scavenge from large carcasses. Females selects the nesting sites and both the partners defend the nesting territory. Nesting site is typically a stretch of shore along tidal rivers and swamps. Nests are open in exposed location, often in mudflats, sometimes covered with vegetation. It is evident that female saltwater crocodile scratches a layer of leaves and other debris around the nest entrance for protecting the eggs. They mate in the wet season, when levels of water is highest. The courtship starts in the months of September-October. Female lays eggs between November to March. Very little information is available on this apex predator in the Sundarbans aquatic ecosystem.

Census

The first census of the estuarine crocodile was conducted for four days in January, 2012 in all the creeks and rivers covering about 1,160 km area. This operation was conducted by the Wildlife Institute of India, Dehradun and the Wildlife Wing of the FD of West Bengal. The field exercise for the census was carried out over three days in January 2012 by 35 groups of enumerators who traveled, along a straight path, for 613 hours, covering 1,164 km. The enumerators found evidence of 240 crocodiles; 141 direct sightings of the reptile and river bank trails of 99. In all, 240 crocodiles were spotted during the period. Presence of 0.12 crocs was found including the breeding population. Five or six of the crocodiles sighted were more than 20 feet long. The eastern part of the Sundarbans National Park, which has higher salinity, has the highest density. The juveniles are concentrated in Ram Ganga area of the estuarine region. A good number of sightings were recorded from the East (28 direct and 15 indirect), West (40 direct and 51 indirect) and Sajnekhali (30 direct and 14 indirect) ranges in the Tiger Reserve, whereas the Ramganga (15 direct and 3 indirect) Range in South 24-Parganas Division. The census for the first time has set a benchmark for the crocodile population in the Sunderbans. During 2020-2021, its population was estimated to be 144 excluding 26 hatchlings.

The crocodile numbers in Sunderbans are "encouraging" and comparable to some of the best crocodile populations in salt water regions of north Australia and Papua New Guinea. This healthy population is an indication that the crocodiles have the capacity to withstand the increasing salinity and surface water temperature in the region. Rapid destruction of breeding grounds, poaching and fishermen's indiscriminate access to rivers and canals of the Sundarbans contribute to rapid fall in the crocodile population of the forest.

Lizards, geckos and skinks

Mangrove areas are also home to a few lizard (Suborder Sauria) species. Water monitors, one of the world's largest species of lizard, are common throughout the coastal areas of the Sundarbans and blend in perfectly with the exposed roots of the mangrove trees. They inhabit forest, river banks, by the side of nullah, marshy land and tidal creeks. They occupy burrows on muddy bunds around ponds, lakes, creeks, marsh lands, dense vegetation, grassland, paddy fields, bamboo, hollows of trees and crevices. They prefer slopes or under exposed roots for burrowing. They are good climbers, runners and also good swimmers. These are predators of different animals such as birds, amphibians and small reptiles. They eat large numbers of insects and other invertebrates. Large predatory lizards, such as monitors may be considered pests because they often prey on farm animals or steal chicken eggs. But lizards are not disease vectors. The monitor lizards have been exploited commercially for their valuable skin and also captured for eating by fisher folk or traditional medicinal use in the Sunderbans. Frequent trafficking of geckos is also reported from the Sundarbans.

Aves

The highly productive ecosystem provides diverse and abundant floral and faunal food resources, having high caloric concentrations, to the birds. All of its terrestrial, arboreal and hydrological habitats are conveniently utilised by the bird communities, a mixture of resident, breeding birds, summer and winter visitors (mostly aquatic) for their survival. There are two types of users in the mangroves- aquatic birds and terrestrial birds. The former ones entirely depend on water for a variety of activities such as foraging, nesting, loafing and moulting, but the latter ones do not entirely depend on water but may visit some time in search of food, shelter and perch. They use multiple habitats in the mangroves, such as forests, mudflats, estuaries and fringe areas, where food resources are available like fish, turtle, snake, amphibian, mammals and invertebrates, e.g. gastropods, bivalves, crabs, prawn, nekton and insects.

Habitat utilisation

Birds of Sundarban use different habitats suitable for roosting, foraging and breeding purposes. Sundarban offers two major terrestrial habitat types (covering about 60% of the area) for avian use:

Littoral forest facing the sea: This coastal habitat is used by the birds under the Orders Charadriiformes, Procellariiformes and Suliformes like Noddies, terns like Sooty and Roseate, gulls (*Larus* spp), Shearwaters, Boobies etc. use this habitat for breeding purposes, when the sea food provides them nutrition.

Tidal swamps: About 40% of Sundarban is under deep water in the form of estuaries and large rivers. This aquatic habitat is used by the birds belonging to six Orders- Anseriformes, Podicipediformes, Pelecaniformes, Ciconiiformes, Gruiformes and Charadriiformes.

The new depositions and intertidal mudflats are characterized by *Avicennia* (Avicenniaceae) and *Sonneratia* (Sonneratiaceae), flanked by foreshore grassland, gradually replaced by the mangrove species. The tidal mudflats provide the much needed benthic food supply for the birds. The inland and foreshore grasslands are also used by many small and large bird species and many make their nest in this habitat. Specifically, the birds such as larks, pipits and buntings use the grass and shrub areas. The birds like Red-vented Bulbul (*Pycnonotus cafer*), Red-whiskered Bulbul (*Pycnonotus jocosus*), Cattle Egret (*Bubulcus ibis*), Indian Pond Heron (*Ardeola grayii*), Little Egret (*Egretta garzetta*) and Purple Sunbird (*Cinnyris asiaticus*) are often seen moving from grasslands to the mangrove areas because they use the grasslands for foraging and the mangroves for perching.

High species richness in the Sundarban mangroves is associated with plant species richness, the density of the understory and food resource distribution. The immensely productive Sundarban estuarine wetlands are constantly fed by nutrients brought in by the north-south flowing freshwater channel and flushed by the ebb and flow of the tides from the BoB to support a diverse plant and animal communities including the terrestrial, arboreal and aquatic birds. Whereas the phyto- and zoo-planktons feed the fish, crabs, prawns, shrimp and mollusks in the shallow intertidal network, these in turn support the wading migratory and resident birds. Mangrove bird species with larger bills comprise both arboreal and ground foragers that feed primarily on crabs or insects, and larger bills are most prominent among passerine species that feed primarily on the ground.

Since the terrestrial type of habitat provides ideal foraging and breeding sites and also shelter for the birds, it is utilized by a large number of terrestrial species, grouped according to their favourite food and technique of capture. The availability of trunks, limbs, and foliage comprising the tree canopy enables a variety of passerines and non-passerine birds, which are not found in other wetland areas, to use mangrove swamps. It also allows extensive breeding activity by a number of tree nesting or ground-dwelling birds. In fact, Sunderbans is a potential breeding ground of immense variety of birds like Heron, Egret, Cormorant, Fishing Eagle, White Bellied Sea Eagle, Seagull, Tern, Kingfisher, Black-tailed Godwit, Little Stint, Eastern Knot, Sandpiper, Golden Plover, Whistling Teal, Pintail, and White-Eyed Pochard etc. because the higher density of mangroves provides them a secure habitat for a nest and the fledglings. Clamorous Reed Warbler (*Acrocephalus stentoreus*), Ashy Prinia (*Prinia socialis*), Plain Prinia (*Prinia inornata*), Scaly-breasted Munia (*Lonchura punctulata*), Baya Weaver (*Ploceus philippinus*) and many others are often seen breeding in the mangroves and associated plants, especially in the monsoon. During monsoons heronaries develop in eastern part of Sundarban. The key breeding areas of the threatened waterbirds are the Haor (backswamp) areas, new accretion in river systems and mangroves of offshore islands.

The mangroves, mudflats, estuaries and adjacent areas are rich in food resources, which include fishes, crabs, turtles, snake, amphibians, small mammals and invertebrates such as gastropods (more dominant), bivalves (less dominant), prawn, nekton and insects. Most of the mangrove-restricted bird species feed primarily on insects in the canopy, on muddy substrate, in water, on trunks and limbs of trees and from air. There are other species which feed on crabs, nectar and fish.

In terms of preference, these mangrove birds are divided into four feeding guilds-

1. Aerial hunters or sallying birds like swifts (Apodidae), wood swallows (Artamidae), swallows (Hirundinidae), fish eagles and kites

(Accipitridae), bee-eaters (Meropidae), kingfishers (Alcedinidae), which used to hover on mudflats and mangrove areas in search of their prey. The raptors (Falconiformes and Strigiformes) prey on fishes, small birds, small mammals, reptiles, amphibians and large invertebrates. They also roost in the mangrove areas.

2. Foliage and bark gleaners: These birds, mostly terrestrial species, prefer using mangrove vegetation, i.e. trees, shrubs, palms, and ferns for foraging, perching, nesting and roosting. The users are-woodpeckers (Picidae), Tailorbirds, Warblers, Flyeaters (Sylviidae), Flycatchers (Muscipcapidae), Thrush, Shyama and Robins (Turdidae), Nuthatch (Sittidae), Sunbirds, Spiderhunters (Nectariniidae), Pigeons (Columbidae), Owls (Strigidae), Cuckoos and Malkohas (Cuculidae), Parrots (Psittacidae), Tits (Paridae), Orioles (Oriolidae), Drongos (Dicruridae), Ioras (Chloropseidae), Cuckooshrikes (Campephagidae) and Pittas (Pittidae). This category includes the frugivores (pigeons and parrots), insectivorous eating the harmful caterpillars, beetles, bugs, and aphids (tailorbirds, shrikes, flycatchers, ioras, woodpeckers, robins, warblers, tits, etc.), nectarivores or pollinators (sunbirds, white-eyes and spider-hunters).

3. Surface and diving foragers: The birds like Pelicans (Pelecanidae), Ducks and Goose (Anatidae) mostly swim on the surface of water in search of small fishes, amphibians, aquatic invertebrates and vegetable matter, while Cormorants (Phalacrocoracidae), Darters (Anhingidae), and Grebes (Podicipedidae) dive into deep water, particularly river beds, in search of food, mainly fishes, amphibians and aquatic invertebrates such as mollusks as well as vegetable matter. Asian Openbill is a molluscivore.

4. Waders: The birds like Egrets, Herons, Bitterns (Ardeidae), Finfoots (Heliornithidae), Plovers (Charadriidae), Oystercatchers (Haematopodidae), Sandpipers, Curlews, Shanks, Stints, Ruffs, Godwits, Knots, Dowitchers, Turnstones, Whimbrel, Snipes, Oystercatchers (Scolopacidae), Stilts and Avocets (Recurvirostridae), Phalaropes (Phalaropidae), Gulls, Terns and Noddys (Laridae), Spoonbills, Ibis, and Storks (Ciconiidae) and Frigate birds (Fregatidae) wade in shallow water to prey on fishes, prawns, mollusks, crustaceans, polychaetes and other invertebrates of the mud-flats during the low tide. These bird species utilize the mangrove areas for

foraging, roosting, nesting for breeding and shelter from inclement weather.

The carnivore is a generalised group composed of Piscivorous, Insectivorous, Avivorous, Crustacivorous and Molluscivorous species. Carrion is also a popular food source for carnivorous birds, particularly vultures.

In terms of diet, the herons (e.g. striated and pond), egrets (cattle, great, intermediate, little), terns (little), kingfishers and cormorants (little) are piscivores. Waterhen (white-breasted), rails (slaty breasted), lapwings (red-wattled), sandpipers (common, green) are examples of crustacivore. The species belonging to Falconiformes and Accipitriformes are carnivores. The gulls, mynas and starlings are omnivorous species. Graminivorous birds include Jungle fowl, sparrow, weaver, silverbill and munia, whereas the columbids are both frugivores and graminivores. The parakeets are frugivores, so also cuculids and barbets. The swifts, picids, most of the passerins and bee eaters are insectivores. Hoopoe is both insectivore and graminivore, whereas bulbuls are both insectivores and frugivores. A few passerins (family Nectariniidae) are nectarivores, whereas the flowerpeckers are both nectarivores and insectivores.

Diet of most of the birds in Sundarbans is insects, which in terms of nutritional value, is adequate because of rich and easily digestible protein and fat. Predominantly, the insects preferred by the birds in Sundarbans were identified as belonging to the Orders Coleoptera (81), Hemiptera (45), Hymenoptera (96), Orthoptera (19), Odonata (27), Lepidoptera (163) and Diptera (133). There are six more orders with a low (2-8) number of species and not preferred by the birds. Beetles are most frequently preyed upon. There seems to be a direct correlation between habitat resource availability and utilization, i.e. the birds choose food opportunistically.

Analysis on the species diversity, abundance and threats

Like the rest of the world, the highest number of species of birds in the study area belongs to a single Order named Passeriformes- passerines. The least number is in the Order Bucerotiformes- hoopoe with a single species, Common Hoopoe (*Upupa epops*), but no hornbill species under Bucerotiformes was recorded in Indian Sundarbans, though they are found in the mangroves of Malaysia. Worse than that, no species of the

Order Otidiformes (Bustards) exists in the Sundarbans today, although one species thrived in the grassland of Sundarban till the beginning of the 20th century. This extirpated bird was Bengal Florican (*Houbaropsis bengalensis*), which is a globally critically endangered species.

In terms of species diversity, the number of species <5 is represented by four Orders: Podicipediformes- Grebes, Phaethontiformes- Tropicbirds, Procellariiformes- Seabirds and Caprimulgiformes- Nightjars. The number of species between 5-10 is found in five Orders: Apodiformes-swifts, Falconiformes- falcons, Ciconiiformes- Storks, Pelecaniformes-Pelicans, etc. and Galliformes- pheasants. The number of species between 11 and 15 is seen in three Orders: Columbiformes- pigeons and doves, Gruiformes- crane-like and Strigiformes- Owls. Species between 16 and 20 are found in three Orders: Psittaciformes- parrots, Coraciiformes-kingfishers, bee-eaters, rollers, motmots and todies and Cuculiformes-Cuckoos. Four Orders with higher diversity are Piciformes- woodpeckers and close relatives, Anseriformes- waterfowls, Accipitriformes- hawks, eagles and kites and Charadriiformes- snipes, gulls, button quails, thick-knees, plover-like waders.

The avifaunal diversity in Sundarban is dynamic and depends on various factors such as the flowing water level, seasonal variations and changes in tide levels twice (two high and two low) a day. At low tide, vast rivers turn into tiny streams that twist along sandy islands, exposing a rich ecosystem, the birds were seen foraging in the open mudflats and resting in the mangroves during a high tide, when large tracts of the forest disappear underwater. During monsoon, many water birds were seen foraging in puddles in the adjoining grasslands.

The post-monsoon period is most productive in Sundarban, when the species diversity and congregation is increased, particularly with the arrival of long-distance migratory (stopover and wintering) birds since September from three flyways- the Central Asian Flyway, the East Asian-Australasian Flyway and the Asian-East African Flyway. Here, they need to feed voraciously, increasing their body mass by 20-60% and energy reserve to undertake the return journey in summer.

Again, low diversity was observed during the monsoon owing to heavy rains [average for pre-monsoon (March-May): 239.4 mm; monsoon (June-September): 1,355 mm; and post-monsoon (October-February): 226.8 mm], increased flow of water, extensive and prolonged inundation, when availability of food and space is limited.

Whereas a huge number of migrants depart Sundarban during March-April, vacating the habitat for limited summer (from March onwards) migrants from their wintering grounds to the mangrove habitat, no severe food crisis takes place.

Overall bird density is often correlated to insect abundance, which is varied between seasons and mangrove patches. Insect abundance is highest, when mangroves are flowering. For example, occurrence of flowering for *Excoecaria agallocha* (Euphorbiaceae) in July, for *Bruguiera sexangula* (Rhizophoraceae) in April-June, for *Ceriops decandra* (Rhizophoraceae) in March-April, but variable for *Heritiera fomes* (Malvaceae) from March to April in the less saline zone, from March to early May in the moderately saline zone and from March to early June in the strongly saline zone. In general, partitioning of the available foraging niches is limited in the extensive habitats, resulting in dominance of the bird assemblage by a few species that are generalists, in terms of feeding, compared to the specialists.

There are historical records of extinction of at least seven globally Endangered resident avian species from Sundarban during early twentieth century- Pink-headed duck (*Rhodonessa caryophyllacea*), White-winged duck (*Asarcornis scutulata*), Indian peafowl (*Pavo cristatus*), Green peafowl (*Pavo muticus*), Greater adjutant (*Leptoptilos dubius*), Red-headed vulture (*Sarcogyps calvus*) and Bengal florican (*Houbaropsis bengalensis*).

Sundarban now harbours some sympatric species like kingfishers (9) and cuckoos (16); several raptors such as, eagles, falcons, vultures, kites, harriers, etc; some terrestrial birds including doves, woodpeckers, pigeons, fly-catchers, oriental magpie robin, red jungle fowl, owls, rose-winged parakeet, etc; aquatic birds like storks, herons, egrets, adjutants, little cormorant, etc; semi-aquatic plovers, red-wattled lapwing, avocet, stint, curlew, sandpiper, common greenshank, gulls, terns, etc.

A few species are recognised as mangrove specialists like Mangrove Pitta (*Pitta megarhyncha*), Mangrove Whistler (*Pachycephala grisola*), Brown-winged Kingfisher (*Halcyon amauroptera*) and Collared Kingfisher (*Todiramphus chloris*). Those species that prefer mangroves, but are not mangrove specialists, include Masked Finfoot (*Heliopais personata*), Lesser Adjutant (*Leptoptilos javanicus*), Buffy Fish Owl (*Ketupa ketupu*),Great Thick-knee (*Esacus recurvirostris*), Streak-breasted Woodpecker (*Picus viridanus*), White-browed Scimitar Babbler (*Pomatorhinus schisticeps*) and Whitebellied Sea Eagle (*Haliaetus leucogaster*). Many species of Sundarban are known to prefer grasslands, viz. Blue-breasted Quail (*Coturnix chinensis*), Red-wattled Lapwing (*Vanellus indicus*), Paddyfield Pipit (*Anthus rufulus*), Zitting Cisticola (*Cisticola juncidis*), Bengal Bushlark (*Mirafra assamica*), Grey Wagtail (*Motacilla cinerea*), etc.

According to the occurrence rate of the extant species during the study period, about 35% was considered rare. These birds include Ferruginous Pochard (*Aythya nyroca*), Ruddy Kingfisher (*Halcyon coromanda*), Chestnut-winged Cuckoo (*Clamator coromandus*), Slaty-legged Crake (*Rallina eurizonoides),* Pintail Snipe (*Gallinago stenura*), Eurasian Thick-knee (*Burhinus oedicnemus*), Grey Plover (*Pluvialis squatarola),* Darter (*Anhinga melanogaster),* Cinnamon Bittern (*Ixobrychus cinnamomeus),* Great White Pelican (*Pelecanus onocrotalus),* Indian Pitta (*Pitta brachyuran)* and so on.

This evaluation of Threatened animals is grouped into four basic categories of conservation concern, as per IUCN Red List- Critically Endangered, Endangered, Vulnerable and Near threatened. Among the globally Critically Endangered birds, Sundarban harbours two resident species- White-rumped Vulture (*Gyps bengalensis*) and Slender-billed Vulture (*Gyps tenuirostris*). Two globally critically endangered species, Spoon-billed Sandpiper (*Eurynorhynchus pygmaea*) and Baer's Pochard (*Aythya baeri*), are winter migrants to Sundarban. Two resident globally endangered species are Masked Finfoot (*Heliopais personata*) and Black-bellied Tern (*Sterna acuticauda*). Globally Endangered species also include winter migrants to Sundarban like Great Knot (*Calidris tenuirostris*), Egyptian Vulture (*Neophron percnopterus*), Steppe Eagle (*Aquila nipalensis*) and Yellow-breasted Bunting (*Schoeniclus aureoles*). Four resident globally Vulnerable species of Sundarban are Lesser Adjutant (*Leptoptilos javanicus*) and Indian Spotted Eagle (*Clanga hastata*), whereas two globally Vulnerable species are breeding migrants to Sundarban. They are Pallas's Fish Eagle (*Haliaeetus leucoryphus*) and Bristled Grassbird (*Chaetornis striata*). The other globally vulnerable winter migrants to Sundarban are Common Pochard (*Aythya ferina*), Woolly-necked Stork (*Ciconia episcopus*), Indian Skimmer (*Rynchops albicollis*), Wood snipe (*Gallinago nemoricola*) and Greater Spotted Eagle (*Clanga clanga*). Among the Globally Near

Threatened birds of Sundarbans, familiar residents are Oriental Darter (*Anhinga melanogaster*), Red-necked Falcon (*Falco chicquera*), Brown-winged Kingfisher (*Pelargopsis amauroptera*), Great Thick-knee (*Esacus recurvirostris*), River Tern (*Sterna aurantia*), Mangrove Pitta (*Pitta megarhyncha*), Red-breasted Parakeet (*Psittacula alexandri*), Alexandrine Parakeet (*Psittacula eupatria*), Blossom-headed Parakeet (*Psittacula roseata*) and Spot-billed Pelican (*Pelecanus philippensis*). Globally Near Threatened winter migrants to Sundarban are two ducks, namely Falcated Duck (*Mareca falcata*) and Ferruginous Duck (*Aythya nyroca*), shorebirds like Eurasian Oystercatcher (*Haematopus ostralegus*), Eurasian Curlew (*Numenius arquata*), Bar-tailed Godwit (*Limosa lapponica*), Blacktailed Godwit (*Limosa limosa*), Red Knot (*Calidris canutus*), Curlew Sandpiper (*Calidris ferruginea*), Red-necked Stint (*Calidris ruficollis*), and Asian Dowitcher (*Limnodromus semipalmatus*).

It was observed that Masked Finfoot, the rare breeding resident, mainly occurs in the eastern half of the Sundarban. They can be seen throughout the year. In the southern Bagerhat and Khulna districts, most of the birds live on the banks of quiet canals, where human traffic is less, but the bird is rarely seen in the north and west of the forest. From 1999 to 2001, a total of 24 birds were seen in the protected forest of Sundarbans. Most of them were found in freshwater habitats in the eastern part of the Shipsa River. The western limit for most of this bird's sightings was the Sarbatkhali Canal in the Khulna Range of the Sundarban West Forest Division, which is a freshwater region east of the Shipsa River. This bird was commonly seen in Katka and Kochikhali canals of Sundarbans East Sanctuary and near Mora Bhola canal in the western entrance (medium saline water zone). Apart from these places, this bird was also found in Big Morgamari Canal and Jongra Canal of Sundarban East Forest Division. The highest sightings were recorded during a trip to Tambulbunia Canal, with a total of eight birds seen at four different locations. In 2002, there were two sightings in the Katka creek. In the last decade, 25 nests have been found from Chandpai to Sharankhola range, covering <100 km², of which five were active. A 110 km waterway survey of the Bangladesh Sundarbans on 16-22 August 2004 showed them near Mirgamari (22°23'N, 89°40'E), Shapla Canal (22°04'N, 89°50'E), Katka Canal (21°51'N, 89°48'E), on the Kachikhali Canal (22°52'N, 89°50'E), on the east bank of the Sela Gang (21°55'N, 89°41'E), east of the Supati Canal (21°57'N, 89°49' E) and near Hiran Point (21°48'N, 89°28' E).

Its population appears to have declined after the devastating cyclones of the 21st century. It was found that between 2007 and 2011 the local fishermen hunted a total of 16 adults, 15 chicks and 4 eggs from their nests. Climate change has had a significant impact on their preferred habitats. Their habitat preferences for breeding have been changed, especially in selecting tree species for nesting, such as from only Sundari in 2004 to a mixture of Genga and Sundari since 2010. This progressive shift in habitat preference is also reflected in their overall breeding area distribution, i.e. favouring the upstream freshwater areas. Differences in nesting habitat preference may account for increasing salinity along the coast. This is probably why the masked finfoots are being forced to move to less saline areas and this is explained by the higher density of their preferred nesting trees. It is possible that they are nesting in environments that hardly meet their requirements, for example, they were seen nesting in less preferred trees along the narrow creeks. Absence of masked finfoots in the coastal habitats of Sundarbans and northward shift of nesting habitat may also be due to several super-tropical cyclones that have occurred in the last decade. However, to ensure their existence in the future, it is very important to provide a safe habitat where no intrusion or disturbances occur.

There are recent sightings of the Spoon-billed Sandpiper on the shore of Sundarban, east or west. Both White-rumped Vulture and Pallas's Fish Eagle are rarely found in the northern Sundarban. The Greater Spotted Eagle is rarely found on the riverbanks or shores of the Sundarban.

About 65% of the species found during the study was common, (not threatened or Least concern category), mostly belonging to Phasianidae (Galliformes), Cuculidae (Cuculiformes), Psittaculidae (Psittaciformes), Caprimulgidae (Caprimulgiformes), Columbidae (Columbiformes), Ardeidae (Pelecaniformes), Scolopacidae, Charadriidae and Laridae (Charadriiformes), Dicruridae, Sturnidae, Hirundinidae, Pycnonotidae and Muscicapidae (Passeriformes), Ciconiidae (Ciconiiformes), Accipitridae (Accipitriformes) and Alcedinidae (Coraciiformes). These include Red Jungle fowl (*Gallus gallus*), Greater coucal (*Centropus sinensis*), Rose-ringed parakeet (*Psittacula krameri*), Large-tailed Nightjar (*Caprimulgus macrurus*), Eurasian Collared Dove (*Streptopelia decaocto*), Whimbrel (*Numenius phaeopus*), Indian Pond Heron (*Ardeola grayii*), Bronzed Drongo (*Dicrurus aeneus*), Oriental Magpie Robin (*Copsychus saularis*), Jungle Myna (*Acridotheres fuscus*), Barn Swallow (*Hirundo rustica*), Red-vented Bulbul (*Pycnonotus cafer*), etc.

Goliath heron (*Ardea goliath*) is one of the world's largest heron species belonging to the order Ciconiiformes and family Ardeidae. Prater (1926) recorded its presence in the Bangladesh Sundarban. The sightings of Goliath heron have been very rare and infrequent in the Indian part of Sundarbans. Sighting of a Goliath Heron was recorded in 2005 in the Indian Sundarbans. As a part of the project on "Monitoring tigers, co-predators, prey and their habitat", a survey was carried out to enumerate signs of prey and predators during February to May 2010 in the Indian Sundarbans. On 24th May 2010, during the beginning of high tide at 12.52 pm, an adult Goliath heron was sighted on the mud bank of a small creek called Bhagavan Bharani, which was identified on the basis of its brown head, yellow eyes, black bill and legs, chestnut throat and brown body (Prabhu *et al.*, 2013). The bird was observed to fly for about 100 m and perched again near shore. The vegetation of the site, where the bird was sighted, composed of *Exoceria agallocha, Ceriops decandra, Avicenia marina* and *Phoenix paludosa*.

Threats

The birds living in mangroves, permanently or temporarily, face a number of environmental challenges not usually present in inland habitat, such as regular tidal inundation and less-predictable storm surges, anaerobic soil conditions, and a saline environment. The birds in Sundarbans are also facing overwhelming pressure due to habitat loss and degradation. Over-exploitation of forest resources due to growing anthropogenic pressures by the surrounding population is alarming. Moreover, coverage and density of larger diameter trees, canopy closure and diversity have declined over the last 100 years or so. Due to variation of freshwater supply and salinity across the Sundarban from west to the east, 0.4 % of the forest area in Bangladesh is replaced by dwarf species every year. Recently, many key mangrove plant species like *Heritiera fomes, Nypa fruticans* and *Phoenix paludosa,* which were very abundant in the Sundarban 50 years ago, have declined relatively as the salinity has increased. This also causes a decline in the habitat for arboreal birds.

In addition, higher salinity in mangroves leads to depletion of nutrients that might influence the population of planktons, benthic organisms and other macro-invertebrates vis-à-vis top-level predators such as birds. Sea level rise by an average of 3 centimeters a year over the past two decades and 12 percent shoreline-loss during the last four decades have also threatened the shorebirds. On the other hand, excessive collections of Tiger Prawn (*Penaeus monodon*) fry from the rivers by thousands of local fishermen since 1980s, is not only restricted to the buffer zone and, in this process, more than hundred times of non-target fish and crustacean fry are destroyed, for which the aquatic birds are also declining.

Oil spills are another recent potential threat to the inland aquatic and seabirds because oil soaked into the plumage damages insulation, water resistance of feathers and floating ability. Consumption of oil also causes health hazards including reproductive dysfunction and even death.

Trapping and hunting birds for food is also a menace in the Sundarban. Existence of a huge forest-dependent population with a density of about 1,000 persons per km^2 in the reclaimed areas, a considerable percentage of which regularly enter the forests, mostly illegally, leads to disturbance during foraging or breeding and resource crunch in the habitats.

As Sundarban is a very popular tourist hotspot (national and international) lakhs of tourists visit the forest areas causing disturbance to the birds and pollute the area and the watercrafts carrying them also pollute the water bodies.

Increasing frequency and severity of the cyclones is causing lots of damage to the habitat and habitants. Of late, the cyclone Amphan, which hit the Sunderbans on May 20, 2020 at a speed of 155-165 kmph, carried the pelagic seabirds deep into the mainland (bird fallout) and different types of frigatebirds, Wilson's storm petrel, sooty tern, lesser crested tern, great crested tern, brown noddy and short-tailed shearwater were sighted, a few of them are new records in the region. The cyclone has also killed and maimed many of these birds. A Great Crested Grebe (*Podiceps cristatus*), a winter migrant, was observed at a river in the Sajnekhali WLS, while conducting the annual biodiversity survey within the STR in November, 2020. Nordmann's greenshank (*Tringa guttifer*), a wader considered endangered, was spotted for the "first time" in the wild in West Bengal and the second concrete record in India after Maharashtra in December 2020. The bird was clicked in the Lothian WLS on 11th February, 2022. There were five-six birds of this species on the shore.

Two-three juveniles were photographed mixed with the grey plovers. The species is a vagrant winter migrant and very rarely seen. It is time for their return and probably they were using the Sundarbans as a transit route.

Prospects

Considering the present threats, the migratory waterbird species are of high priority for future action for conservation in the Sundarbans. Sundarban is also important for the resident waterbirds. These waterbirds need attention for immediate conservation efforts. The research on waterfowls should focus on extensive survey of migratory as well as resident waterbirds in addition to preparation of an updated inventory and assessment of waterbird habitats. There are three more actions to be taken-

(i)   Awareness related to conservation of birds and their habitats;
(ii) Support for capacity building in terms of technical and management aspect; and
(iii) Resource mobilization for conservation of birds.

Mammals (Mallick 2010, 2011, 2013, 2015, 2019)

Large cats including the tigers evolved from a common ancestor probably similar to modern-day leopards or jaguars five million years ago. Probably originated in east Asia, the tiger had a journey from Siberia, across the Caspian, to Korea, and China, through Vietnam, Leos and Cambodia into Thailand and Malaysia, to the Indonesian Islands, and finally into Myanmar and India.

During the evolutionary history of the tiger, it has adapted to a wide range of ecological conditions, from temperate forests of the Russian Far East to the mangroves. It is impossible to tell when the tiger colonised the Sundarbans. However, studies based on models of habitats and climate suggest the tigers occupied most of India around 12,000 years ago possibly through the Eastern Himalayas, somewhere in the Arunachal Pradesh and Upper Assam area.

The Indian tigers diverged from each other around 8,000-9,000 years ago. Models of demographic history of tigers using whole genome sequences suggest north-east Indian tigers are (a) closest to south-east Asian tigers and (b) the most divergent of populations within the Indian tigers. These indicate that the tigers may have entered India through north-east India. The tiger quickly spread throughout India, moving into the mangrove swamps, evergreen forests, dry deciduous forests, and the variety of other vegetation that covered Indian land.

About 6,000 years ago (in the mid-Holocene), the shoreline of the North-eastern Indian Peninsula was situated to the west of the present shoreline, which was comparatively closer to the foothills of the Himalaya. Therefore, the present Sundarban area did not exist for terrestrial life until the shoreline had moved eastward and the tigers could arrive from the Peninsular India. In 1756, when Siraj-ud-Daulah recaptured the City of Kolkata (then Calcutta) from the British, today's Salt-Lake area used to be the main city and the Lower South Circular Road that's now known as Chowringhee used to be the city's southern border. Beyond that were the forests of Sundarbans and there are beliefs that tigers were often sighted in those forests which now house busy localities like Tollygunge and Behala. A shift in the river Ganges brought changes to agricultural land use in the areas adjoining Sundarbans.

In 1819, a news report in 'Samachar Darpan' appeared that described how a woman was killed by a tiger in Gouripur area. After killing the victim, the tiger forcefully entered a neighbour's house. The residents of the house somehow managed to trap the tiger in a room and locked the door from outside. The police was informed, who, with the help of local administrative staff, managed to kill the tiger.

A photograph of a famous man-eater behind bars in the Zoological Gardens at Calcutta was taken by James Ricalton in c. 1903. This image is described by Ricalton in 'India Through the Stereoscope' (1907), "The tablet on top of this animal's cage states that he has devoured two hundred human victims...Higher records than this are claimed for some man-eaters, but it is always difficult to obtain absolute certainty in such matters...There are several reasons given as to why the tiger acquires the habit of killing human beings. Sometimes a mother tigress with cubs finds it difficult to supply her young with food, and this urgent family requirement compels her to make indiscriminate attacks. In other cases, decrepitude, or physical disability renders it necessary to capture something less agile and fleet than common game." In the Sundarbans, tigress lures away the intruding male to protect the young. There are male tigers who kill others' cubs to force the mother to mate again or fight with other males to establish territory.

After that the anthropogenic activities from 600 to 800 years ago started to impact the tiger habitat. All these factors probably caused complete isolation of the Sundarban tiger.

The Sundarban tiger is regarded as the Bengal tiger (*Pantherea tigris tigris*) subspecies and is not a recognized separate subspecies. The scientists and officials of STR are investigating whether the Sundarban tiger is a different sub-species compared to those found in the sub-continent. Some experts suspect that, *prima facie*, the years of evolution, lack of sweet water affecting the height and growth of mangrove forests, the increasing salinity and arduous search for prey in the inhospitable terrain of Sundarban have caused increased physiological stress, triggering a genetic mutation, apparently leading to formation of an entirely new sub-species of the Bengal tiger. The genetic studies, including analysis of blood, hair and scat samples, were conducted at the Centre for Cellular and Molecular Biology, Hyderabad (32 positive scats confirmed), to ascertain whether the Sundarban tigers really differ from the other mainland Bengal tigers but nothing concrete has so far been established. However, considering the mutation rates that led to a genetic change, usually an animal that was isolated for a period of one million years is classified as a different species and one that was genetically isolated for between 20,000 and 50,000 years as a different sub-species. In the case of the Sundarban tiger, it was perhaps separated from the mainland tigers ±1,000 years ago. Some hundred years back, the tigers were found not only in the Sundarbans, but everywhere in the surrounding mainland, where the areas were full of riverain forests and grasslands with prey-base in abundance.

They might have been isolated long enough to become morphologically or genetically distinct. With BoB on the south and surrounding major rivers, the primary concern regarding threats to the genetic heterogeneity of this population is the probable isolation induced by the wide rivers prevalent in the landscape. It is known that the tigers in the Indian Sundarbans rarely crossed channels wider than 400 m (Naha *et al.*, 2016).

Aziz *et al.* (2018) reported a fine-scale genetic structure and clustering within the tiger population of Bangladesh Sundarban which could largely be attributed to the river systems. Though tigers have been observed to occasionally disperse across wide rivers, increased continuous use of these water channels inside the forest as conduit for commercial boat traffic can transform the rivers to barriers to tiger movement. Hence, along with trans-boundary collaboration, regulation of cargo vessel movement to avoid tiger peak activity periods (5 AM to 10 AM) should be enforced (Naha *et al.* 2016).

Since the Sundarban is surrounded by human settlements and agriculture on its three sides, this forest has long been isolated from the nearest tiger occupied habitats by agricultural lands and human settlements, thereby removing any opportunity of gene flow across nearby tiger populations.

The genetic or morphological representation of the tigers from the Sundarbans of Bangladesh and India has been lacking before. Microsatellite markers indicated low genetic variation in Sundarbans tigers (He= 0.58) as compared to other mainland populations, such as northern and Peninsular (*He* between 0.67- 0.70). Molecular data supports migration between mainland and Sundarbans populations until very recent times. Singh *et al.* (2015) attributed this reduction in gene flow to accelerated fragmentation and habitat alteration in the landscape over the past few centuries. It is assumed that the mangrove tigers have diverged recently from the peninsular tiger population within last 2000 years. On the basis of analysing sixteen tiger samples: blood (n = 5), tissue (n = 1), hair (n = 1) and scat (n = 9) from STR and comparing with the tiger samples (n = 73; scat) of the tiger reserves located in the Central and Eastern Ghats (Kanha, Pench, Bandhavgarh, Panna, and Palamau), and (n = 62; tissue, blood and scat) from the tiger reserves located in northern India (Corbett, Rajaji National Park and Dudhwa), the Sundarbans tigers, being the most divergent group of Bengal tigers and ecologically non-exchangeable with other tiger populations, should be managed as a separate "evolutionarily significant unit" (ESU) following the adaptive evolutionary conservation (AEC) concept (*ibid*).

Aziz *et al.* (2022) have generated 1263 bp of mtDNA sequences across 4 mtDNA genes for 33 tiger samples from the Bangladesh Sundarbans (SRF) and compared these with 33 mtDNA haplotypes known from all subspecies of extant tigers. For this purpose non-invasive tiger samples (scat and hairs left by the tigers in their scratch marks on the trees) were collected during 20th November, 2014 to 26th February, 2015 in addition to a blood sample from a rescued tiger, five tissue samples from the skins confiscated from around the Sundarbans and four hair samples taken from the rescued tigers within the Sundarbans). The field sampling sites were- (i) East WLS with additional areas (383 km^2), (ii) West WLS (715 km^2), (iii) Chandpai block (342 km^2) and (iv) Satkhira block (554 km^2).

Combining these haplotypes with previously reported haplotypes from the Indian Sundarbans (Mondol *et al.* 2009), they detected three haplotypes within the Sundarbans tigers, of which one is unique to this population and the remaining two are shared with the tiger populations inhabiting the central Indian landscapes. Haplotype diversity (h) is lower for the Sundarbans population compared to all the regional groups of Bengal tigers, except the Nepal population. At the subspecies level, h is higher for the Sundarban population than for the Amur and Indochinese tiger populations. In terms of nucleotide diversity (π), the Sundarban tigers exhibit moderate values of π, which is broadly similar to other populations in India but higher than those in Nepal.

Phylogenetic analyses using maximum likelihood and Bayesian inferences supported the Sundarbans tigers as being paraphyletic, indicating a close phylogenetic relationship with other populations of Bengal tigers, from which the Sundarbans population diverged around 26,000 years ago. The Sundarbans tiger population, adapted to unique mangrove habitat, has been geographically isolated from the nearest Tiger Conservation Landscape (TCL) in Similipal, Odisha, by just ~200 km of landscapes dominated by human settlements and agricultural land, suggesting little chance of future population movements by dispersal. Considering the phylogenetic analyses, together with evidence of ecological adaptation to the unique mangrove habitat, the Sundarbans population is proposed to be recognised as a separate management unit (MU).

Considering the morphological differences and a lack of gene flow from other populations, it seems that the Sundarbans tigers may be in the early stages of allopatric speciation (Barlow, 2009). The breeding population of Sundarbans tiger in India and Bangladesh is important from the conservation point of view, i.e. the future recovery of tigers. However, this uniquely adapted tiger population is at extinction risk due to the anthropogenic factors like direct poaching and decline of the preferred prey population consequent upon the increasing human-tiger conflict and climatic change. All these anthropogenic factors have caused severe decline of this population in the Bangladesh Sundarbans, from an estimated 300-500 tigers (Barlow, 2009) to only 106 (Dey *et al.*, 2015) in only half a decade (Aziz *et al.*, 2022), whereas the current tiger population in the Indian Sundarbans is 96.

The Sundarbans tiger represents the only population adapted to inhabiting a unique mangrove habitat type, isolated from neighboring tiger populations by hundreds of kilometers of agricultural and urban land, for which they are morphologically distinct in terms of skull morphometrics and body size. While the average weight of an adult Royal Bengal tiger of the mainland is over 180 kg, it has fallen to only ±100 kg in the Indian Sunderbans. Its body length has also come down by at least one-third metre. Loss of weight is also apprehended due to such climatic changes and scarcity of prey. The lustre of its coat is also fading. Tigers used to migrate from the more saline to less saline areas, i.e. southern core area to the northern buffer zones. They used to consume saline water as the number of fresh water ponds inside the core area of Sunderbans has gone down because of rising salinity and lack of maintenance.

A large portion of the Sunderbans gets submerged mostly during the high spring tides of the vernal equinox (March-April) for 3-4 hours and monsoon, when even the high trees get partially submerged in saline water. During the low tide, a few hours' later, the water recedes back exposing the thick clay bed. The short flood-tide (2-3 hours) and long ebb-tide (8-9 hours) occur during the dry season, when the higher land does not get flooded. The daily tides and ebbs have made the topography of Sundarbans hostile to the tigers and associates. As a result, the preferred tiger niches are the thickets of Sundari, Golpata and Hental which are submerged only a number of times during very high tides (*Kotal*). Since the regularly flooded forest floor is muddy and covered with pointed stilt-roots (pneumatophores), projected to a height of 20-30 cm above soil, it is very difficult for the tigers to hunt in this slushy terrain (one-fifth attempt successful). Such adaptive dwarfism of the Sundarban tiger is advantageous as a smaller tiger facing energy crisis in a hostile habitat needs less food. Here, the pugmarks are smaller than those of other parks and shallow in mud in the delta. Being lighter also made it easier to move around in the muddy terrain or negotiate the rivers and creeks of the

Sundarbans with minimum loss of energy. So, it favours the grasslands along the riverbanks or the creeks for hunting.

Unlike the mainlanders, the Sundarban tigers migrate from one island forest block to another and across the international boundary. The animal has been sighted negotiating vast expanse of water; crossing the rivers, creeks and rivulets frequently in search of elevated lands above the inundation level and food as its main prey animals like the cheetal and wild boar are also semi-aquatic. The predator also follows them. In this process, it could have lost the so-called territoriality of the mainlanders owing to obliteration of the urination marks by the tidal waves. Even the cubs have to learn swimming under the guidance of their mother. The Sundarban tigers are capable of swimming with the help of its tail, which is comparatively more muscular and consistently thicker at the base than that of the mainlanders. It is used as a limb for swimming in the winding, sweeping and side-to-side manner of the crocodile. It has developed a technique to keep the entire body submerged in water except for the nostrils. Hence, from a distance I noticed only something black object floating in the water, but when looked through the field glasses, could identify it. I have often noticed it swimming across the bigger streams only when the tide just turns, either from ebb to flow or *vice versa*. It generally chooses high tide to take off and hard ground for landing the opposite bank. The tigers were seen to cross the rivers at right angles parallel to the current once the crossing is decided and a boat coming on the way is often attacked. They also change their course, when they feel disturbed while swimming.

It is very cautious while negotiating from one island to another. First, it observes direction of movement of the floating objects like leaves, fruits, etc. then uses the claw to dig and whirl the muddy water for observing the direction, again turns and puts its tail in the water to physically determine the intensity of current as well as direction. Thereafter, it selects a static object like tree on the opposite bank as a target as well as a suitable starting point often by adjustment, *i.e.* proceeding ahead or going back a bit and ultimately drives into the water. In the mid-stream, if it realises that the target on the opposite bank is lost, it comes back to its original position and follows the same procedure or starts swimming again.

In the Sundarban, the tiger has become an expert climber too (Mallick, 2013). At the time of crisis, it is seen to climb even a straight tree for either safety at the time of straying in a village or for hunting in the forest. At Arbesi of STR, one tiger climbed a long and straight Passur tree at a height of at least 4.5 m above the ground. It was tranquilised and then brought down with the help of a net spread around above the ground. During the cyclonic flooding of the Sundarbans in 1969, many climbed trees and a few were found drowned. During the tidal bores, it clings to the low mangrove branches or is driven to the elevated part of the island. It also climbs into sleeping shelters built on the machans (safety platform or tree house) and often seizes one of the inmates. 20 years ago, Ramen Mistry (55) climbed a tree to pluck some fruits during one of the fishing expeditions. A tiger followed him, climbed up the tree and began pawing at him. It dug its claws on his hip as he bled profusely. As Mistry screamed in pain, his companions rushed to his rescue and shooed away the tiger. He was rushed to a hospital in the South 24- Parganas district, where he remained for three months.

Whereas the tigers of North Bengal are stated to grow bigger in length and lighter in colour(almost yellow with black stripes and white from the throat down to the body), in the Sundarbans, the tiger is much smaller in size, but stockier and thickest; almost brown-red in colour and in the forest, the black stripes are hardly distinguished and movement through the mud and slosh is heavy and slow, where the pugmarks are more splayed out because of loose soil (Bahuguna and Mallick, 2010). Its footprints are smaller than those of other parks and shallow in mud in the delta (Sankhala, 1978).

Two adult female tigers were caught in the south-east SRF of Bangladesh by using modified leg-snares placed near baits and immobilized with 6-8 mg kg-1 of Telazol, administered using a projector and dart (PalmerCap-Chur Inc., Powder Springs, USA). The first female was located at Katka-Kochikali and the second female at Chaprakhali. Based on general teeth condition, the former seemed to be of 12-14 years old; teeth were generally discoloured and worn, one lower pre-molar was broken, and two lower incisors and a right lower canine were missing. The age of the latter was estimated at 10-14 years; teeth were discoloured and one upper canine was broken, as were some upper incisors. There was no evidence to suggest that either female had any dependent offspring. These female tigers had a mean weight of 76.7 kg (Standard Deviation = 2.89, range 75-

80), which is the smallest mean weights of any mainland group. However, recent observations of various tigers over the last few years have cast doubt over this theory. Tigers which have been found and photographed are as big (if not bigger) as those found in the mainland of India. If these images are carefully studied (given the large number of such images), we can come to some conclusion regarding the size of the Sunderbans tigers.

It is suggested that the available number of prey species and their size could be an important influence of tiger morphology. The small skull of the males and lesser body weight of the female tigers in the Sundarbans may be a consequence of having no larger prey available than *Axis axis* and *Sus scrofa*. The tiger's prey in the mainland always includes a large ungulate species. There have been records that the Sundarbans once harboured Javan rhinoceros, wild buffalo, swamp deer, and hog deer, when the tiger population was also high. With their extinction from the tiger land, the tiger perhaps evolved a smaller body size more suitable for the energy requirements of subsisting on smaller prey types.

Both tigers mentioned above were fitted with GPS collars (Advanced Telemetry Systems, Isanti, USA). The two tigers were released at their capture sites after radio-collaring. The former was tracked for 5.5 months until she died of unknown causes. The latter was tracked for 2.5 months until the GPS collar batteries expired. She was then recaptured and released at the capture site after the collar was removed. The GPS locations from the collars were used to construct minimum convex polygon (MCP) and fixed kernel (FK) home ranges using the geographical information system ArcView v.3.3(ESRI, Redlands, California) and the ArcView extension Animal Movement v.1.1 Both tigers were recorded eating from kills within 1-4 days of capture. The former female's collar was set to record locations every four hours, and the latter's collar was set for every 30 minutes. The former was tracked for about six months until she died. The latter was tracked for 2.5 months until the GPS collar batteries expired. She was then re-captured and released after taking the collar off.

Water bodies that were never crossed by the two tigers, or land that lay across from these water bodies, were discounted from the home range estimates. The high frequency of location acquisition by the GPS collars (one location per 4 hours for the former and one location per 30 minutes for the latter) made it unlikely that the tigers could have crossed these water bodies and returned without recording a location.

The GPS collars recorded 679 locations for the former tigress during April–October, 2004 and 1,528 for the other tigress during March–May, 2006. After four months of monitoring the former made a foray to the east of her normal home range, returning after three days. She moved to the same area six weeks later and died c. 9 km from her normal home range. Tracks, judged by their size to be that of a new female, were observed in the former's home range within days of the dead tigress moved out. Female tracks together with those of a large cub were observed in the same area one year later. The poor condition of the former tigress's teeth, her movement pattern and the appearance of a new female, suggests that the dead one may have been unable to defend her territory from a rival. Therefore, the forays to the east were discounted from the calculation of home range size, as they were not representative of her normal movement pattern recorded earlier.

Location-area curves indicated that 95% MCP home ranges were acquired after c. 275 and 910 locations (c. two months) for tigress no.1 and no. 2 respectively. The mean 95 and 50% MCP home ranges were 12.3 km^2 (F1= 14.1 km^2, F2= 10.6 km^2) and 4.23 km^2 (F1= 4.2 km^2, F2= 4.3 km^2), respectively. The mean 95 and 50% FK home ranges sizes were 14.2 km^2 (F1= 16.2 km^2, F2= 12.2 km^2) and 3.0 km^2 (F1= 3.5 km^2, F2= 2.5 km^2) respectively. A mean female home range size of 14.2 km^2 would indicate a density for the south-east Sundarbans of seven adult females per 100 km.

Using one location per 4 hours (F1), the estimated mean straight line distance moved was 2.25 km day^{-1} (range 2.16–2.34 km day^{-1}) and with one location per 30 minutes (F2) the mean daily travel was 3.6 km day^{-1} (range 0.02–10.0 km day^{-1}). Maximum distance moved per day was 11.3 km for F1 and 10.0 km for F2. Both crossed water bodies at a mean frequency of 17 times per month (range 12-21), equivalent to approximately one crossing every two days.

The two 95% MCP home ranges are amongst the smallest recorded for female tigers and are indicative of a relatively high tiger density in SRF. However, considering the small sample size and that both tigers were from the same part of the Forest, these estimates of home range are considered to be preliminary. The two tigers were captured in areas of medium relative tiger abundance, as indicated from a study based on tiger-track frequencies along the creek banks. In areas of lower or higher relative tiger abundance the home ranges may be larger and smaller, respectively, than those known in this study. In the Indian Sundarbans, the home range of one female tiger was estimated by telemetry to be c. 40 km^2 and a camera trap study indicated a relatively low tiger density of 0.8 per 100 km^2.

Further estimates of home range size and movement distances of adult female tigers in both Indian and Bangladesh Sundarbans are required to improve inferences of habitat quality and thus of how many tigers the area can support and to design monitoring approaches for tigers across the whole landscape.

In the Indian Sundarbans, the straying tigers (mostly young adventurer or incapable old) usually cross the creeks or rivers between 50 m and 150 m in width to enter into the villages in Bagna Range (Hingalganj block), but 300-900 m in cases of the villages of Sajnekhali Range (Gosaba block) and cross more than 1 km width of the River Raimangal (Kultali block). Moreover, it is a superb long-distance swimmer. It is recorded to have crossed larger (5 km width) channels as well. Even it has capability of swimming 15–20 km through the tidal rivers. Long back, one swam about 10 km on the River Hooghly. From the sighting records, it appears that though the tiger generally crosses the channels at night for depredation in the fringe villages, it also swims even during the daytime in an undisturbed forest area. Sometimes some of them cross the bordering rivers or creeks (25-900 m wide) at night and stray into the non-forest islands or human habitations on the northwest (54 km^2) in quest of easy domestic prey. A tiger was observed to swim continuously for about three hours crossing two mighty Rivers- Mayadwip and Matla- bordering BoB in an April afternoon. A radio-collared tigress was recorded to cross, on the average, ten channels (mean width 25 m and maximum width 200 m) per day. It moved a distance of 1 km in ten minutes. A tiger swam across about 600 m against the current in seven minutes, eighteen seconds. There is another record of a tiger to swim at a speed of 13-16 km/hour. In a narrow creek near the forest, a tiger can easily outswim a country boat with only one man rowing and leap aboard. It was also seen that on reaching the opposite bank, it first stood up on its hind legs and had a good shake to get rid of the muddy saline water.

The Sundarban tiger often migrates between the two segments of India and Bangladesh. A male tiger was trapped in the Arbeshi forest on 21 May 2010, then tranquillized, radio-collared and released in the Katuajhuri forest bordering Bangladesh on 22 May. This Khatuajhuri male was a blind tiger on the right eye, perhaps indicating signs of territorial fights. The radio collar signals revealed that on the first two days, it was on the hunt and traveled only 6-7 km. But on the third day, the tiger traveled more than double that distance. The signals showed that it crossed the River Harinbhanga and left its command area and moved into a new territory in Bangladesh. Signals were being received from the collar and had located it somewhere in the middle of Talpatty Island, an offshore sandbar landform in the Bay of Bengal, situated 2 kilometers away from the mouth of River Harinbhanga. In recent times, a tiger and a tigress, straying into Shamsernagar village, were also found to have come from Bangladesh by crossing the river.

Barlow (2009) estimated movement of two female tigers in Bangladesh using one location/day as 1.65-1.72 km/day (mean = 1.69 km/day). Using one location/four hours increased the estimates of mean daily movement to 2.16-2.34 km/day (mean = 2.52 km/day). Using one location/30 minutes, mean daily travel was estimated as 0.02-10 km/day (mean = 3.57 km/day. Maximum distance moved/day was 11.3 km for the first female and 10 km for the second female. All distances moved included traversing both terrestrial habitat and waterways. The former crossed waterways at a frequency of 14/month (Standard Deviation or SD= 1.8, range= 12-16), or one crossing every two days. The second female crossed at a frequency of 20/month (SD= 1.5, range= 18-21), or one crossing every two days. In terms of widest water bodies crossed, the first female crossed a 0.6 km wide river on three occasions when she dispersed outside of her normal home range, and the other female had a 0.2 km wide river within her home range that she crossed regularly.

The widest river crossed by a tiger (1.5 km) recorded above is likely to be an underestimate of tiger swimming capabilities. An unverified report, citing records from 1900 to 1922, suggested that tigers swum across 29 km of the Hooghly river, and that one tiger may have crossed 10-56 km of open water, depending on where it started (Garga 1947).

Other than tigers, the medium and small mammals are a significant component of the mangrove ecosystem and a major source of food for a variety of animals such as raptor birds, snakes, crocodiles, etc. Frugivore mammals such as monkeys, squirrels, and bats are also important as seed dispersal agents. Herbivorous mammals browse on young shoots of trees, shrubs and other vegetation; hence, they control the growth of shrubs and bushes that may compete for nutrition with trees.

The Spotted deer are found throughout the Sundarbans, especially in the feeding areas like the meadow (40%) to feed on grasses, followed by forest edge (35%) and water-holes (14%). The mean density (number/km^2) was maximum in *S. apetala-E. agallocha-* open grasslands (44-195) followed by *C. decandra-E. agallocha-S. apetala* associations (43-55), *X. mekongensis-B. gymnorrhiza-A. officinalis* (14-18), *E. agallocha-C. decandra-X. mekongensis* (15-16), *H. fomes- X. mekongensis-B. gymnorrhiza-E. agallocha* (12-14), *H. fomes- E. agallocha* (10-13), *E. agallocha-H. fomes* (6-7) and least in *H. fomes* (3-4), whereas maximum number of the Barking deer, found only in the north and north-east was observed in the forests of *H. fomes- X. mekongensis-B. gymnorrhiza* (3-5), *E. agallocha-H. fomes* (4-5), *H. fomes-E. agallocha* and *X. mekongensis-B. gymnorrhiza-S. apetala* (2-3), and only *H. fomes* (1-2)(Dey, 2007). Therefore, the least preferred habitat of both the living deer in the Sundarbans is *H. fomes*. During the breeding season in April-September the Spotted deer congregates and during the non-breeding season (October-January) they disperse, although they used to breed throughout the year. During the breeding season their home range is multiplied and a male share its range with two-three females. The Spotted deer mostly forages during the day, particularly morning and evening, also at night, often in herd and the Barking deer, solitary or pair, are seen during the night. Three main activities are observed-

Mobility: Direct and indirect movement from one site to another;

Feeding: Manipulation, ingestion and mastication of the food items in the open areas or meadows;

Resting: Leaning against any object, lying down or sleeping.

The food habit of these two deer (browser/grazer) species is almost similar and all the plant (tree, shrub, climber, herb or sedges) parts are consumed- leaves, stem, flowers, fruits, bark, shoot, seedling and pneumetaphores. These species are- 20 Trees: *S. apetala, S. caseolaris, B. gymnorhiza, B. parviflora, A. alba, A. officinalis, A. marina, X. mekongensis, X.*

granatum, A. cucullata, L. racemosa, R. mucronata, A. rotundifolia, B. racemosa, C. manghas, E. fruticosa, P. pinnata, K. candel, S. chinensis, T. populnea; 12 Shrubs: *A. corniculatum, C. crista, C. nuga, C. decandra, D. roxburghii, F. virosa, I. pescaprae, M. repandus, N. fruticans, P. corymbosa, P. mucronata, P. roxburghii;* 9 Climbers/creepers: *T. grandiflora, D. trifoliata, D. scandens, S. globosus, Flagellaria indica, D. spinosa, T. bracteolatum, Finlaysonia, S. carinatus;* 44 Herbs/grasses: *I. cylindrical, C. dactylon, P. repens, M. wightiana, E. procera, F. polytrichoides, C. javanicus, C. cyperides, C. procerus, P. karka, T. elephantiana, C. exaltatus, C. jukkubga, O. rufipogon, P. paludosum, C. aciculatus, D. bipinata, S. spontaneum, S. bengalense, V. zizaniodes, P. coarctata, A. donax, C. nardus, C. citratus, B. intermedia, L. chinensis, L. hexandra, A. lanata, H. parasitica, H. curassavicum, F. halophila, F. sub-bispicata, E. roxburghii, T. tristis, S. maritima, V. loteola, H. plomoides, P. corymbosa, S. trilobatum, S. surattensis, C. roxburghii, N. porphyrocoma, F. indica.*

The wild boars are mostly seen moving around (73%), feeding (23%) and resting (4%), whereas the rhesus macaques spend 46% of time in moving, 41% in feeding and 13% resting.

New records of mammalian species have been recorded in both Bangladeshi and Indian Sundarbans. For example, carcass of a sperm whale (*Physeter macrocephalus* Linnaeus, 1758) was found on 22nd November, 2007 at Selar char. *Balaenoptera brydei* (Olsen, 1913), *Stenella longirostris* (Gray, 1828), *Steno bredanensis* G. Cuvier (Lesson, 1828), *Tursiops aduncus* (Ehrenberg, 1833) and *Pseudorca crassidens* (Owen, 1846) were observed in Swatch of No Ground MPA. The Fawn-coloured mouse (*Mus cervicolor* Hodgson, 1845) was recorded from Hemnagar (Mandal, 1988). The Hoary-bellied Squirrel (*Callosciurus pygerythrus* I. Geoffroy Saint Hilaire, 1832) is reported from the Bangladesh Sundarban (Adhri *et al.*, 2015). In 2019, an adult male wolf (*Canis lupus pallipes*) was killed by the locals in a remote village in the Sundarbans named Taltoli, a suboptimal habitat for the wolves, in Barguna district. A photograph published in a news report in early June showed a canine-like animal, beaten and dead, legs splayed, hanging from the makeshift posts. The dead animal had a white patch around the cheek and throat indicating that it was a wolf. There had not been a record of wolf in the region since independence. On reaching the spot a researcher found that its body was buried near the local branch office of the Bangladesh Forest Department. However, the animal was dug out and its skull appeared to be of a wolf.

Samples of hair and tissue were taken from one of the legs for DNA test, which later proved to be of a wolf. The researcher doubted that the wolf got caught up in the cyclone and was swept into this remote region of the Sundarbans. It survived by lying low in a nearby Tengragiri WLS. It lived there but hunted in the village because there is no prey like the deer in that PA, till it was killed by the locals. This is not the first wolf found recently in the dense, muddy, watery mangroves. On 14th April, 2017, a wildlife photographer Riddhi Mukherjee took a photo of an adult male wolf at Jyotirampur in the buffer zone of the Indian Sundarbans. But there is little possibility for that animal to cross the border to reach beyond the Bangladesh Sundarban's border after a gap of two years.

Poncins (1935) observed an unidentified species of tree-shrew at Bungsipore, Northern Sundarbans, during 1892; the field characters given exclude any confusion with a squirrel. However, the occurrence of tree-shrews has not earlier been confirmed for the Bangladesh or Indian part of it. Hence, it was assumed that the tree-shrews are absent from the Sundarbans. Definitely more field work was necessary to confirm the presence or absence of tree-shrews in the Sundarbans. At last on 18th December 2021 at about 11.29 am, a Madras treeshrew, also referred to as the Indian Tree Shrew [*Anathana ellioti* Waterhouse 1850 (Family: Tupaiidae, Order: Scandentia)], was discovered in the Indian Sundarbans too when conducting the All-India tiger estimation exercise. The camera trap, installed beside the Bagmara channel in Bagmara-4 compartment (21.644 N 89.049 E), has clicked the animal in this core area of STR (Justin *et al.*, 2022). The habitat is surrounded by Sundari (*Heritiera* fomes), Gewa (*Exocoecaria agallocha*), Pasur (*Xylocarpus mekongensis*) and Garan (*Ceriops decandra*). The site is in close proximity to the Indo-Bangladesh International border, separated by a river. Hence, there is possibility that this species may still be found in the Bangladesh Sundarbans, as was recorded by Poncins (1935).

The Malayan or Himalayan porcupine (*Hystrix brachyura* Linnaeus, 1758) was sighted at Supati Forest Camp (22.047°N and 89.827°E), Eastern Sundarbans, Bangladesh, on 22nd May 2018 and was photographed at around 7.35 pm on 24 May (Hasan and Neha, 2019).

CHAPTER 5
TAXONOMIC ENUMERATION AND SPECIES EXTINCTION

Biodiversity profile

Biodiversity may be defined as the variability among living organisms from all sources including *inter alia* terrestrial, marine and other aquatic ecosystem and ecological complexes, of which they are essential components. There are three components of the biological diversity-

(1) Genetic diversity : The allelic variation within a species, such as different food crop varieties, different floral and faunal. The enriched mangrove flora has diverse genera, viz. *Acanthus, Acrostichum, Aegialitis, Aegiceras, Aglaia, Avicennia, Bruguiera, Camptostemon, Ceriops, Conocarpus, Cynometra, Dolichandrone, Diospyros, Excoecaria, Heritiera, Kandelia, Laguncularia, Lumnitzera, Mora, Nypa, Osbornia, Pelliciera, Pemphis, Rhizophora, Scyphiphora, Sonneratia, Tabebuia, Xylocarpus*, etc. The distinctive adaptive features of individual mangrove species are salt glands in leaves (*Aegiceras, Avicennia, Acanthus, and Aegialitis*), thickening of leaves, viviparous seeds, and waxy epidermis, which add to the ecological success in addition to the species diversity.

From the conservation point of view, due to different barriers to free movement of the tigers living in the Sundarbans, inbreeding occurs among this small population. Inbreeding among the tigers with similar genes increases the morbidity and mortality rate of the tigers as well as birth of diseased and weak cubs. Transferring of the tigers from one population to another inside the Sundarbans can increase their genetic variety, possibility to survive and, as a result, the number of tigers will rise forming a viable population.

(2) Species diversity: This is the commonest method of measuring biological diversity which indicates total number of species and their dominance. This also includes variation of different taxa at various levels. Total number of terrestrial species are much more than the aquatic species. Within the floral genera, there are rich species diversity, comprising a total of 73 true mangrove species and hybrids.

The generic and specific diversity of the Sundarban mangroves varies in three (northern, central and southern) identified zones (Mallick, 2013). The tiger mostly uses the low diversity areas above the general tidal level (Zone A i.e. northern). At Zone C (southern), both the generic and specific diversity are the highest amongst the three zones, but the generic diversity of Zone A (above the tide level) is more than that of Zone B (central), which is frequently inundated, while the reverse relationship exists between Zones A and B so far as specific diversity is concerned. But, according to timber and fuel resources, Zone A is richer than Zone B, which, in turn, is richer than Zone C. So, the richness of resource is inversely related to the generic and specific diversity of the vegetation. Again, it may be seen that Zone C, which records the highest generic and specific diversity, does not form an ideal and permanent abode of the tigers as the tiger possibly avoids the constant tidal fluctuations and, hence, the increased salinity. Salinity indexes are very high in Zone C followed by Zone B and then by Zone A. A unique example of a very dense but closed vegetation matrix is evidenced from the low figures of both generic and specific diversity. Hence, Zone A is the mostly used habitat of the tigers and its prey animals.

There are threats within the same species. For example, in Bangladesh Sundarbans, about 49 tigers died from 2001 to 2021. Of them, 22 tigers died in Sundarban's eastern division and 16 in western division. Besides, members of law enforcement forces recovered 11 tiger skins and body parts from different areas. In such case, the birth and survival rate of the cubs are important for conservation of the species for recovering the loss of breeding individuals due to anthropological conflicts. Research has shown that the eastern Sundarban has the capacity to harbour 200 tigers. But, according to a survey conducted in 2018 by camera trapping, there were 114 tigers. The forest department is implementing various initiatives to increase the number of tigers from the present number. In order to ensure unimpeded breeding of all kinds of wildlife including the tigers during the monsoon season, entry of the tourists and fishermen into the Sundarbans has been prohibited during the three months of June, July and August. The activity of robbers in the Sundarbans has declined in the past few years as they have surrendered to law enforcers and returned to normal life. Along with that the activity of poachers also decreased. Apart from the regular operations, Smart Patrol is being conducted in the Sundarbans to stop the activities of tiger poachers and timber mafias. Along with the Forest Department, CMC, CPG and BTRC regularly patrol the Sundarbans boundary area. Sequel to these positive effects, recently, the tourists and forest staff have spotted tigers of different sizes in various parts of the Sundarbans. At the same time, tiger footprints have been observed in different areas of the forest, suggesting an increase in the number of tigers in the Sundarbans.

(3) Eco-system diversity: The forest eco-systems are mostly highly diverse and within the forest eco-system the rain forests are more diverse than the low rain fall areas or mangrove. The mangrove ecosystems are highly productive in terms of forest biomass and nutrient contribution, especially through detritus-based food webs, to support rich biodiversity in the wetlands and adjacent estuaries. The mangroves also play vital role in atmospheric CO_2 sequestration, sediment trapping and nutrient recycling.

As already stated before, the Sundarban region originally had two eco-regions, swamps and mangroves, but in the recent years, these fresh water swamps became extinct due to speedy expansion of agricultural lands and fisheries. Frequent destructions of mangrove forests were evidenced due to devastating cyclones like Aila in 2009, Amphan in 2020 and Yaas in 2021 and the recurrent tropical cyclones continuously damage the Sundarban coast and significantly reduced the provisional, regulatory, supporting and cultural ecosystem services of the Sundarban mangrove ecosystem. Recently, drastic land use change and rapid modification of forest area can significantly decrease the vegetation cover as well as carbon stock in this ecosystem rich forest cover area. With the ecosystems being lost continuously, rare species are now left with fragmented and fragile habitats among the mangrove, and their movement or dispersal are now full of obstacles.

The Sundarbans is gradually degrading due to the age-old massive extraction of its resources. The biodiversity of terrestrial ecosystem of this mangrove forest is tremendously declining and there has been massive degradation of forest coverage and its density in last two and half centuries. The aquatic resources are also declining mainly due to the exaggerated extraction of resources by unsustainable means accelerated by profit making impulses of people. As a result, both aquatic floral and faunal species significantly have declined in terms of quality and quantity. The inappropriate expropriation has triggered rapid destruction and has failed to take into consideration, sustainability and conservation of the forest resources.

On the other hand, the sea level rise is engulfing a huge portion of the mangroves, while the associated salinity increase is posing immense threats to the biodiversity and economic losses. Climate-mediated changes in the riverine discharge, tides, temperature, rainfall and evaporation have negative impact on the wetland nutrient variations, influencing the physiological and ecological processes, thus biodiversity and productivity of Sundarban mangroves. Hydrological changes in wetland ecosystems through increased salinity and cyclones lower the food security and also induce human vulnerabilities to waterborne diseases.

To sum up, the Sundarban is an ecologically vulnerable area due to the degradation of biodiversity resources triggered by both human interventions and climatic changes. The outbreak of the COVID-19 pandemic and the recent trail of destruction left behind by cyclone Amphan have had a catastrophic impact on the forest and millions dependent on the Sundarbans for their livelihood.

The biodiversity of Sundarban includes numerous species of phytoplankton, zooplankton, micro-organisms, benthic invertebrates, mollusks, amphibians, reptiles, aves and mammals. As per the present enumeration in the Sundarbans and reclaimed areas, the biodiversity includes 7,571 species- Bacteria (255), Fungi (243), Algae (745), Protozoa (294), Rotifer (62), Sponge (8), Lichens (319), Vascular plants (1,340), Gymnosperm (1), Non-vascular plants (10), Moss (11), Zooplanktons (159), Ciliates (5), Pelagic larvae (indeterminate), Polychaete larvae (67), Crustacean larvae and Fish eggs and larvae (indeterminate), Molluscs (247- Gastropods: 152 and Bivalves: 95), Cephalopods (10), Scaphopod (1), Arthropods (326), Insects: Thrips (1), Butterflies (170), Moth (250), Flies (212), Wasps (86), Bees (23), Ants (65), Firefly (2), Beetles (176), Grasshoppers, Crickets and Allies (108), Dragonflies and Damselflies (57), Cockroaches (3), Mantis (10), Termites (14), Spiders (266), Ticks and mites (163), Flatworms (1), Nematodes (77), Spoon worms (4), Annelids (55), Aphid or plant lice (34), Echinoderm (18), Ostracods (3), Collembola or springtails (78), Water fleas (25), True bugs (119), Fish (646); Amphibians (27), Turtles (19), Snakes (56), Crocodiles (1 extant, 1 very rare and 1 extinct), Lizards, geckos and skinks (20), Birds (603) and Mammals (74).

The fauna in the reclaimed areas is represented by practically every group of animals although the higher vertebrates, specially the mammals, are not well represented.

Hunting, poaching, extinction, fossils

Since the Matla-Bidyadhari river system was cut off from the sweet water sources completely leading to change in the salinity regime, this has resulted in loss of the riverine grassland and disappearance of four large herbivorous prey species of the tiger from the Indian Sundarbans during last two centuries:

(i)   Water buffalo (*Bubalus bubalis*)(1885)[800-1,200 kg],
(ii) Javan Rhino (*Rhinoceros sondaicus inermis*)(1888)[900 kg],
(iii) One Horned Rhino (*R. unicornis*)[1,600-2,200 kg],
(iv) Swamp deer (*Cervus duvaucelli*)(1930)[170-180 kg] and
(v)   Hog deer (*Axis porcinus*)(1945) [20-30 kg].
(vi) Barking deer (*Muntiacus muntjack*) [20-30 kg] population vanished from the southern portion (1976) except Halliday, Bulcherry and some other seaface islands.

The Cheetal deer survives due to excess salt excretion in the form of lachrymal secretion and the tiger has adapted to drinking waters of the saline creeks, yet the sweet water is an attraction and preferred to the saline water as evident from the pugmarks found near the water holes dug artificially all over the forests (Chaudhuri and Choudhury, 1994). When the area is swept over by severe storm along with high current and the wild animals have to cling hard to the forest floor for survival.

An estimated 80,000 of tigers were killed from 1875-1925 and probably more till 1971 when hunting tigers was totally banned in India. On the north of present Sundarbans there were extensive swamp lands inhabited by the mega-herbivores, all of which have become extent during the early 20th century. One of the most tiger-infested jungles, the Sundarban was stretched up to Govindapore before the Fort William II was constructed. It was told that Warren Hastings had a luck to shoot a Royal Bengal tiger on the spot. where the St. Paul Cathedral stands today (Cotton, 1907). Earlier, in 1756, the plains of Govindapore were occupied by native huts and by salt marshes, which afforded fine sport to the buffalo-hunters. Cotton (p. 24) described the surroundings as follows:

"...Chowringhee, which was in 1717 a hamlet of isolated hovels, surrounded by water-logged paddy fields and bamboo groves and separated from Govindpore by a tiger-haunted jungle where expands the grassy level of the maidan". The jungle, presumably, had been once a part of the Great Soonderband (Sundarban). Many traces of trees were found at a considerable depth below the surface of the ground. These remains are thought to be those of the soondrie forest that covered the site of Calcutta when newly emerged from the waters of the Gangetic Delta.

Hence, during the early 18th century, the swamps, salt marshes and grasslands of the old Calcutta was known as the tiger habitat. The habitat was reclaimed and the tiger, the buffalo and other wetland species inhabiting there were either hunted or wiped out.

As far as the hunting of the tigers inhabiting the mangroves of sundarbans was concerned, the colonial government thought it safe to engage the indigenous elite shikaris in the destruction of man-eaters of the mangroves. The government also adopted a policy of endowing monetary reward to encourage indigenous shikaris in the mission destruction of man eaters. Notification published in the Calcutta Gazette 16th November 1883, mentioned that the foresters and rangers were authorised to pay rewards for killing tigers. Initially the value of the reward was fixed at Rupees Fifty for each full-grown tiger, which was later enhanced after periodical review. In 1906, the reward was raised to Rupees One Hundred, which was again raised to Rupees Two Hundred in 1909. However, there were also numerous human casualties, which took place during this mission of destruction of the man-eaters. The indigenous companions of the elites were perhaps the worst affected (Chakrabarti, 2001: 67).

Though the area teemed with large game like tigers, buffaloes, single horned rhinoceros, four species of deer, etc, they were extremely difficult to get at and only to be obtained with the aid of staunch and skilful local shikari.

As referred to above, a government policy for reward to induce the indigenous shikaris (hunters) into killing the tigers, led to large-scale slaughter of the tigers in the Sundarbans between 1881 and 1912, when more than 2,400 adult tigers were killed in the region. Curtis (1933) recorded hunting of 452 tigers between 1912 and 1921. Hunting was banned in India as per the WL(P)A (1972) and in Bangladesh according to the provisions of the WLPA (1973), still 19 tigers (poaching 15 and 4 shot as man-eaters officially) were killed between 1976 and 1992.

A local correspondent from the village of Maheswar Pasha (approx. 22°53'20"N, 89°30'22") in Khulna wrote about the traditional process the local shikaris resorted to for killing the leopards. Their machinery consisted of several pieces of nets made of jute ropes interwoven with each other. These nets were usually eight to nine feet in height, propped up by bamboo pieces, the base being stuck to the earth by bamboo hooks closely set. A stretch of jungle was encircled with it with the object of driving the animals into it. Once entangled, they used to try to break through them by making a hole downwards. The shikaris then assembled round the tightly enclosed net and spear the leopard. Such practice confirms presence of the leopards in the northern belt of the erstwhile Sundarbans.

The saltwater crocodile is the top predator of aquatic ecosystems of the Sundarbans mangrove landscapes. Two deer-feeding incidences were recorded from Dobanki and Bagna areas of STR. Even a tiger, the terrestrial predator, was hunted by this aquatic predator. Additionally, most of crocodiles detected were basking on the exposed muddy bank of rivers between 0900 in the morning and1600 hrs in the afternoon during the low tide period.

In the past, the saltwater crocodile was abundant across the southern coast but large-scale exploitation for skins until 1970s greatly depleted its population. Though it is legally protected, anthropogenic threats are still prevailing. One crocodile was found dead in the eastern part of the Sundarbans but the cause of its death could not be known.

Occasionally, the crocodiles stray into the village ponds and are rescued frequently in September and October, for example, in Patharpratima block. A team from the South 24-Parganas division and trained personnel from the Bhagabatpur crocodile centre in Patharpratima arrived at the village and initiated a rescue operation along the muddy banks of the pond to capture about 10ft-long adult female crocodile, that got swept into a pond of the village because of the storm surge during the cyclone Yaas when the pond was inundated with water from the creek during the high tide and in all possibility the crocodile got swept in. Crocodiles swimming into ponds in the Sunderbans are now uncommon after the fringe villages are secured with nylon fencing. But the fencing got damaged in several areas because of the cyclone. The rescue team erected a bamboo barricade around the pond to prevent curious onlookers from getting too close to the crocodile. As the crowds kept shouting to draw the reptile's attention, it was mostly swimming below the surface. The team could spot it only when its nostrils surfaced to breathe. Thereafter, they lowered a nylon net at one end of the pond and started pulling it towards the bank. The crocodile managed to slip away several times before a team member managed to get a noose around its snout. The crocodiles are incredibly powerful, so several men were engaged to pull the net as well as the rope. Once near the bank, the team used bamboo poles to "steer" the reptile head towards them. A team member threw a gunny bag connected with a rope over its head and caught hold of its snout. Another got on top of its back to restrict its movement. Once the crocodile failed to see, it became easier for the staff to tie it with ropes. The reptile was later checked by a veterinary doctor and released into a creek near the Lothian island. The Forest Department staff often translocate the straying crocodile from the civil areas and reportedly, about a dozen individuals were translocated from the fringe areas and released in a suitable habitat.

In the Bangladesh Sundarbans, the number of crocodiles in the rivers and canals is on the rise in recent years because after a survey in the 2019-20 fiscal year, reporting only about 200 crocodiles across 1,874 km^2 of water areas of the Sundarbans, 206 saltwater crocodiles, born at the Karamjal Wildlife Breeding Centre, have been released into the rivers and canals of the Sundarbans in four ranges of the Sundarbans last December on the occasion of the birth centenary of the Father of the Nation Bangabandhu Sheikh Mujibur Rahman. At present, there are 92 crocodiles of different ages at the centre, which is a tourist destination. A 10 feet long saltwater crocodile was rescued from a pond in a village Srirambha in Rampal upazila of Bagerhat on 7th January, 2022 and released it back into the wild. The crocodile, aged around 15, reached there through the rivers and canals of the Sundarbans. On March 11, another saltwater crocodile was captured by the fishermen, while they were netting in the Bhaga area of Rampal and released it in the Sundarbans. In April, a crocodile was rescued from a shrimp enclosure at Bururdanaga village in Mongla upazila of Bagerhat district. The 5-feet long crocodile is suspected to be seven years old and it came to the enclosure through crossing canals and rivers in the area. So far about 10 such rescues have been made. In the

Sundarbans, one crocodile rules a particular stretch of water area. When another crocodile enters there, a fight breaks out between them. A rise in crocodile number in the Sundarbans and territorial fights might lead the reptiles to move towards the localities.

At present, hunting of crocodile may be insignificant, but the main threat to surval of the crocodiles is extensive fishing, fewer available prey species and increasing crocodile mortality. A second threat is pollution, which may reduce reproductivity and prey availability.

In Bangladesh Sundarbans, a crocodile conservation plan has been drafted by the Sundarban Biodiversity Conservation Project (ADB), proposing the following conservation objectives: improved protection; ranching and restocking; feasibility study on reintroduction of the marsh crocodile; and an improved understanding of population dynamics, distribution and breeding biology, and public awareness. The habitat has been subdivided into two types of conservation zones: wildlife sanctuary and production zone, where sustainable resource use is allowed. Protection measures for crocodiles target fishing, mainly by reinforcing its interdiction in sanctuaries and by closing smaller canals for fishing in the production zone. Measures are particularly useful for nesting areas, which are still remain to be identified.

A new crocodile rearing station has been constructed in the Karamjal Visitors Center with a capacity of 110-120 crocodiles. Hatchlings will be obtained from the wild, but the collection of eggs from the wild is dissuaded, as this may be too risky for the current endangered population. A key issue is the development of a financial mechanism, using revenues from the visitor center, to cover the operation costs Building experience in crocodile rearing for conservation could simultaneously contribute to the future development of commercial crocodile farming in Bangladesh.

Establishment of extensive human settlements and agricultural land by deforestation, habitat loss and degradation as well as unregulated hunting and poaching caused extirpation of many species (local extinction) and contraction of habitat for several species, during the period- 18th-20th century. The species now known to be extirpated from the region are: the water buffalo (*Bubalus bubalis*); the swamp deer (*Cervus duvaucelli*); the Javan rhinoceros (*Rhinoceros sondaicus*); the gharial (*Gavialis gangeticus*); and the chitra turtle (*Chitra indica*). The barking deer (*Muntiacus muntjak*), last sighted in the 1970s in the Indian Sundarbans is also locally extinct, although it is still surviving in the Bangladesh Sundarban.

There are historical records of extinction of at least seven globally Endangered resident avian species from Sundarban during early twentieth century- Pink-headed duck (*Rhodonessa caryophyllacea*), White-winged duck (*Asarcornis scutulata*), Indian peafowl (*Pavo cristatus*), Green peafowl (*Pavo muticus*), Greater adjutant (*Leptoptilos dubius*), Red-headed vulture (*Sarcogyps calvus*) and Bengal florican (*Houbaropsis bengalensis*).

The French pharmacist and explorer Christoph Augustin Lamare-Picquot made an excursion to the Bangladesh Sundarbans, where on 17 November 1828 his local hunters shot a female rhinoceros and caught her young one the next day, just south of Khulna, for some 20-40 km to the area around Mongla (22°50′ N, 89°60′ E). Both animals (female) were completely hornless. The total weight of the adult female was estimated to be about 3400 pounds ("livres") and the female calf, about 4 months old, was weighing 300 pounds. The experience allowed him to taste the milk of the adult cow, sweeter than the domestic cow's milk and to try the meat of the young animal prepared as a steak and the liver of the mother, all said to be tasty.

Lamare-Picquot must have remained in the area for about another three weeks. His zoological collections at the end of the trip included two rhinos, one tiger, three axis deer, five crocodiles (two species), four tiger cats (two species), one Ganges dolphin, two wild pigs, six monkeys (two species), 10 monitor lizards (two species) and a variety of other reptiles, molluscs, 133 birds-of-prey and herons.

He returned to France in the spring of 1830, where his zoological specimens were assessed by Georges Cuvier. After he published an account of the hunting expedition in an obscure work of 1835 (Lamare-Picquot 1835), the animals were described as a new species *Rhinoceros inermis* by René-Primivère Lesson, not in 1838 or 1840 as has been assumed, but in 1836, as shown later. After transfer to Berlin, the specimens were again examined by Wilhelm Peters, who upheld the validity of the species as a rhinoceros without horns. The type specimens of R. inermis are still preserved in the Museum für Naturkunde in Berlin.

The adult female (ZMB_Mam_1957) was selected as the lectotype. But the tubercles on the skin, the number and direction of skin-folds and the nasal bones exactly resembled the species described as "Rhinocéros de Java" by G. Cuvier.

Edward Blythe, the then Curator of the Asiatic Society, in a letter to Edward Newman (Calcutta March 1, 1862), highlighted four morphological differences based on examination of museum specimens of the large Indian and small Sundarban rhinoceros:

1. The tubercles of the hide (of their stuffed Sundarban rhino) are much smaller than *R. indicus* (now synonym of *R. unicornis*) ;
2. Length of skull, from middle of occiput to tip of united nasals, in *R. indicus* 2 feet (±half an inch); in *R. sondaicus* the length is 1¾ foot at most;
3. Height of condyle of lower jaw in *R. indicus* 1 foot+; in *R. sondaicus* 9 inches;
4. Breadth of bony interspace between the tusks of the lower jaw in *R. indicus* 1½-1¾ inch; in *R. sondaicus* ¾-1 inch.

Rainey (1878) described the physical features of the species:

"The largest of the *feræ naturæ* is the one-horned rhinoceros, which is identical with that of Java, *Rh. Sondaicus,* and differs from the other Indian one-horned species (*Rh. Indicus,* sic. *unicornis*) in being shorter in height, but not in length, and the female only possessing the nasal protuberance, which is not really a bony structure, but merely agglomerated hair, that is, it is similar to the horns of hollow-horned ruminants."

Pennant (1798), an employee of the East India Company, first wrote about the presence of both tiger and rhinoceros in the islands of Sundarbans:

"The one-horned rhinoceros is very common in these islands. It loves forests and swampy places, and is a frequent concomitant of the tiger."

He also wrote about the aggressive behaviour of the rhinoceros:

"... the rhinoceros repairs to wet places out of love of rolling itself like a hog in the mire. The Rhinoceros, when provoked, is a most dangerous enemy, and extremely swift. A gentleman of my acquaintance once in the service of the Company, had landed on one of these islands, and routed a Rhinoceros, which rushed on him, flung him down, and ripped open his belly; the animal proceeded without doing him any farther injury; the gentleman survived the wound, and lived to a very advanced age."

The Morrison brothers (M.S.S. Field Books of Lieut. Hugh Morrieson, of the 4th Regiment Bengal Native Infantry, and Lieut. W.E. Morrieson, Bengal Engineers; Surveyors of the Soonderbuns) recorded their hunting expedition:

"Having come to an anchor we saw a rhinoceros on the opposite side of the river drinking. I crossed in a pausway, he allowed me to approach to within 30 or 40 yards, I fired at his head and put the ball through his cheek, he ran off into the jungle before I could get a second aim at him. On reaching the pinnace I learnt from the party I had sent on shore that they had been successful in finding a tank of good water under the cocoanut trees, it was however surrounded by long grass and other jungle, the haunts of many rhinoceroces, they had made a regular bed in it. Being anxious to save a trip all the way to Chandealley (in the North) for fresh water, I went on shore with an armed party, carrying fire-brands with which we soon set the whole place in a blaze. I left it to burn out meaning to return in a day or two to try and fill our casks (Calcutta Review 32: 19)."

Poncins (1935) wrote about the ecological adaptation of the Javan rhino in the Sundarbans:

"The foot decidedly broader than the girth of the leg as even in sinking, as is the rule, over 10 inches (25.4 cm) in mud, the leg itself does not touch the walls of the foot hole. The foot is lifted, and comes out absolutely neat and clean without any drag of any sort. Inspection of the lower part of the hole shows that the weight of the body does not make it expand as would appear likely under the pressure. The animal must be extremely tough ... The hind feet steps exactly into the print of the forefoot. It is a little smaller and more oval. It seems also that the hind foot rises on the ground at a little outward slant. The beast goes at a quite slow walk, under the big trees and through the heaviest thickets, without paying the slightest heed to obstruction. The spoor is made in a regular zigzag line like a cat's track on snow."

Poncins also described the food habits of the species:

"A rhino browses very steadily… he seizes the tips of branches in his stride without altering the position of his feet in any way. He does not trouble to move his head right or left to inspect his food. He does not pull at a branch sideways… the branch is crushed and cut sharp without a move of the head. His meal lasts until late in the morning (c. 9 o'clock)."

He also informed that the rhino used to drink river water of the Rimangal and seen to swim well. He described its other habits:

"When he goes into a thorny or leafy thicket, he does it as a needle piercing a cushion. The sort of tunnel left behind is hardly visible and, at most, 3' high, as the branches spring back almost exactly… the rhinos traveled singly, only occasionally two together for a short time. They usually move in a straight line, generally east-west or *vice versa*."

Baker (1887) confirmed mud-bath of the rhinos:

"On the margin of a mud-hole twenty or thirty feet in diameter stood a huge rhinoceros in deep contemplation of two shapeless slate-coloured lumps just showing above the muddy water; in other words, two companions enjoying a mud-bath, while he, having had his, as his well-plastered hide testified, was basking in the sun half asleep, working his ears and stamping with a foot now and then as flies pestered him".

Baker (1887) also described his hunting activities when he shot three rhinoceros in the Sundarbans:

"The mud-hole was near the jungle on the north, and fully two hundred yards from our ambuscade; too far for a shot at so tough a customer, and there was no cover between us beyond a few rushes and a little scrub, too thin and low to afford concealment. Backing out, therefore, a little distance, I entered the bushes, which formed a fringe of the forest all round, and charily making a little sweep, not altogether unmindful of the possibility of a tiger being an interested spectator of my evolutions, I wriggled my way to a position within sixty paces of the pool, the wind favourable and the enemy's broadside bearing almost directly on me. Looking about me while recovering my breadth after the stalk, I observed that the water in the hollow was rising by the influx of the tide along a rill issuing from the woods, so that the patriarch's fore-feet were now immersed, but the sedges and rushes in which he stood barely reaching his belly, his whole right side was fully exposed to view. As soon as my breathing had settled down to its normal state, the big rifle was directed to the neck, but on drawing a sight, it was somewhat covered by the huge bulging shoulder, the head being turned a little away from me, making the shot an uncertain one at the angle presented; the aim, therefore, was rapidly changed to a point a trifle behind the shoulder, and the heavy bullet told truly with a loud smack. On feeling the wound the great creature threw up his head with a grunt, and glared round for the enemy who had struck him, and before his position was changed a second bullet hit him on the same spot, but a little more forward, and brought him on his knees with a wheezy sort of a groan. He was up immediately, and dashed into the woods with blood spurting from his mouth. At the report of the first barrel the other two rose from the mire in a mighty hurry, but paused on failing to discover aught on which to vent their wrath, and then seeing or scenting the smoke, galloped off after their leader, the larger of the two receiving from the second rifle one ball in the fore-ribs, and a second in the head, fired at short range, and driven home with four and a half drams of powder… followed as fast as I could on the broad trail left by the flying monsters, and before I had gone fifty yards, a loud crash and a long-drawn groan announced the fall of one at least, and soon after I almost stumbled over its huge carcase lying in the death-agony. Dashing on upon the bloody trail for another hundred yards I came upon the bank of a narrow creek, just as one of the animals was disappearing in the overhanging wood on the opposite bank, and the other was rising out of the water, exposing its broad back as it struggled with mighty efforts to extricate itself out of the sticky mud. A shot, planted in the middle of the back over the loins, followed by another just behind the head, caused the stricken beast to plunge forward stone dead; its fore-parts on land, and the hind-quarters and legs in the tide now near the full."

Later in 1850, a hunting expedition was organised in the Piyali island (Simson,1886):

"...Peering about, presently I saw that something was moving in the dry fern-bush, which was about ten feet high, so I stood by the trunk of a tree about twenty-five yards from the fern... soon I saw a nose poked out, then the eye, and then the ear of a rhinoceros; as soon as this came out I let fly, and you can scarcely conceive the row which followed- something between the roar of an elephant and the neigh of a horse, but far stronger. The smoke hung, and as it passed I saw a rhinoceros standing, looking directly towards me. He stood a minute. I knew it was no use to fire at his lowered head, there is no vulnerable place there and the ball would have glanced...the animal turned round, and as he did so I shot him high in the shoulder and he bolted. I followed, but was instantly stopped, for there lay a rhinoceros stone dead. There had been two in the bush, and my ball behind the ear had killed the first. After the judge and I had made sure that the beast was really dead, we went after the other. The blood had gushed out on the trees and was frothy... " I

Rookmaaker (1997) recorded presence of the Javan rhino in various parts of Sundarbans, mostly from eastern parts falling in Bangladesh, including Jessore [89º10' E. long. and 23º10' N. lat.] towards the north, Barisal [90º20' E. long. and 22º45' N. lat.] in the east and Xavaspur Island in the south. The Piyali river side near Baruipur [88º32' E. long. and 22º20' N. lat.] in the north and Sagar [sic.] Island in the south [88º05' E. long. and 21º75' N. lat.] of western Sundarbans (West Bengal) were also having rhino-records. These records of capture, shooting and sighting of animal, and recovery of skull include three records in seventeenth century, 17 records mainly in the later half of nineteenth century and two records in the beginning of twentieth century. Out of these, 11 specimens were sent to museums of Calcutta, Berlin and London.

The remains of recent fossils of *Rhinoceros unicornis* have been collected from Bakkhali. Fossils of sweet water tortoise Chitra turtle and jaws of Gharial (*Gavialis gangeticus*) were found in the excavations of Dum Dum along with great stumps of Sundari (*Heritiera* spp.) in association with fruits of *Derris* and *Ceriops*. Sweet water enabled the survival of *Rhinoceros sondaicus, R. unicornis, Bubalus bubalis* and *Rucervus duvaucelii* within mangrove forests of Sundarbans until 19th century. Skeletal remains of *R. sondaicus* were found at Mollakhali, Gosaba on 19th April, 2000. The skeletal remains (skull, ribs, legs, etc) at Mollakhali were found only at 2.7 m below the surface, by the side of island. The bones were sent to the ZSI, Kolkata for proper identification. The scientists of ZSI reported that the bones belong to a Javan Rhinoceros. Again, on 14th May, 2000, few bones were recovered from a pond in Tentulia village near Pathankhali within the Goasaba Police Station of SBR, where people were excavating a pond with depth of about 20 feet. The bones were also sent to ZSI, Kolkata and were identified as the bones of *Rhinoceros sondaicus*. Moreover, on 3rd March, 2001, some bones were recovered from Netidhopani (Compartment 1) within STR, which was badly eroded by the storm surge. The bones were sent to ZSI for identification and were identified as the bones of the Wild Buffalo. Another skull was found opposite to Bidya village in Pirkhali forest block in an eroded area on 4th March, 2002. The skull was sent to ZSI for identification and was reported to be that of a Wild Buffalo.

Along with the fauna, some floral species got extinct from the Sundarbans. In case of flora, Prain (1903) recorded 13 species of orchid, but during a survey of 21st century, only seven species were recorded. Since the orchids are very sensitive to salinity, increasing salinity in the habitat caused extinction of the other six species during the last century.

CHAPTER 6
MANGROVE EXPLOITATION

Population and forest dependence

Till the 1st half of the 20th century the population growth in SBR was very low. From 1872 to 1951 the population was increased from 296045 to 1159559 (74.46%), whereas the increase was 73.80% during the period 1951-2011 (from 1159559 in 1951 to 4426259). In 2011, the population density was 1089.2 as against 925 (per km²) in 2001, particularly due to significant migration after the partition of Bengal in 1947 and the Bangladesh liberation war in 1971) live in the reclaimed transitional zone of SBR with mostly non-forest and agricultural areas. The population (2011) was 3309526 in 13 blocks of South 24-Parganas district and 1116733 in 6 blocks of North 24-Parganas district.

In Bangladesh, population was 2,306,550 (2011) with density of 560.3 (per km²) as against 537.33 (per km²) in 2001 live within a 20 km wide radius surrounding the periphery of SRF, known as Sundarban Impact Zone.

Most of the people living in the fringe areas are directly dependent on the natural resources of Sundarban, but population pressure on the Indian Sundarbans is almost double than that of the Bangladesh counterpart because of higher reclamation areas and population density as well as less forests in SBR. In India, the livelihood of nearly 2 million people is linked with the Sundarban, which mainly include fishing, crab collection, honey and beeswax collection and allied activities (Singh et al, 2010). Over 32,000 households in the Sundarban have at least one member exploit the forest regularly for various purposes, such as collecting fuel wood, sustenance, cash income (from the sale of honey), medicinal requirements, and harvesting timber for construction of houses and boats. More than one lakh families belong to the fisher community and also involved in crab collection. Around 150,000 men, women and children are engaged in estuarine and riverside shrimp fry collection, which has caused biodiversity loss in the Sundarbans. Out of estimated entrants of 45,000 into the forests of Tiger Reserve per day, about 6,000 were legal permit holders and approximately 38,000 illegal entrants.

In Bangladesh, the Sundarban provides livelihood for about 740,000 people who work as fishermen, woodcutters, honey gatherers, shrimp fry collectors and nypa-leaf (*Nypa fruticans*) and thatching grass (*Imperata* spp.) collectors (Islam, 2010). Of them, 80% are 'collectors'. About 95% of the collectors work for others. With regard to the distribution of the total number of collectors across the districts, Khulna occupies the highest position (48.7 percent) and Satkhira the lowest (4.1 percent). After the initial modest beginning during the late 1970s, the shrimp farming rapidly gained popularity and by early 2000 reaches its peak, covering about 200,000 ha of Satkhira, Khulna and Bagerhat districts in 2005. The largest number of persons (estimated 423,000) is engaged in *Macrobrachium rosenbergii* or galda prawn fry (24.3 percent) collection, followed by collection of *Penaeus monodon* or bagda shrimp fry; 40% of these collectors are women and children. Some of them get involved in collecting molluscs, shells and oysters. There is usually a ban on crab collection in specific months of the year but the ban is often not adhered to. Households (70 percent) located in the vicinity of the SRF were mostly dependent on fishing and forest related activities than the distant households. About 56 percent of fishing took place within the 0 to 5 km zone. 20,000-22,000 people are involved in dry fish processing. Approximately 5% of the population is engaged in wood collection and about 3% of the population is engaged in the seasonal honey collection.

CHAPTER 7
MANGROVE RESTORATION

Niche-based afforestation programmes

There are many pocketed degraded areas in the Sundarbans. Mangroves are degraded along the shorelines, channel banks and around the salt affected surface by the physical processes of erosion, shifting channel banks, over wash sand deposits and increased soil salinity. Several factors and sub factors are estimated as diversity of mangrove degradation (over uses by humans, development of fisheries, hypersalinity, sediment movement, storminess or storm effects, shoreline erosion, and mangrove regeneration problems) in the Sundarban coast predominantly influenced by hydrodynamic and morphodynamic changes intensified from time to time by the super cyclones during the 21st century. Low density in forest cover is led by low canopy coverage, immature growth of mangroves and isolation of vegetation patches and has caused changes in the mangrove dominated wetlands of Sundarbans.

Mitra and Pal (2002), the degraded areas in Sundarbans are divided into three categories, viz.

(I) Areas devoid of any vegetation: It is characterised by saucer-shaped areas with no vegetation. Flood water remains in the depressions from March to December. Salt accumulates on the top soil and no plantation can be undertaken. Salinity can be reduced by improved drainage of these areas and the soil may be improved through use of gypsum, sulphur and natural grasses.

(II) Areas with scrubby growth of mangrove species like *Ceriops decandra*: This area requires silvicultural improvement through planting of upper storey species. Some experiments have been carried out where strips are cleared and upper storey species have been planted.

(III) Areas with scrubby growth of mangrove species like *Excoecaria agallocha*: There is no need of naturally transported or artificially dibbled germination on the hard soil of this area.

Restoration of mangrove wetlands is a challenge to the SBR authorities. The management has adopted a two-pronged strategy for conservation and restoration of mangroves in the Indian Sundarban-

1. **Prevention of degradation of the mangroves:** The objective is to protect the existing mangrove patches (preventive management), and
2. **Ameliorative management:** The objective is to restore the degraded mangroves by planting and bringing newly emerged mudflats under mangrove afforestation before any encroachment takes place.

Objectives

The objectives of such programme are-

- To stabilise the mudflats.
- To stabilise the embankments.
- To regenerate the degraded biodiversity resources of the wetland ecosystem in the Sunderbans for acting as a biological defense against the natural calamities and protecting the age-old earthen embankments.
- To ensure continued existence of the community in this vulnerable ecosystem through increased access to wetland resources and reduced exposure of local populace to the hazards of tidal surges.
- To involve the village women folk in the plantation programme for enhancing their financial status as well as social security.
- To reduce carbon footprints and get carbon credits in the process of restoration of green belts.
- To create a community participative model and develop a sense of ownership in the community for the mangrove, the lifeline of Sundarban.

Plantation technique

Seed collection

- From mid-February, the seedlings of *Rhizophora* sp first starts coming up followed by *Excoecaria* sp in late June, *Avicennia marina*

in June-July, *Bruguiera* sp in July-August, *Aegiceras* sp in July/August, *Avicennia alba* from August till October, *Sonneretia* sp in September and so on. These seedlings are collected during the monsoon according to the variable phonological pattern.

- Seeds suitable for collection are commonly found along the high tide line. The seeds are swept off and carried away with the tide water. Since their size and shape are different they float different in the water. Small propagules such as *Avicennia* spp and small mangrove fruits such as *Sonneratia* spp float far and wide on normal tidal currents. On reaching new site, if the soil condition is suitable, it is established and known as colonisers. These are collected from the tide water or mud flats. The collected seeds are planted in the forest or village nurseries for germination and growth of new saplings. If seeds are not naturally available in an area of plantation, the seeds are procured from other areas.

- The *Avicennia* sp is quite common and for all other species, the seeds that drifted along with the tidal waters into the village sides were collected by the community and plantation took place by direct seed rowing (*Avicennia* and in some cases *Bruguiera*). Other species were maintained through nurseries. However, *Avicennia* was also maintained in the nurseries as the mortality for direct seed plantation was almost 48-52%. In case of plantation through nursery-raised saplings, the mortality rate was calculated to be 75-80%.

Processing of the seeds/propagules

After collection, the propagules are kept in a jute bag and placed under shade where tidal waters keep them wet for 15 days in order to prevent loss during seedling establishment.

Nursery and germination techniques

The nursery site is selected when it is inundated twice a day during high tide and the water is drained off during the low tide. Nurseries are required for the species like Baen (*Avicennia marina, Avicennia alba*), Kankra (*Bruguiera gymnorhiza*), Bakul Kankra (*Bruguiera parviflora*), Garjan (*Rhizophora apiculata*), Garan (*Ceriops decandra, Ceriops tagal*), Khalsi (*Aegiceras corniculatum*), Tara (*Aegialitis rotundifolia*), Keora (*Sonneratia apetala, Sonneratia caseolaris*), Dhundul (*Xylocarpus granatum*), Pashur

(*Xylocarpus mekognensis*), Hental (*Phoenix paludosa*) and Golpata (*Nypa fruticans*).

However, some of the above plants can be sown or planted directly on the spot. Baen (*Avicennia marina, Avicennia alba*), Kankra (*Bruguiera* sp), Garjan (*Rhizophora* sp) and Garan (*Ceriops* sp). These are planted in the mother bed. When leaves and roots are sprouted, they are ready for plantation. The saplings of Keora (*Sonneratia caseolaris)* are planted directly in the plantation area after trimming the roots and stem. During lowest tide, washed Keora seeds are usually sown on the silt accumulated near the embankments. When the sprouts are two inch long, they are planted at the site during low tide. The seeds of Kankra, Garan and Garjan may also be sown directly on the silt after cutting trench. In case of Hental (*Phoenix paludosa*), the seedlings are first prepared like the Date palm before being planted at the site. In case of Golpata (*Nypa fruticans*), the mature dark brown seeds are packed in a jute bag and are left on the diurnal tidal bed for several days for germination. The sprouting seeds are planted on the mother bed. When the saplings are 6 inch long, they are taken out of the mother bed for plantation.

A seedling is fit for plantation at the age of about 6-8 months, when its height would be 35-40 cm, bear 8-10 leaves, stem stand upright with profuse root system and healthy with no sign of disease or pest attack.

Site selection and preparation of ground for plantation

There are large strips of newly created mudflats and char lands as well as nearly 3,500 km of embankments, which have been afforested. Many emerged bars and inter tidal flats are also utilised for seedling mangroves in the Sundarban. An attempt is also made to protect mangroves along the margins of embankments, roads, ponds and fish farm plots for achieving surface stability of tidal sediments through social forestry. For this purpose, several mangrove nurseries of SBR provide quality propagules for an afforestation programme in different parts of the Sundarban, which is being monitored through an independent mechanism to evaluate its success.

The mangrove afforestation programme has been undertaken to protect the coast from natural catastrophes and to conserve rare and endangered floral species such as *H. fomes*. A mangrove rehabilitation project was launched for afforestation of embankments with mangrove species. NGOs have taken active role in afforestation programme involving the local population. For example, WWF-India and S.D. Marine Biological Research Institute organised a plantation programme in the northern portion of Sagar Island since 2001, where mangrove species were planted in the experimental plot and initiation was taken to measure their growth rate with respect to the environmental parameters.

After identifying the planting site and species, the site was cleared off the ferns and weeds, dead standing wood and debris. An inverted V-shaped spacing with the point of the V facing the sea deflects wave impact. Spacing was less than 0.5 meter. Planting was done in triangle formation with one of the corners of the triangle pointing seaward. Spacing in between triangles was less than 1 m and cluster planting was also done to act as a wave break. To maximize survival, spacing was made much closer (25 x 25 cm). After 2-3 years, when the clusters were fully established, the gaps could be created for proper growth through a wise management. Even temporary inverted V-shaped bamboo structures were put up to dissipate the wave energy.

The same principle was applied in strip planting. Strips (10 or 20 x 100 or 150 m) were established at very close spacing to withstand strong waves. Once established, the open areas between strips and shoreline might then be planted at a wider spacing, like in most of the planted sites of Bidya-Raimongol, Saptamukhi-Thakuran and Matla.

In some areas, where the upper shoreline had become stiff and unsuitable for plantation, the strip plantation was done with *Avicennia* sp. in lower shoreline and silt thus trapped in the upper shoreline made it possible for plantation in the coming year.

Channels of size 1 ft wide and 1 ft deep and 5 ft apart parallel to river flow was dug in the month of July for collection of silt to enable the seeds/plants to get new alluvium, after one and a half month of which the mudflats became ideal for plantation of seedlings.

Those areas which are chars and low-lying, dug-channel management was not considered. In the initial stage, plantation was directed to obtain a plantation density of 4,000 per ha. Considering the mortality, this spacing was estimated to produce a plantation of density 2,500-3,000 plants per ha, an ideal for optimizing carbon sequestration rate.

A common practice responsible for high mortality rates in mangrove planting is the sowing of propagules more than half their length in the soil. Generally, propagules were sown one third of their length on firm substrate and one half of the length on soft substrate. The first two years after establishment were the most caring phase of plantation. Generally, from the third through the fourth year the level of the care was somewhat less, when spread of disease was checked. The sixth through the fifteenth years was a period of relatively low maintenance. The sixteenth through nineteenth were typically on lower maintenance. By the twentieth year the growth was in most cases over.

Regular visits especially at low tides and intense guarding of the area with community mobilization was an integral part of the maintenance process. Checks were made so that if a large number of green algae would float into the area, regular visit would make its removal before much damage. Other debris that might adversely affect the seedlings included pieces of driftwood, fishing nets and other heavy naturals or excess silt deposition that could knock over the seedlings or damage them.

Plantation projects

During the 20th century, there was no intense afforestation programme in the non-forest areas, for limited funds and human resources. Plantation of freshwater as well as mangrove plants were done in some of the degraded area. However, artificial planting had since then been discontinued within PA and natural mangroves had covered the areas. Artificial regeneration had however, been taken up on the mudflats, riverbanks and char-lands in adjacent areas outside RFs though at a small scale and is being continued.

Under the IUCN LGP Project "Alternative Livelihood Options for Vulnerable Mangrove Resource Users in the Sundarban Biosphere Reserve" a study entitled "Benchmark Studies on the status of Sundarban Mangrove Forests" was conducted by an NGO "Nature Environment & Wildlife Society" (Kolkata) during 2012-14 in 12 forest blocks including Lothian Island WLS- eight in STR (Bagmara, Chamta, Gona, Haldibari, Mayadwip, Netidhopani, Bagna and Pirkhali) and four in South 24-

Parganas Division (Ajmalmari, Dulibhasani, Chulkati and Lothian) and prescribed a guideline for plantation in SBR.

Extent of plantation

Mangrove afforestation programme was carried on in SBR since late 20th century. During a period of 27 years (1989-2015), plantation was raised over about 160 km² area covering mudflats, degraded mangrove forests and river embankments (1989: 636; 1990: 280; 1991: 645; 1992: 677; 1993: 715; 1994: 845; 1995: 347; 1996: 360; 1997: 440; 1998: 830; 1999: 400; 2000: 850; 2001: 1146; 2002: 770; 2003: 750; 2004: 830; 2005: 760; 2006: 780; 2007: 800; 2008: 400; 2009: 500; 2010: 500; 2011: 285; 2012: 453; 2013: 220; 2014: 320; 2015: 430).

Species planted

The species that have been most extensively planted under this programme include *Brugeria gymnorhiza* (Kankra), *Heritiera minor* (Sundari), *Rhizophora apiculata* (Garjan), *Sonneratia apetala* (Keora), *Nypa fruticans* (Golpata) and *Xylocarpus granatum* (Dhundul), as naked propagules and *Avicennia* spp. (Baen), *Excoecaria agallocha* (Genwa), *Ceriops* spp. (Garan).

Monitoring

Continuous monitoring of plantation sites has been ensured through a forest department team, who protect the plantation sites from trampling by the Tiger Prawn seed collectors. Trenches that get damaged due to wave action are repaired immediately when detected by the watchers. As the seedlings enter their second year, the watchers protect the plantations against illicit felling, damage, etc. This on-site monitoring of mangrove plantations is discontinued after two years.

An independent monitoring wing of the Forest Directorate, Government of West Bengal undertakes assessment of the status of the planted sites every year. Additionally, the GIS Cell of the Forest Department examines the changes in the mudflats and erosion patterns over the years using satellite imagery. Coastal shelter-belt planting of non-mangrove species has been carried out to stabilize the bunds in the SBR. Planting was also carried out on a few islands of the Sundarban, especially on some sand heads and sea-facing islands, where there are natural beach formations. *Acacia* and *Casuarina* are also planted along the embankments to stabilise the access ways and to protect the aqua farms and crop lands by involving self-help groups.

For example, in the plantation programme of Nayachar Island, species such as *Sonneratia apetela, Rhizophora mucronata, Excoecaria agallocha* and *Bruguiera gymnorhiza* were considered in the afforestation scheme. Species like *S. apetala* attained about 12 m mean height within a period of 3 years. These are all locally available plants and worked very well in context to erosion protection of Nayachar Island.

The nurseries were maintained by the villagers, training was imparted and intense campaigns were organised. Simultaneously, livelihood promotional activities were undertaken in the adjacent villages to create a sustainable 'care and protection' initiative for the mangroves. However, a sizeable portion of the plantation was lost immediately after cyclone, but in the adjacent mudflats, the plantation had grown into thick forest. The mangroves, especially *Avicennia* spreads cluster of roots, appreciable larger than the shoot in the first three months to stabilise themselves in the tidal wave energy system.

Actions against common threats

- Removal of encrusting organisms like barnacles
- Removal of insects and moth larvae eating leaves or damaging roots
- Removal of dead and dying plants
- Removal of plants entangled in green algae or other debris.
- Natural (storm) and anthropogenic (prawn seed collection, grazing of cows, goats and irregular ishing and boating activity) threats.

CHAPTER 8
JOINT FOREST MANAGEMENT

Indian Sundarban

Initiated in 1995, Forest Protection Committees (FPC) had been formed to manage a dedicated forest area with the technical and minor financial assistance from SBR. They are allowed to harvest non-timber forest products (NTFPs) like honey, medicinal plants, fruits etc. The major forest produce is harvested at the end of a felling cycle and 25% of usufructs are distributed to the FPCs. In case of management of PAs, the dedicated forest area is managed by Eco-Development Committees (EDCs). Whereas up to 1998, 10 FPCs and 12 EDCs consisting of 10,000 families have been formed in TR and 21 FPCs consisting of 8,300 families in undivided 24-Parganas Forest Division, at present, 51 FPCs and 14 EDCs have been registered in SBR.

In TR, 8,548 families of 14 EDCs and 11 FPCs protect about 252 km² of forests, falling under WLS and RFs (buffer area). In 24-Parganas (South) Division, 40 FPCs composed of 26,519 families protect about 390 km² of forests. Development of Self Help Groups (110) involved 800 women in STR and 234 groups in 24-Parganas Division involving 2,808 women. Various JFM Support activities and Eco-Development activities are being implemented in SBR.

Bangladesh Sundarban

While community based forest management have been quite successful in some areas of Bangladesh, in the Sundarban the Forest Department has yet to introduce effective participatory forest management. A gazette notification for implementing a co-management approach for managing the Sundarban involving key stakeholders stipulates the establishment of co-management councils and committees. Two Co-management Committees have been functioning in the Chandpai and Sarankhola Forest Ranges of the Sundarban East Division and two more are being formed.

After implementation of JFM in the Indian Sundarbans, human-wildlife (mainly tiger) conflicts have been reduced considerably in the villages. Prior to the 1990s and in the 1990s due to the frequent problem of tiger straying into the villages adjacent to TR, problem of cattle-lifting by the strayed tigers, many tigers were killed by the villagers. To ease frustration and reduce retaliatory killings, the Indian government provides compensation to owners of cattle attacked by tigers, but the process is long and tedious, impacting its effectiveness. In the absence of immediate interim relief, villagers affected by attacks on their cattle would continue to poison tigers in revenge.

In July 2001, a strayed tiger was killed by local villagers at Pakhiralaya. The big cat was hiding in a bush. About a thousand of people gathered within 500 m from the Sajnekhali Range Office, located just opposite to Pakhiralaya to kill the tiger. Its body was mutilated into pieces, put into gunnysacks and thrown into a river. The Forest Department staff was scared and none of them tried to prevent the tiger-hunting in anticipating physical assaults by the mob. Formation of EDCs definitely helped to curb retaliatory killings of tiger by the villagers and it also helped spreading awareness among the people regarding the conservation of tiger. Rather than killing the straying tiger, villagers scare it back into the forest by brandishing flaming torches and setting off firecrackers. If this fails, they have a number to call to get a swat team on the ground with a tranquiliser to sedate the tiger so that it can be released back to the forest.

But it has been alleged that manipulation, corruption, etc. by the village elites and politicisation of JFM benefits are depriving the real needy and deserving villagers from enjoying the usufructs.

The data available for rescue of the strayed wild animals in Sundarbans since 2001 has indicated a major success of JFM programme in Sundarban as large numbers of wild animals have been saved by people every year.

EDCs around TR are also eligible for 25% of the total revenues earned from tourism, which is then used for development works in the villages, such as construction of irrigation canals, jetties, community halls to organize any public meeting, digging tube-wells to supply drinking water, building of brick paths, and distribution of van-rickshaws.

Various JFM Support activities and Eco-Development activities that are being implemented, are as follows:

Community Development Projects :

- Construction of irrigation canals for rain-water harvesting.
- Construction of sweet water ponds for irrigation as well as fresh water pisciculture.
- Construction of village brick paths to improve communication.
- Construction of jetties.
- Digging of tube-wells for drinking water supply.
- Supply of solar lights.
- Organising regular medical camps in remote areas.
- Construction of embankments for protection of villages

Individual beneficiaries oriented schemes

- Apiary, mushroom cultivation.
- Piggery, Goat rearing, Duck rearing, Poultry.
- Vocational training for Cottage industries like boutique printing etc.
- Development of Self Help Groups.
- Training and input for vocations/professions like tailoring, cycle repairing etc.

CHAPTER 9
CAPTIVE BREEDING OF ENDANGERED SPECIES AND REHABILITATION

Indian Sundarban

Captive breeding of saltwater crocodile

Bhagabatpur, South 24-Parganas Forest Division

The saltwater crocodile conservation breeding programme started in 1976 at Bhagabatpur. The centre has six breeding pools of different sizes and 14 enclosures with tanks to keep hatchlings of different age groups. In the beginning, 32 eggs were collected from a creek (Dabarkhana Khal). Since 1982, eggs were being collected from the project site itself. The crocodiles usually mate in March and each female lays 25-90 eggs in their natural nests in May. The eggs laid by female crocodiles are collected within a fortnight by the forest staff, who keep them in a temperature and humidity-controlled incubator. The temperature maintains 29^0 to 32^0C thrice a day and maintains desired humidity. The hatchlings come out within 80-85 days. Up to 2021, 629 captive-bred crocodiles at Bhagabatpur were released into the wild, i.e. rivers and creeks of the TR and South 24-Parganas Division. Captive breeding of saltwater crocodiles at Bhagabatpur since 1976: Up to 2021, 629 captive-bred crocodiles at Bhagabatpur were released into the wild, i.e. rivers and creeks of the TR and South 24-Parganas Division.

Karamjal, Bangladesh

In 1997, after a saltwater crocodile was caught in a fisherman's net at Dublar Char, the initiative of crocodile breeding was taken up. In 2002, a crocodile breeding center was built up on eight acres of land at the Karamjal Tourist Center in the Chandpai range of East Sundarbans. The crocodile gave birth after 3 years in 2005. On 15 November 2006, Super Cyclone Sidre washed away 75 crocodiles from the centre. Already 72 crocs have been released in the Sundarbans. Besides, five more were sent to Dulhazra Safari Park in Chittagong and 3 to Bhola. One male named Ramiet, aged 25 and two females named Juliet and Pilpil and 206 baby crocs were reared at the Karamjal Crocodile Breeding Center. One night, a leopard cat broke through the fencing of the breeding center and allegedly killed and ate up 62 baby crocs. Later, the forest guards killed the leopard.

During the breeding season, Juliet and Pilpil laid eggs in their own nests dug in the ground or built in the bushes on the bank of pond. The eggs are picked up and hatched through a specified process with the help of a machine. The eggs, which were hatched at a temperature of 34.5°C, produced all males and those eggs hatched at a temperature of 27°C produced all females. They are reared in cages for fixed period of time and allowed to eat fish and meat. When the young are two meters in length, they are released into the creek. This release programme has been interrupted during the lock down period.

Captive breeding of Olive Ridley turtle at Sajnekhali

In the 1980s and 1990s the West Bengal forest department was collecting eggs of olive Ridley turtles along BoB for a breeding programme. The eggs were collected from Mechua beach, Bagmara beach (Bagmara 4, 5, 6, and 8) adjoining Bangladesh Sundarban area of National Park (East) Range and Chaimari beach of Mayadwip-2, Narayantala sea beach of Goasaba-4, Mayadwip area of Mayadwip-3 of National Park (West) Range. As the egg laying season starts from November, so special patrolling team is created during this period, which keeps vigil in these areas. During early 21st century, 1,928 eggs had been collected from Mechua, out of which 190 hatchlings were again released in nature. The point where the up and down tracks of the female meet is the point of laid eggs, where sand surface is also not even and these areas are then very safely dug and eggs are normally found below 5-10 inches. The average dimension of the nests was found to be 7" X 10" X 8", however the biggest size was 12" X 13" X 10". The smaller nests are of the size 6"X 8' X 6.5". These cluster of eggs are very properly stacked so that the sand does not penetrate between the eggs and the inter-spaces between the eggs are very clean. The number of layers in different nests has been found to be 3-6 and the distance between the sea and the nest varies from 50 ft. - 400 ft. However, normally they are found at a distance between 100 ft. and 120 ft. from the sea. Number of eggs found in each nest varies from 60 to 180, however, normally about 100-130 numbers of eggs were found in a pit. It is normally found that an Olive Ridley lays maximum eggs during 7lh and 8th day of both the Lunar phases and eggs are collected, the day after in the early morning before 8 o'clock to save them from predators.

Due to high predation of nests in different nesting beaches, the State Forest Department has developed two Hatchery Centres for marine turtle at (i) Sajnakhali in STR and (ii) Bhagabatpur of 24-Parganas (South) Division. The first measure is collection of eggs from the nesting beaches like Mechua and Kalash and those eggs have been brought to hatchery centres. Generally, it takes 60-70 days for the hatching of the eggs. When the hatchings are two months old, these are released in the nearby creeks, where the nestings are recorded and found. Whenever a nest is found, departmental staffs dig it and collect eggs by putting identification marks on each egg, layerwise. After that, the eggs are placed carefully on plastic trays filled with local dry sand. After cruising for about 6-7 hours, the eggs are brought to Bhagabatpur and these are placed in sand pit maintaining the layers of original nest, and filled up tightly.

Hatchings come out after 60 to 70 days. They are fed with mashed prawn. The weight of the hatchings varied between 15 and 18 grams. The average length is 6.5 cm. The average length of the front and back flippers are 4 cm and 2 cm respectively. Some white dots are seen on the body of the hatchlings and then the dots are spread all over the body and some infection is also seen in their mouth. Due to this infection, good percentage of the hatchlings dies every year. In such cases, Terramycin anti-germ is used for treating the disease, which is generally mixed with the food. However, the Bhagabatpur hatchery was not functioning and no eggs had been collected from the islands. But the hatchery of Sajnekhali was operating and during the early first decade of 2000, 1,928 eggs had been collected from Mechua, out of which 190 hatchlings were again released in the nature.

Captive breeding of *Batagur baska* at Sajnekhali

First breeding occurred in 2012 and continued successfully. As the numbers are increasing there gradually, six more assurance colonies (Chamta, Jhingakhali, Netidhopani, Dobanki, Jhilla and Harikhali) have been set up by the year 2020.

The eggs were collected (May-June) during 2012-2017 were:

2008= 7♂+5♀= 12; 2012 (June)= 32♀+1 (unidentified)= 33; 2013= 20♂+36♀= 56; 2014= 4♂+50♀+1 (unidentified)= 55; 2016= 96 (sexing not done); 2017= 74; 2019= 50 (Grand total= 362). 10 acclimatised sub-adults (3-4 years) fitted with acoustic transmitters for post-release monitoring, were released in the wild and 25 and 34 were soft-released in Chamta and Netidhopani. One of them has been spotted 400 km away- in Karamjal, Bangladesh. On 19th January 2022, the Turtle Survival Alliance and STR, released 10 subadult northern river terrapins *Batagur baska* into a tidal river in STR. This group of terrapins, c. 9 years old and comprising seven females and three males, selected from <370 individuals hatched through the captive breeding programme, is the first monitored return of the species to the wild in India. The terrapins are offspring of 12 founder animals discovered in 2008 in a pond at the Sajnekhali Range Station.

Bangladesh Sundarban

Forest Department has taken some small-scale captive breeding programmes for Estuarine Crocodile, *Batagur baska* and Spotted Deer at Karamjal, under the Chandpai range of East Sundarbans, while special measures have been taken to conserve the habitats of the Estuarine Crocodile in Mrigamari. On March 11, 2019 this terrapin laid 32 eggs at the breeding center, which started to hatch on May 16, through natural incubation (keeping them in sand) process. The Batagur laid eggs for the third time in this breeding centre. Earlier, in 2017, two turtles laid 63 eggs, of which 57 offspring were born. In 2018, two turtles laid 46 eggs, of which 24 hatched. In the year 2014, the eight main Batagur and 94 offspring were brought to Karamjal wildlife breeding center. They have been pushed towards extinction owing to the hunting threats, habitat degradation and changing climate conditions. The wild terrapin population precipitously declined due to unsustainable collection of adults and eggs for food.

It was reported that the olive ridley turtle used to nest on the sandy beaches of Dimer char (opposite Kochikhali), Kachikhali, Dimer char (the northern side of Alor khol) and the Mandarbaria areas in February and March. These turtles visiting the Sundarbans during the winter season are often caught in the fishing nets and brought to shore either dead or exhausted. Nests are often destroyed by the predators like *Sus scrofa* or *Varanus* spp.

CHAPTER 10
CONSERVATION CHALLENGES

The conservation challenges are summarised below-

Climate change- vulnerability to floods, earthquakes, cyclones, increasing temperature and salinity, sea-level rise and coastline erosion; biotic pressure and unsustainable exploitation of forest resources causing degradation of the natural habitat and resulting in loss of biodiversity; extension of non-forestry land use into mangrove forest; human-wildlife (tiger, crocodile, sharks) conflict; poaching of tiger, spotted deer, wild boar, marine turtles, horse shoe crab, etc.; uncontrolled collection of prawn seedlings; uncontrolled fishing in the waters of Reserve Forests; rapidly expanding shrimp farming; brick making; mangrove plant diseases; encroachment of forest areas, forest fire, sedimentation, chemical pollution; oil Spill; adverse impact of tourism; organisational and infrastructure deficiencies.

Reclamation

The entire process of reclamation of Sundarbans depended on erection of embankments (presently about 3,500 km in length) and closing of the mouths of the creeks so as to destroy the entire mangrove forests within the island area. This has altered the tidal inundation regimes, sediment accretion and geomorphic character of the deltaic inlets. Embankments have impacted on the biodiversity and physiological adaptations of mangroves within the sphere of tidal ingression, habitat fragmentation and seedling establishment. Earlier, Hunter (1875) also predicted severe consequences of such fast and unwise reclamation, i.e. absence of the guarding green wall to protect this vast reclaimed area from the high devastating waves due to super cyclones, as is evident from the impact of recent devastations.

Land cover change

In the Bangladesh Sundarbans, Mondal (2017) showed changes in the land cover of SRF from 1973 to 2010. Areas occupied by the water bodies

in 1973, 1978, 1989, 2001 and 2010 were 5,488.04 km², 5,456.85 km², 5,341.20 km², 5,218.19 km² and 5,084.92 km², respectively. The percentage occupied by water bodies to total area decreased from 58.79% in 1973 to 54.52% in 2010. Water bodies decreased by about 7.35% over the last 37 years. On the other hand, area covered by the barren land in 1973 was 8.78%, which was decreased in 2010 and made up 4.43% to total area. Significantly, the grassland made up around 5.54% of the total area in 1973 but increased to 18.51% by 2010, whereas the dense vegetation (both mangroves and non-mangroves) decreased notably during this period from 2,501.05 km² to 2102.94 km². During this period, the forest vegetation was more agglomerated in the western part than the eastern part. This change has affected the existence of the flora and fauna in the two regions.

Degraded ecosystems

Construction of the Farakka Barrage and other dams has resulted in the reduction of fresh water flows to the Sundarbans resulting in increased salinity and alkalinity, damaged vegetation, agricultural cropping pattern and changed landscapes in the Sundarban region. The impact of soil starts with the destruction of surface organic matter and of soil fertility required for the mangrove plants. The changes alter basic soil characteristics related to aerations, temperature, moisture and the organisms that live in the soil.

While erosion of the estuary margins and the sea facing coastline- up to 40 m/yr- was continuing for decades in the southern islands, intervening channels between the northern islands were getting silted up, especially in the western sector, resulting in land gain. Area change in the central sector mostly tended to be small and erosional. Erosion prevails in the seaboard islands and increases away from the sediment sources whereas accretion predominates the landward islands due to waning of the intervening creeks. Sediment starvation and landward rise in tidal range and asymmetry cause the changes.

During the period 1980 to 2019, the river erosion rates were significantly higher for Passur (2,139 ha, 55 ha/yr), Baleshwar (2,122 ha, 54 ha/yr) and Shibsa (1,809, 46.5 ha/yr) rivers compared to the other three rivers Arpangasia (995 ha, 26 ha/yr), Malancha (774 ha, 20 ha/yr) and Shella (332 ha, 8.5 ha/yr)(Sohel *et al.*, 2021). The spatial pattern of lateral erosion was not uniform for all major rivers. The southern parts of all the rivers were prone to severe erosion compared to the central and northern parts. Degradation of the water quality of about 60% of the rivers in the region, especially in the dry season when the high salinity intrusion occurs, also affects the quality of the three other core elements (soil, vegetation and wildlife) of the ecosystems.

Floristic changes

The degree of salinity has a strong influence on the regional distribution of species and stand-height of tress. Salinity in the Sundarbans increases from the eastern Sundarbans in Bangladesh to the western Sundarbans in India and, correspondingly, the density of vegetation growth and canopy closure decreases from the eastern part to the western part. Height and growth of different mangrove species in the Sundarbans are correlated to the changing salinity level. For example, Sundri (*Heritiera fomes*) growing in the low saline zone (5-10 ppt), was once a dominant tree species of the Sundarbans, but failed to survive in SBR due to high salinity (17 to 32 ppt), whereas Gewa survives well in the moderate saline zone (10-25 ppt) and Goran in high saline zone (over 25 ppt). In fact, germination of seeds of Sundari is highest at 0 to 5 ppt. salinity. For the growth, survival and pigment function of the species, salinity less than 15 ppt. is required, which is available only during the peak monsoon months. It is also predicted that top-dying disease of Sundari trees is likely to be the consequence of slow increase of salinity over a long period of time. Almost 265 km^2 *Heritiera fomes* type of forest are moderately and 210 km^2 areas are severely affected in SRF, which is one of the main threats for a sustainable mangrove forest management and its ecosystems.

As a consequence, the salt tolerant *Avicennia* sp. and *Excoecaria* sp. now dominate the species-assemblage by replacing the less tolerant *Heritiera* sp, *Nypa* sp., etc. Moreover, due to increasing salinity some non-woody shrubs and bushes replace the tree species, reducing the forest productivity and habitat quality for valuable wildlife.

Thus, due to hydrological modifications, with altered flooding and soil salinity conditions, above floristic changes are taking place, which, in turn, influence the entire food chain within the Sundarban mangrove forests. Furthermore, the increased runoff reduces soil salinity, and, owing to higher erosion, high siltation causes lower transparency and thus lowers photosynthetic activity of the phytoplankton with adverse consequences on food chains and oxygen production.

On the other hand, unavailability of required depths due to inadequate freshwater flow and siltation restricts as well as swelling salinity, the some aquatic species like *Tenualosa ilisha* and *Platanista gangetica* within the riverine systems of SBR are disappearing.

Due to sea level rise, nearly 75 km² of mangrove forest in SBW is projected to be flooded.

Increase of cyclones is destroying the mangrove forests to a large extent, changing the mangrove structure as well as the coastline.

Biological invasion (Islam *et al.*, 2019)

One of the cyclonic impacts is colonization of the invasive species. 23 invasive species, belonging to 18 families and 23 genera have been identified in the Sundarbans. These species have been categorised as-

Potentially invasive (14 species)

Arundo donax L- Gramineae; *Clerodendrum inerme* L- Gaertn Verbenaceae; *Cryptocoryne ciliata* (Roxb) Fischer ex Waylder- Araeceae; *Dendropthoe falcata* L.f.- Etting Loranthaceae; *Flagillaria indica* L- Flagillariaceae; *Hibiscus tilliaceus* L- Malvaceae; *Hoya parasitica* (Roxb.) wall. ex Wight- Asclepiadaceae; *Imperata cylindrica* L- Raeuschel Gramineae; *Ipoemea fistulosa* Mart. Ex Choisy- Convolvulaceae; *Pongamia pinnata* L- Pierre Leguminosae; *Saccahrum spontaneum* L- Poaceae; *Salacia prinoides* DC- Celastraceae; *Sarcolobus globosus* Wall- Asclepiadaceae; *Typha angustata* Borry f- Typhaceae;

Moderately invastive (Six species)

Acrosticum aureum L- Polypodiaceae; *Entada rheedii* Spreng- Leguminosae; *Excoecaria indica* (Wild.) Muell.-Arg.- Euphorbiaceae; *Micania scandens* Willd- Compositae; *Syzygium fruticosum* (Roxb.) DC- Myrtaceae; *Tamarix indica* L.- Tamaricaceae; and

Highly invasive (three species)

Derris trifoliata Lour- Leguminosae; *Eichhornia crassipes*- Pontedteriaceae; *Eupatorium odoratum* L.- Compositae.

Climbers (6 out of 23) were the most frequently encountered invasive species followed by trees (5 out of 23) and shrubs (4 out of 23). The impact of climate change and anthropogenic influences are causing development of secondary plant species and reducing the native plant species. These species affect the mangrove ecosystem through different ways. These negative impacts are ecological, economic and environmental. The ecological effects include replacement of native plant species and reduction in ground cover, which leads to loss of biodiversity, forage, habitat and scenic quality, and even soil productivity. The economic impact includes loss/reduction of revenue.

The native plant species *Heritera fomes* was dominating 34% of the Sundarbans forest area and at present it is representing only 21%. In addition, the agricultural land and mangrove forest areas are being used for shrimp cultivation.

A study was conducted in the Chandpai Range of Bangladesh Sundarbans (Rahman and Vacik, 2016), which was classified into `less affected' (LAA) and `high affected' (HAA) areas due to the tropical cyclone `Aila'. In total, 23 invasive plant species were identified in the Chandpai Range of Sundarbans, out of them 19 species were indigenous and the rest were alien. All species were found in the HAA, where only 09 species were found in the LAA. The abundance, diversity and rate of invasion were higher in the HAA than that of LAA. Proper management must be adopted to control the invasion to protect the endemic species.

Human-animal conflict

Human settlements are located at the northern and north-eastern periphery of the SRF and northern and north-western fringe areas in SBR. The encounter rate of human sign and sighting is higher near the edges of the forest and overall higher in Bangladesh Sundarban, which is further exacerbated by the usage of river channels for transportation of commercial vehicles.

Records of man-tiger conflict are available since 1670. A very limited number of the tiger is man-eater in the Sundarbans and they are officially hunted by proclamation. Besides the permit holders, a large number of non-permit holders also enter into the forests and die due to tiger attack and these go unrecorded because they could not claim compensation from the Government because of their illegal entries.

The number of man-tiger conflict is mostly reported from the western Sundarbans because the populations of the spotted deer and the wild boar are not so abundant here compared to the eastern Sundarbans. Retaliatory killings of the tiger were also on record. Curtis (1933) recorded 427 human killing during 1912-1921 as against 452 tigers. Chaudhury and Choudhury (1994) recorded a toll of about 1,000 human beings during last 50 years. 401 human casualty and 41 tiger poaching cases were reported during 1984-2000 in Bangladesh (Dey, 2001). Most of the (humans entering with permit or illegally) killings takes place in the forest involving fishermen (44%), woodcutters (36%) and honey collectors (18%). Current levels of human and tiger deaths in the Sundarbans are relatively low compared to mean levels recorded in the last 140 years. Killing tigers in the Sundarbans may have reduced tiger attacks on human in the past; Curtis (1933) attributed the sharp decline in human deaths in the early 1900s to the high levels of tiger hunting (43 tiger deaths/year) during that time. Hunting problem animals, however, is not currently politically acceptable or in line with conservation objectives to preserve the tiger population. Furthermore, the 2-3 tigers killed each year in the Sundarbans in addition to an unknown number poached, could threaten the long-term viability of the tiger population. Reducing both tiger and human deaths is, therefore, needed to improve conservation prospects for tigers in the Sundarbans.

In the Indian Sundarbans, the then (1970s) Divisional Forest Officer (DFO), 24-Parganas Division, AB Chaudhuri with whom I worked later on in the Wildlife Wing (Headquarters), opined that although water salinity increases as one proceeds from the northern zone to the central zone and then to the southern zone, similar pattern was also noticed so far soil salinity index is concerned, but the frequency of human casualties does not vary according to such zoning. Therefore, it is evident that the man-eaters of the Sundarbans are concentrated in some habitat formations only, which also incidentally records significantly more salinity. The man-eating nature of the tigers of Bangladeshi part of Sundarban was authoritatively studied by Hendricks (1975), who, reported 392 casualties between 1956 and 1970, out of which 365 were known by place of incidence with 198 (54.2%) occurring in Satkhira

Range. He has given possible reasons behind such occurrences, among other factors, the saline estuarine water. Subsequent data (2002, 2003, 2007) prove that the portion of casualties in the Satkhira Range has always been high and has increased slowly since the investigations by Hendrichs (1975). During the period 1974-1983, the differences in casualties between the low and high salinity zones became insignificant and the casualties in the medium salinity zone were significantly higher than those of the other two zones. This trend does not justify the hypothesis that the salinity of the water causes tigers to develop man-eating behaviour. In fact, the frequency of man-killing is highest in areas and at times of heaviest concentration of people representing various occupations, suggesting that the man-killing and the frequency of man-tiger contacts are directly correlated. But there are some uncertainties in calculating the number of people entering forests for the purpose of extraction including the illegal entrants (Neumann-Denzau and Denzau, 2010) to establish this interrelationship statistically because-

(a) it remains unknown how many members of the primary dependent households enter the forest for how many days per year;

(b) the group of secondary resource users, whose main income is generated outside the forest but who enter the forest occasionally, was not analysed in the SBCP Baseline Study (it might be the bulk);

(c) the number of illegal entries is not available; and

(d) the number of persons who enter the forest from outside the Impact Zone is not recorded.

It is assumed that this region is under intense human pressure with around 3.5 million people living within 20 km of its northern and eastern borders and depending upon the forests for livelihood resources. Annually around 35,330 people enter the forests of Sundarbans to collect timber, fish, honey and other products. This is an estimation based on the permits issued by the forest department. The actual figure may multiply but it is beyond calculation.

The poachers of tiger or deer include the amateurs, semi-professionals and professionals- all illegal and timber mafias. A scarcity of herbivorous prey, caused either by poaching or environmental factors, might increase the tiger's interest in human prey (*ibid*). An unknown number of Bangladeshi nationals are crossing the border for illegal forest resource extraction in the Indian Sundarbans. This has been proven for the honey collectors. In a case witnessed in April 2009, a Bangladeshi honey collector was killed by a tiger on the Indian side. Unequal management realities may give rise to increased legal or illegal resource exploitation and thus create higher disturbance levels in certain forest areas (*ibid*). Reducing the number of illegal entrants into the forests by SMART patrolling, as has been conducted during the last few years in both India and Bangladesh, is a viable solution to release high extraction activities in the Sundarbans.

The man-eating tigers in the Sundarban are divided, according to their behaviour pattern, into three broad categories-

- Category I: This includes two sub-categories, of which one is inherent and designed maneater (25%). This sub-category is again subdivided into two micro-groups: one is aggressive maneater (80%) and the other lusty and adventurous (20%).

- Category II: Undesignated man-eater (15%), and

- Category III: Circumstantial man-eater(60%).

There is absence of concrete evidence to prove the theory that the tigers in Sundarbans are man-eaters by nature and it may be a behavioural characteristic manifested in some tigers (M. Monirul Khan). Mallick (2007) categorised the problem of human-tiger conflict in STR into two types:

(i) Conflict outside the forest area: This is caused when the tiger stray out of the forests and enter the villages on the opposite side by crossing the channels. In the past, these fringe villages were established on the reclaimed forestland, where the interface is mostly demarcated by a small creek, e.g. Kalitala and Samsernagar on the boundary of Arbesi-1.

(ii) Conflict within the forest areas: This is caused due to intrusion (with or without legal permits) of people in the tiger habitat. Mostly the honey collectors and fishermen fall prey to the tigers inside the forest. The honey collectors penetrate deep inside the forest in search of beehives and, in the process, mostly get isolated from the other group members because of the thick and almost impenetrable forest. As a result, they become easy prey for the tigers. Sometimes these groups accidentally and unknowingly get close to the tiger habitat and the disturbed tiger attack them. The honey collectors also burn the beehives by using the Golpata (*Nypa fruticans*) and Hental (*Phoenix paludosa*) leaves, which creates heavy smoke and causes irritation and disturbance for the tiger and they are attacked by the big cat.

Most human kills took place in the forests of north-eastern part. Most of the victims (59%) were residents of Gosaba block followed by Hingalganj (14.96%), Basanti (9.99%), Hasnabad (3.8%), Canning II (2.54%), Pathar Patima (2.54%), Kultali (2.03%) and others (5.14%). In case of the fishermen and crab collectors, the incidents mainly occur in the small creeks deep inside the forest and very close to the tiger habitat, where the tiger may hide behind the Hental bushes on the bank and they are attacked by the tiger. Crab collection is higher in the eastern part of the study area, while fishing is more in the western part and honey collection is concentrated in the north-eastern side. The low mud flats on both sides of the creeks are potentially best areas to find the crabs and a time consuming activity and these areas are also the most probable sites for the tiger.

Incidences of the tiger straying in the villages of STR are not recorded or heard of before 1974. The first case of tranquilisation of a problem tiger was reported in 1974 (Seidensticker *et al.*, 1976). In August 1974, a young male tiger moved near the village of Jharkhali, a mosaic of paddy fields and a large central mangrove marsh used as a fishery and killed one woman (not eaten) and a number of livestock in the early morning. The tiger was sighted repeatedly over the next few days and people were very alarmed. The area had been part of reserve forest lands until about 1955, when it was reassigned for the resettlement of refugees. The remaining reserve forests on the southern end of the island were at a distance of about 3 km from the village where the tiger was observed. A narrow belt of mangroves along the Matla River on the west side and the central fishery area provided the tiger with shelter and cover but no large mammalian prey. Further south, in the reserve forests and in the tiger reserve, axis deer (*Axis axis*) and wild pigs (*Sus scrofa*) occur in good numbers.

Rather than eliminating the animal, the Forest Directorate decided to capture it, using immobilising drugs, and release it in STR. This was successfully done, but less than a week later, it was found dead from wounds evidently inflicted by another tiger. There were numerous tiger tracks around both the translocation cage and the carcass, and the freshest tracks, which had been made in the few hours before the last high tide and many of the older ones were larger than those of the dead tiger. The carcass was already badly decomposed. The right side and abdomen had numerous maggot-infested puncture wounds, the thoracic cavity had puncture wounds and there were multiple puncture wounds in the hips and right shoulder. Only the extremities had been fed on by scavengers.

Only three straying cases in 1970s (2nd August 1974, 14th August 1974 and 13th July 1977. A successful tranquilisation of a tiger is recorded in July 1979 near Dayapur village and the animal was translocated to the Alipure Zoo, Calcutta. During the period 1986-March 2010, 288 cases of tiger straying in STR were recorded.

The Indian Sundarbans has recorded the highest number of tiger straying incidents in 2021 after witnessing a decline over the past few years- three cases in 2014-15; no cases of straying for four years between 2015-16 and 2018-19; two such cases in 2019-20, only one case in 2020-21 and the number of straying cases shot up to seven in 2021-22. The villagers, however, said that this number may be much more, as on several occasions the tigers have strayed into the fringe villages by crossing the rivers and later returned to the forest of their own. The forest department records only those cases where tigers had to be trapped after they strayed and released back into the wild. Experts said that the rising number of tigers to 96 could be one of the reasons for the increase in straying incidents. The male tigers always move out of their mother's territory to establish their own territory. It is a part of their evolutionary process and inherent behavioural need. They would also move out if the area has reached its carrying capacity. Also, the Sunderbans has been hit by two major cyclones in two consecutive years- Amphan in May, 2020 and Yaas in May, 2021, which took a heavy toll on the Sunderbans. Nearly one-third of the mangrove forest suffered extensive damage. The nylon net fencing set up by the forest department along the forest edge to prevent tiger straying also suffered extensive damage. This could also be one of the reasons, as gaps might have been created in the fencing through which the tigers may have come out and strayed into human territory.

There could be multiple reasons behind the rise in tiger straying. But with the sea-level rising and erosion of land going on simultaneously, the Sunderbans is under real threat. However, the joint forest management initiatives taken by the authorities along with the locals, tigers have not been killed by the villagers, when they stray into the villages. But if the tigers repeatedly enter the villages and kill the domestic animals regularly, it may reverse the situation to retaliatory killing of the intruder. The moment a tiger enters a village, chances are there that it could be harmed.

Another dimension of the tiger population in the Sundarbans is skewed sex ratio. For example, in the Bangladesh counterpart the Tiger census in 2018 was conducted on 1,656 km² of forest and used camera traps to count the big cats. A poaching crackdown by authorities in the Bangladeshi part of the Sundarbans mangroves saw an increase in the big cat population from 106 to 114 four years ago, according to a census published in May. But closer analysis of the data found the number of male tigers was lower than the typical ratio of one male for every three tigresses, with the figure now at one male for every five females. The standard ratio should be one male for every three female tigers. In the Sharonkhola range, only two males were detected against 19 female tigers, adding the future of the tiger population in the Sundarbans could be at stake if the forest continues to lose male tigers. Authorities are planning to capture male tigers from other parts of the forest and introduce them in the Sharonkhola range, which is the densest tiger habitat in the mangroves. Male tigers are aggressive and they cover bigger territory, which make them vulnerable to poaching.

During the late 19th century, the rivers in the Sundarbans were infested with both *C. porosus* and *C. palustris*, which were equally destructive to man and beast venturing into the stream. On one occasion, when the prisoners of the Khulna jail were drawn up on the banks of the river to be mustered before being sent to the Presidency Jail, a crocodile suddenly swept one of them with its tail into the stream and carried him away. Another species of the fresh-water rivers of the northern Sundarbans, known as *Gavialis gangeticus*, which were not known to prey upon the man or beast but feed almost exclusively on fish, though they would seize and devour any dead of wounded bird that might be dropped into the stream on being shot.

The crocodiles are often found in the rivers of forest-fringe areas of SBR, such as Kumirmari, Satjelia, Jharkhali, Harbhangi, G-plot, Banashyamnagar, Ghanashyampur, K-plot, etc. Mainly fishermen and crab collectors often enter the crocodile's habitat, while the riverside residents also go inside the forests for fishing, when many of them loss their lives due to crocodile attacks. This also happens when the crocodiles rest on the riverbanks and creeks, especially in the evenings during the rainy season when the fishermen, crab hunters and other gatherers return home. A 10-year survey (2007-2017) has recorded a total of 460 crocodile attacks in 10 villages (Satjelia- 48, Sudhanshupur- 46, Laxbagan- 41, Lahiripur- 51, Dayapur- 38, Sadhupur- 35, Hamilton abad- 21, Pakhiraloi- 53, Rangabelia- 62 and Dulki- 65). The crocodiles are also caught in the nets of poachers and fishermen and since the crocodiles eat fish, the fishermen beat them to death.

In June, 2021, Sekendar Ali, a fisherman from Phatikpur village in Kakdwip, along with other neighboring fishermen, went fishing in the Tinpuri river creek near Barpara in the evening. When he went ahead leaving his companions behind, a full-grown crocodile suddenly came out of the bush and attacked him. The fisherman resisted by grabbing the crocodile's rostrum. This fight ended when the fisherman diligently wrapped his fishing net round the crocodile's body. The crocodile tried to jump, but its mouth, teeth, legs, tail were all caught in the net. Hearing loud noises all the fishermen came there and captured this 1.5 long animal. Later, it was handed over to the forest department's staff, who took it to the Bhagabatput Crocodile Farm.

Human-shark conflict often takes place in low water in the fringe areas. The shark is said to be the most ferocious animal in the sea that roams around in search of prey all day long. A shark over two meters in length may be dangerous to the human beings because of large size, well-formed jaws and strong teeth. Although all sharks are not deliberate man-eaters, a few species do attack humans when they wander around in search of food. Most of the sharks prefer to stay away from the humans, except the great white and bull shark. The sharks sometimes enter the interior Sundarbans during the high tide. Smaller sharks even move into the knee-deep water, especially during the rainy season, when the tide rises high. When the tide recedes, they remain confined there and attack people with great speed leading to injury and amputation of limbs.

But the old records of conflict between the people and the sharks in Sundarbans are not available. Therefore, it is assumed that this type of conflict has been associated with a recent influx of shrimp (*Penaeus monodon*)-collectors of all ages and sex. But death due to shark depredation is rare. Only one boy is known to have died due to shark's attack near Mollakhali. As these attacks occur mostly in fringe lagoons, the incidents are mostly not recorded in the forest department. However, out of 3,082 families living in Lahiripur and Satjelia villages of Gosaba block along the rivers like Gomar, Melmel, Sajna and Datta adjacent to the Indian Sundarbans, a total of 8 (two males and six females) shark-bite incidents is recorded from August 2004 to March 2005. But this figure accounts for 7.2 percent of the region's overall human-animal conflict (tiger 82 percent and crocodile 10.8 percent).

Snakes are bitten to death in retaliation leading to decline of their population in the Sundarbans. Thousands of people are bitten and die from snakebite every year. The two most commonly observed poisonous species are the common krait, *Bungarus caeruleus* (51 %) and common cobra, *Naja naja* (40 %), and that of non-poisonous varieties are the *Ptyas mucosus* (41 %), *Typhlina bramina* (34 %), *Xenochrophis piscator* (12 %), and *Amphiesma stolata* (10 %), whereas apart from killing of snakes out of fear; habitat loss, unscientific handling of snakes by snake catchers and charmers, and netting by fisherman contributes to snake mortality to a large extent; 72 % killed snakes are of poisonous varieties, 60 % of which are *B. caeruleus*, the most venomous snake in Sundarban (Das, 2013).

Poaching methods

Poaching of the tiger and its prey species is the vital threat for their survival in the Sundarbans. The only tiger poaching method detected in the Bangladesh Sundarbans was poisoned baits (Aziz *et al.*, 2017). First, the poachers use twigs and leaves of *Sonneratia apetala* and epiphytes as bait on either side of a snare set up in order to lure the ungulates. The prey visiting the site is hunted with snares and by shooting (metal pellets found during the post-mortem). The baits are attached to a tree trimmed to the approximate height of a tiger and placed next to tiger trails (indicated by tiger tracks). A single spotted deer is used to create two bait stations, with body parts being prepared by removing the intestines, dismembering, skinning and coating in poison, usually Carbofuran pesticide.

A Bengal tiger was found dead with a wire snare around its waist in Kultali area in April, 2019. It is being speculated that the tiger was not the intended and that it got entrapped in a deer-poaching trap (snare made of galvanized iron). The incident took place inside compartment 1 of Ajmalmari forests under South 24-Parganas Forest Division. Earlier in 2013, deer poachers were held in the Lothian WLS with 10 traps made of nylon wires. In the same year, a group of villagers were caught red-handed while they were feasting on deer meat. While a kilo of deer meat is sold in the Sunderban's villages for more than Rs 400, a skin can fetch as much as Rs 50,000 in the international market.

In the Indian forests, instances of poisoning tiger are also available. A male tiger, age approximately 14 years, entered into the Samsernagar village, adjoining Jhingekhali Station of Basirhat range on 28th August, 1998. The tiger's carcass was later found floating in a village pond. The analysis of the visceral parts confirmed the presence of endosulfan which is a commonly used insecticide for farm crops.

During the dry, hot summer months small forest pools or creeks are also poisoned by the poachers. There is a sophisticated and well organised supply route operated by the major traders, to distribute poison and collect tiger bones from the remotest villages. Poison fishing also puts freshwater dolphins and other globally important piscivorous species such as turtles, otters, fishing cats, and crocodiles at risk due to prey depletion and potential toxic effects. Firearms are used where hunting can be carried out with little hindrance.

In the wild the wild boars run very fast and also change their direction sharply to hide behind the heavy- foliage and respiratory roots. Hence, it is very difficult for the poachers to shoot them down. They are poached mainly with the help of traps made of nylon wire. The culprits are the fringe villagers of Arbesi, Khatuajhuri and Jhilla in Basirhat Range and Pirkhali in SWLS Range. The villagers adjoining the Duttar, Jhingekhali, Bidya and Lahiripur Stations are also involved in this offence and the meat is locally sold in the villages.

At present, poaching of the Water Monitors has reduced considerably, however, stray incidences have been reported wherein the crabs are used as bait for trapping them. Earlier, in the mid-seventies, poaching for its skin was rampant and this has now reduced due to adequate protection measures.

In Indian Sundarbans, local people from Sagar Island reported to exterminate the fishing cat from the island. In STR from 2003 to 2009, 11 fishing cats faced natural death and within the same period 20 fishing cats were rescued. In February 2019, an adult fishing cat was rescued by the local people and was handed over to forest department.

Entanglement in fishing nets

During the period February, 2007 to December, 2013 52 mortality incidents (40 Ganges dolphins and 12 Irrawaddy river dolphins) were recorded in the Sundarbans. Among the 40 carcasses of Ganges river dolphins, 13 were killed due to entanglement, while in case of eight carcasses of the Irrawaddy dolphin, five had died due to entanglement, constituting about 62.5% of the cases. In some cases, the entangled dolphins are rescued from by-catch or accidental entanglement in and around the dolphin sanctuaries. Dolphins are also killed for illegal trade and consumption of meat. The locations of dolphin trade and consumption was reported from a relatively small geographical area along the Passur River between Chandpai, Khulna, which includes two of the dolphin sanctuaries. Due to inadequate monitoring and enforcement, fisheries regulations are sometimes inconsistently applied.

Human-crocodile conflict is common in SBR. The problem is on the increase since the late 20th century since the people are mostly operating deep into the forests illegally for higher returns. Though the crocodiles are reported to attack more or less throughout the year, but there are peak periods too. The crocs mostly attack during daytime, particularly winter (November to February) and pre-monsoon (May-July) because the number of people operating in the water is highest during these seasons than the monsoon with higher frequency of natural hazards. Fewer incidents occur during the post-monsoon period (August-October). The lowest number of attacks takes place in September. The prawn seed collectors are mostly attacked during hawling the nets in waste deep water of the rivers and creeks in the forest fringe areas, e.g. Gosaba (Pirkhali, Gomdi, Korankhali, Rangabeliya, and Bidya), Patharpratima (Nuchara, Thakuran), Namkhana (Saptamukhi, Muriganga), Basanti (Matla, Bidya) and Kultali (Kaikalmari, Ajmalmari) areas. There are other victims who are professionally dependent on the aquatic animals resources like the crab collectors and fishermen. The victims are mostly women than the males because women are involved more in those aquatic occupations. The crocodiles also attack the boats plying in the rivers and creeks. Width of the water-bodies, where the incidents occur, varies from 100 m to 800 m. Creeks or small rivers are more vulnerable than the large rivers, as this apex aquatic predator prefers muddy banks during low tide for thermoregulation. There is very little medical facilities in the region for saving the victims. Occasionally, crocodiles stray into village ponds and are rescued, particularly during September and October, mostly in Patharpratima block. World Wildlife Fund (WWF) in a survey in 2006 found 30 persons died of crocodile attacks in three months in Patharpratima.

Indiscriminate fishing

Another important threatening issue is the change in the pattern of tidal currents, human interventions and the negative effect on the breeding patterns and on the lifecycle of the fish. In practice, depth of the water column remains low (on the average >30 cm) due to increased sedimentation for most of the year (average 280-300 days). Due to low water depth and stagnancy of water, most species are on the verge of losing their natural habitat and are becoming vulnerable. Due to change in the tidal bed and abrupt decrease in the water level, (i) the water birds are consuming fish; (ii) the bigger fish are consuming more smaller fish; (iii) fishermen catching the fish that have come to breed; and (iv) the people catch local fish varieties, juvenile prawn, crabs and other molluscs together, all in their eagerness to obtain the juvenile prawn. Since change in tidal patterns exposes the fish, the predation pressure on the population of the smaller fish has increased multifold and due to such indiscriminate fishing, there is a noticeable squeeze in the edible biomass and fish diversity. This has a very destructive effect because it abruptly terminates the lifecycle of the fish.

Using monofilament nets

In addition, excessive use of monofilament nets to catch surface feeders has also led to unnecessary netting of small-sized fish. Intensive misuse of trawl nets for bottom trawling results in change in turbidity and that in turn, adversely affects the fish habitat.

By-catch of fish fingerlings and non-target crustacean fry in post-larvae (PL) collection nets

This practice is a major cause for declining fish and crustacean populations in the Sundarbans.

Large-scale collection of shrimp fry

Collection of seeds of wild tiger prawn (*Penaeus monodon* Burkenroad, 1959), a fast-growing and largest shrimp in the world, however, started only from the late eighties of 20th century with the introduction of highly profitable scientific brackish water prawn culture with direct or indirect support of multinational companies. Before that period fish and prawn culture in the 'bheries' was dependent on the natural sources, in which fingerlings of fishes and prawn seeds used to take entry into the *bheries* with the influx of river water during hightide of new or full moon days. During that period, tiger prawn seed stocking (concentration) was not more than five per m². With increasing export value of tiger prawn the seed stocking density increased at a very high speed beyond its natural carrying -capacity and the density rises to 8-10 times higher than the traditional farming. Natural resource of tiger prawn seed was not sufficient to fulfill the demand of this growing trade and to meet up the demand, collection of tiger prawn seeds from the rivers started from the late eighties. After invading the surrounding area, from 1993 tiger prawn seed collectors started to trespass into the protected areas slowly and from 1994 it became almost a common practice.

About 33,000 ha in North 24-Parganas and 12,000 ha in South 24-Parganas are being used for shrimp farming. More than one lakh rural people of the area, especially from the local fishing communities are now engaged in this lucrative occupation. This profession is becoming so attractive in the region that the whole family, from 6 year to 60 year old members, involved in search of tiger prawn from the early morning till sunset.

Lakhs of tiger prawn seed collectors invade almost all the surrounding riverine system of SBR except the portion nearer to sea. The creeks, sandy mudflats, etc are the main area from where the tiger prawn seeds are being collected. However, they have been driven out, but for small portion of Jhilla inside the TR, the main affected rivers were Jhilla and

Dattar. With the availability of tiger prawn seeds the number of seeds collectors naturally fluctuates. Diurnal rhythm varies with the tidal system, as this collection is made mostly during high tide and attains the highest peak during new moon or full-moon session. Seasonally, the number of collectors goes down to its bottom during the winter (November-January) and attains the peak during the rainy season. However, heavy showers minimise the number of collectors as seed mortality is higher during that period.

In practice, the seed collection process involves forward or backward dragging or spreading of a net, mainly hand-operated small nets, which they drag through the thick sticky mud and salty water (approximately 6.5 km/day/collector) spending about 4 hours in a day. Aluminium pan is used for storing the seeds.

But this practice affects very adversely on the ecology, as well as, on the natural resources of the area. Together with the tiger prawn shrimps, other shell and finfish and planktons are caught in the net. The seed collector keeps the tiger prawns and the rest of the catch is thrown on the embankment. Thus many other varieties of aquatic fauna are destroyed and the biodiversity is lost.

From those samples collected in the field, fingerlings of 79 fish species, 10 species of prawn seeds other than seeds, 4 species of crabs, 4 other crustaceans and 2 annclidan species have been recorded. It has been calculated that with every single tiger prawn seed, 46.7 other prawn seeds, 4.1 fingerlings of fishes and 0.3 other aquatic fauna (crustaceans and annelids) get trapped and most of them do not survive later on.

This large-scale tiger prawn seed collection plays against the normal health of this ecosystem. This practice is not only diminishing the tiger prawn population but also a large number of fingerlings and seeds of other prawn and fish species get trapped and are being vanished from the nature. Naturally, it affects negatively those animals (big fish species and even crocodiles) who take these small creatures as food. Step by step the food chain of this ecosystem is breaking up and its deleterious effect may be more harmful than it is estimated today. Almost 90% of engaged women reported of skin disease and attacks by crocodile and small river shark. The dragging of nets by the fry catchers along the intertidal mudflats often uproots the mangrove seedlings and salt marsh grass

(*Porteresia coarctata*) . Such process also destroys the saplings planted on the riverbanks by trampling.

It causes havoc to the mullet population and subsequent reducing the fish population and diversity, even leading to extinction of some vulnerable species. The by-catch aquatic fauna observed during collection of the tiger prawn seeds are-

Fish: *Setipinna phasa* (Hamilton 1822), *Setipinna taty* (Valenciennes 1848), *Coilia reynaldi* (Valenciennes 1848), *Lates calcarifer* (Bloch 1790), *Daysciaena albida* (Cuvier 1830), *Cynoglossus arel* (Bloch & Schneider, 1801), *Cynoglossus lingua* (Hamilton, 1822), *Scatophagus argus* (Linnaeus, 1766), *Strongylura strongylura* (van Hasselt, 1823), *Sillaginopsis panijus* (Hamilton, 1822), *Chirocentrus dorab* (Whitehead *et al*, 1966), *Liza parsia* (Hamilton, 1822), *Liza tade* (Forsskål, 1775), *Anguilla bengalensis* (Gray, 1831), *Terapon jarbua* (Forsskål, 1775), *Pomadasys argenteus* (Forsskål, 1775), *Epinephelus coioides* (Hamilton, 1822), *Epinephelus tauvina* (Forsskål, 1775), *Lutjanus johnii* (Bloch, 1792), *Harpodon nehereus* (Hamilton, 1822), *Mugil cephalus* (Linnaeus, 1758), *Osteomugil cunnesius* (Valenciennes, 1836), *Liza parsia* (Hamilton, 1822), *Rhinomugil corsula* (Hamilton, 1822), *Eleutheronema tetradactylum* (Shaw, 1804), *Polynemus indicus* (Shaw, 1804), *Polynemus sextarius* (Bloch & Schneider, 1801), *Trichiurus* sp., *Gudusia chapra* (Hamilton, 1822), *Tenualosa ilisha* (Hamilton, 1822), *Anodontostoma chacunda* (Hamilton, 1822), *Escualosa thoracata* (Valenciennes, 1847), *Stolephorus baganensis* (Hardenberg, 1933), *Stolephorus commersonnii* (Lacepède, 1803), *Coilia dussumieri* (Valenciennes, 1848), *Epinephelus diacanthus* (Valenciennes, 1828), *Parastromateus niger* (Bloch, 1795), *Nandus nandus* (Hamilton, 1822), *Polynemus paradiseus* (Linnaeus, 1758), *Photopectoralis bindus* (Valenciennes, 1835), *Otolithoides pama* (Hamilton, 1822), *Harpadon nehereus* (Hamilton, 1822), *Periophthalmus* sp., *Glossogobius giuris* (Hamilton-Buchanan, 1822), *Pseudapocryptes elongatus* (Cuvier, 1816), *Odontamblyopus rubicundus* (Hamilton, 1822), *Platycephalus indicus* (Linnaeus, 1758), *Eleotris senegalensis* (Steindachner, 1870), *Lepturacanthus savala* (Cuvier, 1829), *Lepidocephalichthys guntea* (Hamilton, 1822), *Mystus gulio* (Hamilton, 1822), *Fenneropenaeus indicus* (H. Milne Edwards, 1837 [in Milne Edwards, 1834-1840]), *Thryssa baelama* (Forsskål, 1775), *Thryssa kammalensoides* (Wongratana, 1983), *Torquigener oblongus* (Bloch, 1786), *Thryssa hamiltonii* (Gray, 1835), *Sillago sihama* (Forsskål, 1775), *Sillago soringa* (Dutt & Sujatha, 1982), *Zenarchopterus dispar* (Valenciennes, 1847), *Glossogobius giuris* (Hamilton, 1822), *Pseudapocryptes lanceolatus* (Bloch &

Schneider, 1801), *Eupleurogrammus glossodon* (Bleeker, 1860), *Pseudorhombus* sp., *Pisodonophis boro* (Hamilton, 1822), *Lagocephalus lunaris* (Bloch & Schneider, 1801), *Leiognathus blochii* (Valenciennes, 1835), *Leiognathus elongatus* (Günther, 1874), *Setipinna taty* (Valenciennes, 1848), *Sphyraena* sp., *Sardinella longiceps* (Valenciennes, 1847), *Macrognathus* sp., *Tetraodon cutcutia* (Hamilton, 1822), *Bregmaceros mcclellandi* (Thompson, 1840), *Ichthyocampus carce* (Hamilton, 1822), *Stigmatogobius* sp., *Channa* sp., *Kurtus indicus* (Bloch, 1786), *Suggrundus rodricensis* (Cuvier, in Cuv. & Val., 1829), *Moringua raitaborua* (Hamilton, 1822), etc.

Prawn: *Penaeus indicus* (H. Milne Edwards, 1837), *Metapenaeus brevicornis* (H. Milne Edwards, 1837), *Metapenaeus monoceros* (Fabricius, 1798), *Metapenaeus ensis* (De Haan, 1844), *Macrobrachium rosenbergii* (de Man, 1879), *Macrobrachium malcolmsonii* (Milne-Edwards, 1844), *Palaemon styliferus* (H. Milne Edwards, 1840), *Parapenaeopsis sculptilis* (Heller, 1862), *Acetes indicus* H. Milne-Edwards, 1830, *Exopalaemon styliferus* (H. Milne-Edwards, 1840).

Crab: *Scylla serrata* (Forsskål, 1775), two unidentified sp., *Megalopa* larval stage.

Illegal fishing activities

Fishermen enter into the STR area for fishing after getting official permits for a specific period in a month for a specific area. These fishermen usually extend their period of stay in the forest area and also sublet their fishing permits to other fishermen even to fishermen from the neighbouring districts, who take big boats with all facilities of storage etc. and tend to enter in the prohibited areas for fishing. The fishermen with legal permits from neighboring 24-Parganas (South) Division also illegally enter into the waters of STR for good catches. Big trawlers sometimes even Bangladesh also illegally enter the prohibited areas of STR along the sea during winter. In spite of the prohibitions against fishing in the rivers during the rainy season which coincides with breeding season of a major number of fish species, in the absence of any effective regulatory or even monitoring mechanism, the fishermen violate such provision.

As per the Hon'ble Supreme Court's judgment in December 1996, shrimp culture industry/shrimp ponds are covered by the prohibition contained in para 2(1) of the CRZ Notification 1991 and no shrimp culture pond can be constructed or set up within the coastal regulation zones as defined in the CRZ Notification. It also directed that all aquaculture industries/ shrimp culture industries/ shrimp culture ponds operating/set up in the coastal regulation zone as defined under the CRZ Notification was to be demolished and removed from the said area before March 1997. Audit observed that in response to the order of National Green Tribunal regarding violation of CRZ norms in Sundarban, Fisheries Department had furnished (February 2015) a report that 2098 brackish water farms were registered with Coastal Aquaculture Authority, out of which only 1068 farms were under active registration. As such, without registration, the remaining farms were operating unregulated. WBSCZMA had not taken any action to curb these unauthorised farms, despite availability of information about these unauthorised shrimp farms.

Fishermen are resorting to shark hunting as the hilsa and other commercial fish are not abundant in the BoB adjacent to the Sundarbans. Fishermen are catching different types of sharks indiscriminately. Sharks are not usually harvested singly. Sharks along with other fish are extracted from the sea and estuaries using Hilsa nets, Lakshma nets, Behundi nets and fishing nets. Most fishermen hunt sharks at night. The price of sharks varies depending on the size. They get good price in the market. So, the fishermen always seek for capturing more sharks in the net. A big shark weighs about 4-6 maunds (about 37 kg/maund), which can be sold for 30-50 thousand Bangladeshi taka (Tk). Besides, a small shark is sold for Tk 20-150, depending on the size. In the BoB, about five hundred small sharks are caught in a week. A decade ago, an average of 5,000 to 6,000 tons of sharks were harvested every year. In recent times, this extraction has reduced to an average of 3,000-4,000 tonnes per year. Although shark hunting is prohibited by law, the sharks are on the verge of extinction in the world's best wetland ecosystem as thousands of fishermen used to hunt sharks for selling at high prices.

Illegal crab hunting

Two peak abundance for crab larvae were recorded in the present study. Winter (December-February) peak was observed in freshwater zone, whereas monsoon (June-August) was in saline zone and crab larvae was less abundant in post-monsoon period. When the fishing is discontinued in the Sundarbans in October, the villagers go for crab-hunting in a small sized country boat with two others twice a month. The fishers generally catch the mud crabs (*Scylla serrata*) for their demand in the market. Fishers continue catching crabs throughout the winter until March-April. Though it remains in its traditional form even today, there is increasing demand for live crabs in the local market as well as export market and hence there is increasing catchment, livelihood dependency, decreasing catch per unit effort and clandestine catchment of mud crabs (*Scylla* spp.) including adult and undersize crabs. Expansion of crab aquaculture aims to target all size of the crab population (megalopa, juveniles, sub-adults and broods) and affect the traditional capture fishery as well as the sustainability of the aquaculture itself.

Ilegal exploitation of molluscan shells for lime manufacture

Although it is not permitted in the Sundarbans, some local people illegally collect the molluscan shells from the wild. This prevents calcium recycling into the system and affects the ecological balance.

Illegal brick kilns

Records showed that in response to the order of National Green Tribunal regarding violation of CRZ norms in Sundarban, District Magistrate, South 24-Parganas had reported (February 2015) to DoE, GoWB that in Sundarban area there were 88 unauthorised illegal brick kilns which were operational without consent of WBSCZMA resulting in violation of CRZ. WBSCZMA, however, did not initiate any action to curb this illegal activity affecting the coastal ecosystem. WBSCZMA stated (December 2016) that the matter was subjudice at National Green Tribunal. However, the fact remains that National Green Tribunal had directed in September 2014 that WBPCB should take action to stop all brick kilns operating in Sundarban.

Pilferage of timber, fuelwood and habitat destruction

The biggest peril to this landscape is direct habitat destruction by humans. There is no departmental provision for meeting the local needs except issuing fuel wood permits to local fishermen entering the forest with permits for fishing. Hence, illicit fellings of timber and fuelwood are rampant in the areas adjoining the villages, as a result of which the good sized trees of locally valuable species like *Xylocarpus granatum*, *Xylocarpus mekongensis* or *Heritiera fomes* were not available in the entire fringe forest areas and illegal timber markets grew up at Satjelia, Basirambazar, Kumirmari, Basanti and Canning of SBR.

The Sundarbans mangrove forest is a frequent victim of tropical cyclones. Cyclonic damage in the Sundarbans forest is caused to the delicate stem diameter of the species at breast height (dbh) and their spatial position in the forest stand on the degree of cyclonic thrust. For example, *Heritiera fomes* was more vulnerable to cyclonic damage than *Excoecaria agallocha*. In *Heritiera fomes,* the intensity of wind damage during cyclone increase with increasing dbh. In Sundarbans, cyclonic damage also depended on stand factors such as proximity to the riverbank or forest edges.

Usage of water channels inside forest as conduit for commercial boat traffic

For example, over 200 vessels ply everyday through the Sela river and Passur river located in and near the Chandpai-Sarankhola range of Bangladesh Sundarban respectively. This constant movement of boats acts as potential barrier to the isolated tiger's dispersal between islands within Sundarban. A male tiger was seen to cross the Gaji River at noon from Pirkhali 6 to 5 on 3rd December, 2016. It transpires that the radio-collared tigers in general avoids in crossing channels wider than 600 m and are most active between 5 and 10 in the morning. Their movement corridors that minimise crossings of wide water channels need to be identified so that the various islands harbouring tigers remain genetically connected. Boat traffic during the active phase of tigers needs to be controlled.

Fire

A fire broke out at noon on 3rd May, 2021 in the Das Varani area of Sharankhola Range under the Sundarbans East Forest Division, which has been extinguished after 30 hours through many attempts. At least 3 acres (may be more up to 5 acres) of forest had been burnt. 3% of forest land including several large trees was burnt in a fire in Chandpai Range in the eastern part of the Sundarbans on 8 February. These fires do not just burn the trees, various animals are also affected severely.

There have been about 22 fires incidents so far in Sundarbans over the last 19 years. On 22nd March, 2002 a fire broke out at Katka in Sharankhola Range and about one acre of forest was burnt. Besides, two fires broke out in Nangli and Mandarbaria under the same range. On March 25, 2004, a fire engulfed three acres of forest in Madrashachhila area of Nangli Camp in Chandpai Range and on 27 December 2004 Aryarber area was burnt. On April 6, 2005, fire destroyed two and half acres of forest in Kalamteji area of Dhansagar Station under Chandpai Range. On 13th April of the same year, four acres of forest were burnt at Tulatala in Chandpai Range. On 09th March, 2006 one acre of forest was completely razed in Terabekay area in Sharankhola Range. The same year on 11th April about 0.5 acre of forest in Amurbunia Patrol Outpost area of Chandpai Range was burnt. On 12 April one and half acres of forest in Khutabaria area under Kalamteji was burnt to ashes. Three fire incidents were reported in 2007. On January 15, five acres of forest was burnt in Dumuria Camp area of Sharankhola, on 19th March two acres in Nangli area of Chandpai Range and on 28th March about eight acres in the same area were burnt. On March 20, 2010, five acres of forest was burnt in Gulishakhali area of Dhansagar Station of Chandpai Range. In 2011, three fire incidents took place. On 01st March about two acres of forest were burnt in Nangli area and on 08th March three acres of forest were damaged in Aaryarber area. 10 acres of forest were destroyed on 25th March in Gulishakhali under Chandpai range. From March 27 to April 27, 2016 the miscreants set fire to the mangrove forest of Pachakoralia, Tengra and Tulatuli in this Nangli Camp of Dhansagar Station. On May 26, 2017, about five acres of forest in Abdullahharchilla area of Nangli Camp at Dhansagar Station in Chandpai Range were burnt down. During the Corona epidemic, a fire broke out at Dhansagar Station on February 8, 2021 and lastly, on May 3, 2021, a fire broke out in the Das Bharani area of Sharankhola Range. No legal action has been taken against the miscreants so far.

Negative impact of tourism

Ecotourism is considered a potential management tool for conserving the Sundarbans mangrove ecosystem. Although the growing tourism business benefits the local economy, it has proven detrimental to the natural environment of the Sundarbans due to habitat destruction for hotel construction, air pollution, pollution by garbage disposal, poor sanitation, and noise caused by mechanized boats.

Variable carbon stock

The total forest CO_2 and soil CO_2 stocks in the SRF are 204.8 and 90.4 Megatons respectively [Collaborative REDD+IFM Sundarbans Project (CRISP) in Bangladesh (2011)]. Overall the forests CO_2 stocks are increasing (1997-2010) at an annual rate of 4.81 ton/ha in over nearly two-third of the forest area against decrease in rest one-third (mainly on the northern periphery, adjacent to 210 fringe villages, where biotic pressure on the forests is comparatively high).

Air pollution

At present in Sunderbans region the suspended Particulate matter (SPM) level varies from 71 to 178 Kg/Cu.m i.e. within the permissible limit of residential area. The RPM, SO2, NOx levels are also fairly low i.e. almost coinciding with standard or sensitive locations. Due to the construction process, a short term and localised adverse impact is caused due to dust emission. But due to increase in number of tourists, smoke emission from transport vehicles, motor launches and mechanised boat and from domestic activities also affects the air quality to some extent.

Noise pollution

The noise level during daytime ranging from 48 to 78dB(A) and at night ranging from 42 to 62 dB (A). It appears that due to night traffic at some places noise is bit higher. During peak tourist season time, noise from mechanised boats and launches are tremendous in terms of calmness of the forests. During this time, considerable number of boats and launches enter the sanctuary and ply through different parts of the forests and causes disturbances to the animals and birds.

Water pollution

Thousands of international ships, mechanised boats and trawlers move in the Sundarbans. The vessels plying inside Sundarban often carry cargo like oil, fly-ash, cement, fertilizer, etc. Oil spills from these watercrafts make the natural habitat in the world's most luxuriant mangrove forests vulnerable to water and air pollution as was revealed by the massive oil spillage in December, 2014, when the ship Southern Star-7, ran aground and dumped 358,000 liters of Heavy Fuel Oil in the Sela river. Earlier, two such ships had run aground and emptied their entire cargo in the highly sensitive mangrove system and similar incident by one ship also took place after the incident of 2014. Such occurrences threatened the trees, plankton and vast populations of small fish and dolphins of the Sundarbans. Even after such a disaster, the government did not restrict movement of the vessels through Shela River and other linked channels of the Sundarbans. The brackish water fishery, fishing and shrimp farming are also contributing to the environmental pollution and adversely affecting the mangrove ecosystem. The coastal zone was also affected by the major consecutive super-cyclones since 2007.

The human settlements in this region generate domestic/ municipal sewage and industrial effluents, which find their way into the sea. The agricultural run-off also adds to the pollution load. The existence of port and shipping through the estuary further complicates the situation in the Hooghly - Matla estuarine complex. The overall pollution on biota in the estuary reflects a poor biological quality near the sources of pollution indicating a general deterioration in the ecological conditions.

Heavy metals are the normal constituents in the marine and estuarine environment. Pollution of the estuary with trace metals, particularly lead, cadmium, chromium and manganese, is on the rise because heavy metals are gradually contaminating the Sundarban estuaries. Their concentration is moderately high in sediments in all the estuaries of Sundarban. Most of the heavy metals in estuarine sediments are either precipitated in insoluble forms by alkaline water or absorbed into bottom sediment and low quantities are liberated into water phase by soil-water exchange. Hence, the sedentary organisms are adversely affected by the trace metal pollution.

These are detailed below.

- The industries, which are causing pollution from the point sources include paper, textiles, chemicals, pharmaceuticals, plastics, shellac, food, leather, jute, pesticides, oil etc. The domestic/ municipal sewage contributes maximum pollution to the estuary.
- Oil spillage damages beaches, marshlands and fragile aquatic ecosystems, destroy the habitat and breeding grounds; kill the users like birds, marine mammals, fish, etc.

- Heavy metals like Cadmium, chromium, copper, nickel, lead, mercury, and arsenic, which may have changed the mangrove ecosystem's biogeochemistry and physiology of mangrove species, death of mangrove species, contamination through the food chain, significant environmental risk for the future generations. Mercury and cadmium are of greatest concern for living organisms, including mangrove wetland species, which concentrate heavy metals in their tissues, becoming highly contaminated in the process.

Two resultant effects of increased pollution and soil salinisation on the mangroves of SBR are decrease in canopy closure and prevalence of a top-dying disease affecting the formerly dominant and economically most important tree species of *Heritiera* sp. Such degradation will also have strong negative effects on different interrelated components of the biodiversity in SBR.

- Agrochemical pollution through excessive and improper use of pesticides (aldrin, dieldrin, endrin, and heptachlor), herbicides, fungicides, insecticides, fertilisers in the crop fields pose a significant risk to coastal waters with surface soil erosion and wash-out of the agricultural field by rainfall. Such use changes the physical and chemical properties of mangrove sediment, the patterns of sedimentation and shoreline configuration, loss of biodiversity, increased fish mortalities, contamination through food chain and health risk.

- Nutrient Pollution: Nitrogen and phosphorus compounds enter the coastal waters from point and non-point sources. Eutrophication may cause algal blooms, change the aquatic community structure, decreased biological diversity, fish kills and oxygen depletion.

High levels of pollution, sediment load, and salinity also cause negative effects on the faunal reproduction and growth, as well disturbing their composition and distribution patterns. In the central SBR, more saline part, opportunistic trash fish have been increased significantly, while the abundance of commercially important taxa has been decreased. In addition, low water quality may also cause a reduction of commercially important fish-larvae due to the negative impacts on the quality of their nursery grounds within the forests.

Reduction of upstream freshwater flow and dominance of brackish water wetlands

This problem is more acute in the southern Sundarbans due to construction of dams and barrages, which, in turn, has been choking the rivers and enhancing the salinity and siltation of the streams, rivers, canals and creeks inside the habitat. The total freshwater impoundments, i.e. freshwater bodies, in the Indian Sundarbans is 34,810.65 ha while brackish water bodies occupy an area of 39,874.7 ha. Thus, altogether 74,685.35 ha of inland water bodies are found in this region, out of which 46.61% is freshwater in nature, while 53.39% water bodies are brackish in nature. Hence, brackish water bodies are dominant in the Sundarbans even within the inland water bodies. This is particularly true for the part of Sundarbans under North 24-Parganas, where the brackish water bodies occupy a total area covering six times of that of freshwater areas. It is, therefore, evident that freshwater bodies are comparatively scarce and for the purpose of daily needs like potable water, irrigation, domestic chores, livestock feeding etc. more freshwater bodies are required.

Salt pans

These are formed within the forest areas because of lesser tide inundation leading to destruction of the vegetation and larger blanks are emerging within.

Soil erosion

Soil erosion along the river system has increased the sand deposition, especially in the seaface.

Erection of high embankments along the riverbanks in the human settlements

This was done to protect entry of the saline water during the high tides for development of agriculture, initiation of the brackish water fishery but ultimately affected the tiger and its ecosystem.

Increasing threat of super-cyclones

In the last 23 years, the Sundarban has witnessed 13 supercyclones. The occurrences of cyclones increased by 26% between 1881 and 2001. Most importantly, there has been a significant increase in the frequency of very severe cyclones in the post-monsoon season during the period 2000 and 2018. As much as 73.5% of the entire Sundarban mangroves witnessed a decline in the NDVI values (plant health) after Amphan and about 3.45 km² of the Sundarban mangrove forest (roughly 0.05%) had a considerably lower NDVI value almost a year after Amphan, which likely reflects the damage caused by the cyclone. The southern and western regions of the Sundarbans in India and Bangladesh showed the most damage to the mangrove cover following the cyclone. The shorter mangrove stands were more widely affected in the western Sundarbans, because close to 40% of the mangroves here supports short-statured mangroves, whereas the range of extreme damage decreased in case of relatively taller mangrove trees in the eastern Sundarbans. The northern and eastern parts of Bangladesh Sundarbans seem to have escaped most of the damage from Amphan because the cyclone track stayed away from this subregion. While mangroves are resilient to the impact of cyclones, anthropogenic pressure is changing their behaviour, making them vulnerable.

Effective implementation of a joint long-term tiger conservation strategy and action plan

It is prerequisite for effective and sustainable management and conservation of the tiger and its habitat.

Management strategy

The conversion, management, and conservation efforts for the entire Sundarbans can be grouped into four distinct time periods based on the outlook of resources use e.g.-

1. Conversion for agriculture from 1780 to 1875;
2. Timber production for revenue and control for theft from 1876 to 1951;
3. Inventory based management from 1952 to 1992;

4. Integrated management and co-management as well as project-based overlapping management during 1993-2020 (Mahmood *et al.*, 2021).

This is the roadmap of 240 years' journey towards the conservation of the nationally and globally important ecosystem of the Sundarbans.

After independence in 1971, Presidential Order No. 23 of 1973, known as the Bangladesh Wild Life (Preservation) Order, was issued in 1973. In 2012, parliament passed another Act called The Wildlife (Preservation and Security) Act, 2012 to adapt new legal requirements for coping with shifting scenarios in order to ensure the protection of forests, wildlife, and biodiversity. The latter nullified the former. The poaching of endangered species like the tiger is now a cognizable and non-bailable offense. In India, an offence committed inside the core area of a Tiger Reserve, attracts a mandatory prison term of three years, extendable to seven years and a fine of Rs. 50,000 extendable to Rs. 2 lakhs. In case of a subsequent conviction of this nature, there is an imprisonment term of at least seven years and a fine of Rs. 5 lakhs which may extend to Rs. 50 lakhs. Despite the penalties, the laws are difficult to enforce.

Although the PAs have been declared, the forest department is neither well-equipped nor possesses adequate man-power to monitor (i) the long coastal stretch (marine zone), (ii) the protection zone (PAs), where all types of resource collection is prohibited, including the sensitive international border between India and Bangladesh, (iii) the production zone, subjected to sustainable harvest and (iv) the buffer and transitional zone in the periphery of the forests. India shares about 70 kms of its riverine border with Bangladesh in this region. The border winds through both land and water here.

Management strengths of selected PAs

1. Halliday Island is a transitional ecotone that supports a unique and diverse flora and fauna- mangrove forests and a littoral or supra-littoral forest floral–faunal assemblage within the smallest island (5.95 km²) of the Sundarbans.
2. It provides shelter and protection to various species of wildlife, particularly birds, included those listed in the Red Data Book (R.D.B.) of the IUCN and the appendices of CITES. Besides, the

WLS is visited by a number of animals of Schedule-I of the Wildlife (Protection) Act 1972, including the Gangetic Dolphin, Estuarine Crocodile, Fishing Cat and Tiger.

3. Haliday Island is a small, remote island of the Sundarbans. It is a protected area bounded by the river Matla and BoB on the south. It is well protected from external threats such as illicit felling, poaching and encroachment since villagers do not have easy access to the protected area.

4. Training has been imparted regularly to the division on legal aspects, the use of arms and tranquilising.

5. The forest department has good co-ordination with the animal resource development department of the protected area.

Management weaknesses of selected PAs

1. Potable water is not available on Haliday Island. Wildlife is solely dependent on estuarine water.

2. A large number of mechanised boats and launches contribute to the chemical pollution of the mangrove ecosystem.

3. There is no inter-agency co-ordination and co-operation between various departments such as the tourism, forest, I&CA and education departments with regard to development of tourism.

4. The infrastructure is inadequate. There is no permanent camp on Halliday Island. Patrolling is carried out by the staff of Kalash Camp, of this area.

5. Abiotic factors such as cyclones, of varying intensity, usually accompanied by tidal waves, cause damage to the area, including soil erosion of the land facing the sea.

6. The banks of the river Matla and the coastal area are getting continuously eroded, and a portion of the land is being lost every year.

7. The staff have not been imparted training in monitoring and in technical aspects of natural resource management and eco-tourism.

Immediate actionable points

1. One watchtower is needed to improve the surveillance in the PA.

2. It is essential to dig a few fresh water bodies at suitable locations to conserve water for the wild animals.

3. A palisade with a porcupine structure of bamboo piling will be an effective measure to arrest soil erosion.
4. In some places, specifically on the western side of the PA, embankment protection work of constructing spurs, RCC walls, etc. will be effective in controlling soil erosion.
5. Specialised training specific to PA management and eco-tourism should be imparted to the staff.

Measures to curb human-tiger conflict

- Digging of sweet-water ponds within the forest to change the drinking water habit of tigers in 1975.
- Stoppage of permit for collection of Phoenix (leaf) in 1979.
- Releasing farm-bred wild pigs in the micro-localities of the prey-depleted buffer zone to reduce tiger straying inside the habitation areas
- Tranquilisation and capture of tigers which stray out in the villages and their translocation into the core area or zoo garden
- Introduction of human dummies equipped with electric wires in 1983. In this method, electrified clay models dressed as honey collector or wood cutter are kept in the jungle "charged to 230 volts by an energizer and a 12-volt battery source". When the man-eaters attack these dummies they get the electric shock.
- Human face-mask was introduced in 1986. In this method, honey collector, fishermen, and wood cutters enter in the buffer zone wearing a rubber mask "at the back of the head", resembling a human face. As the tiger usually attacks from behind, they think they are being watched by the man and thus they become reluctant to attack the person. About 2500 rubber masks were distributed among locals between November 1986 to October 1987, who were permitted to enter in the deep forest and not a single man wearing mask was attacked by the tiger. The average number of victims reduced from 45 per year (1975-1982) to 21 per year (1983-1985) since the method has been introduced in 1983.
- Both dummy and face-mask discontinued in 1990, resumed again in 1993 but used in a limited scale.
- In 1994, the Forest Department introduced a headgear made of fiber-glass covering head, neck, and chest. At present this type of headgears are only available to Forest Department's staff and

provide a higher degree of protection as compared to rubber masks.

A series of operation from trapping of the problem tiger/tigress followed by darting, sedation, radio-collaring, release in the wild, to intermittently or continuously monitoring of its mobility and activities by radio telemetry is also being done since 2007. Since both the young and old adventurous or aberrant tigers in Sundarbans are accustomed to the conventional cage traps, this technique was used for the first time to radio-collar an adult female in December 2007 for conducting a study through satellite tracking, when she was fitted with a customised (species-specific, i.e. a device for the tiger does not fit a penguin) imported US-made radio-collar (GPS/ARGOS/VHF) and tiny radio transmitter attached to a little harness or strap on the neck (the device well-tested before by placing it at different mangrove forest locations with variable canopy density). There were a number of receivers high above ground level, which pick up the signal of the transmitter carrier via satellites and convey this information to the researcher stationed elsewhere. When the information is being conveyed from the animal to the receiver, it is called an uplink. When it is sent from the satellite back to the researcher, it is called a downlink.

GPS (Global Positioning System) tracking is a process to (i) remotely observe relatively fine-scale movement or migratory pattern of the radio-collared animal from island to island, (ii) calculate home range (normally 10-12 km in the mainland) and (iii) get an idea about habitat use and behaviour. The receiver is placed on a mechanised boat. The device records and stores the incoming data at a pre-determined interval (duty-cycle) by an environmental sensor. The tiger's locations are plotted against a map or chart.

GPS collars are programmed to acquire locations at set time intervals either in advance or in the field as per requirements, whereas the radio transmitter allows monitoring of the tiger's movements on a daily basis. The battery in the collar allows for a set number of possible GPS locations. The collar's VHF signal is programmed to be transmitted for a set period of time each day, normally 8 hours. Its battery life in the field is determined primarily by the time interval between attempts to fix a GPS location. If it is set for every 2 hours, the field life would be approximately 8 months, whereas at an interval of less than 2 hours its field life would decrease and *vice versa*. Furthermore, battery life varies depending on the ratio of successful and unsuccessful GPS locations because it is for a longer period in case of the unsuccessful locations.

The collar locations are stored within the collar which has to be retrieved either by recapture, automatic release after the batteries run down, or by remote release using the equipment that transmits a signal to activate the collar release system. After the limit of GPS locations are acquired, the collar retains some battery life for radio telemetry and activation of the self-release mechanism. Even if the collar is fallen off subsequently, the still active radio signal helps retrieve it.

Moreover, it is vital to track the stationed or moving tiger every day using radio-telemetry. The GPS locations by themselves do not divulge information regarding the activities of tiger at each location. By radio-tracking, it is sometimes possible to determine the tiger's behaviour, for example, when the tiger hunts or if it is associating with other tigers.

GPS collar locations are analysed to construct Minimum Convex Polygon (MCP) and Fixed Kernel (FK) home ranges. Sign deposition rate is evaluated as the number of creeks/channels crossed by the radio-collared tiger per day using Google Earth. If a tiger remains at a particular location within a periphery of only 100 m for more than 12 hours during the daytime, it is considered that it must have hunted. Similarly, if it is localized for more than 8 hours at night, the same conclusion is arrived at. When the transmission is temporarily cut off in case the tiger enters deep inside the forests beyond the signal receiving range (2-5 km), the monitoring team tries to retrack the animal by to and fro movements for hours. Therefore, it needs calculative use, otherwise frequent readings will drain limited battery power of the collar more rapidly, whereas longer intervals between readings may provide lower resolution but over a longer deployment.

Before conducting the first radio-collaring experiment in Sundarban Tiger Reserve in 2007, a preliminary ground survey was carried out in the protected areas during 2006. Camouflaged cage traps were placed in six different locations, at least 5 km apart (to capture maximum number of different individuals), three each in two distinct (separated by at least 25 km aerial distance) zones National Park (Core area closed to human use other than management purposes) and Wildlife Sanctuary (Buffer area open to tourism and regulated fishing and honey collection by the fringe people), 100-200 metres inside the forest (from the edge of a river channel), mostly near the confluence of two or more river channels or near the rainfed freshwater ponds as these places are visited by the tigers frequently. In addition, camera traps Trailmasters™ 1550 (Goodson Associates Inc., Lenexa, KS, USA) were deployed at each cage trap to monitor the visitation rates and the behaviour of individual tigers at cage traps. Initially, the traps were kept non-operative so that the tigers could enter and leave the cage freely so that they are habituated to such practice in future when the traps are made operative. The traps were deployed for two months (30 days x 2 months x 3 sites x 2 zones = 360 trap-days in total). The trap sites were monitored each day to check the signs of tiger's movement and whether the live bait is hunted. Tiger visits were recorded only at two of the three sampled locations in the core area. One of the camera trap locations recorded a tiger visit only once, while the second location registered three tiger visits at an interval of 25-30 days, but the visitors could not be identified because the camera trap did not trigger on the first occasion and on two subsequent visits the tiger stayed away from the infrared beam necessary to trigger the cameras. Whereas the individual tigers visited all the three trap locations in the Sanctuary 5, 7 and 12 times respectively at a 9-15 days' cycle. Fresh pugmark sets and photographs from camera traps placed at the mouth of the traps confirmed that the tigers picked up the bait from near the traps. Hence, the live baits were kept inside the cage traps and tigers did not seem wary of picking the baits from right inside the traps.

In the field, it appeared that the tiger's visit rate was low in the core but high in the buffer area. For example, a trap was laid with goat-bait at far-flung Chamta block (National Park area, centrally located south of Panchamukhani, in between Chandkhali and Netidhopani) because earlier a number of aberrant tigers had been released here for rehabilitation, but it was observed that none of them was encouraged to enter the trap and lift the bait. A few pugmarks were, however, seen within 200 m radius of the trap, which indicated that a tiger came to the spot at night but it was very clever (or recollected the bitter experience when captured earlier) and, naturally, being suspicious, did not take the risk of further confinement, instead made a few rounds surrounding the spot for inspection of the trap cage and then left the place to utter dismay of the field staff. They did not attempt to lift the bait, even after prolonged baiting or when it was tied outside the trap at the same location, except on one occasion. However, fresh pugmarks were found on the mudflat to indicate that they used to pass by ignoring lure of the easy prey.

On the contrary, in case of the sanctuary, the tigers used to visit all the trap locations. Individual tigers followed a cycle (gap) of 9-15 days to revisit the area. Here, the visiting tigers were not at all shy or hesitant like those of the core area. Most surprisingly, there were a few instances when the intelligent tiger(s) frequently took away the bait from within the cage without being trapped. Fantastic, a jewel of the cunning hunters! You may be eager to know the innovative tactics of such bold tigers. First, it started moving all around the cage, but instinctively did not enter at the time of its first visit, not to speak of any attempt to take the bait away. It appeared that the females were comparatively more intelligent. However, after a few days' visit and observations, it had taken away the bait without stepping on the lever; lest it would lead to closure of the cage. In another trapping site, the animal avoided the main cage door, instead hit the cage repeatedly from the back till one of the nut bolts was slackened and fallen off and it was able to secure a small opening. It then somehow squeezed its paw inside the cage through that small opening to pull out the bait in such a manner that, as a result of the tremendous tug, half of the goat's body was left inside the cage. However, all the residents are not equally cautious or innovative in securing their prey and a few of them entered through the main cage-door in lure of the live bait and could be trapped.

Considering that majority of the tigers were sighted in Pirkhali and Panchamukhani blocks of Sajnekhali Wildlife Sanctuary and, as per the annual monitoring reports of the previous years, highest frequency of sighting during the four months from January to April and the sightings started reducing from May onwards during the monsoon, recovering only in December, indicating a periodical fluctuation of tiger population in these northern blocks, Sajnekhali Wildlife Sanctuary seemed to be the most effective site for capturing and radio collaring of the tigers, particularly from December onwards. This hypothesis proved to be true, when fully operational traps were deployed in December, 2007.

On the 4th, a healthy tigress, approximately five years old, later named 'Priyadarshini', could be captured at Panchamukhani-3. It was tranquilised and radio-collared on the 5th. After recovery from drowsiness, the tigress was released in the same block, i.e. her original habitat, at Choragajikhali location next day (6th) at 8.20 a.m. It was being tracked for the whole day, when she was found either resting or moving intermittently and, till the evening, covered only a distance of about 2.5 km from the place of release. The continued monitoring conducted later also revealed that it usually rested during the day and became active at dusk and dawn and once observed to kill a deer.

The satellite link stopped working within a day of radio-collaring the tigress. The GPS also failed within a week. However, the VHF beacon continued to work, providing a range of over 2-3 km that allowed us to monitor the tigress from a boat. During the period from 7th to 18th, Priyadarshini remained within the surrounding areas of the Panchamukhani forests because signals received confirmed her local movement. As usual, she was mostly resting. Perhaps she might have managed to hunt in the meantime and consumed food sufficient to meet her appetite for a few days.

Continuous monitoring began since 20th afternoon, when Priyadarshini was staying at Pirkhali forest near the Deulbharani khal/khanri (narrow creek), opposite the river Gosaba. She was located at almost the same habitat for the last few days. On 21st, she moved towards Panchamukhani by crossing the khal. But, she did not proceed for long and, ultimately, came back to her earlier refuge at Deulbharani-Gosaba junction. On 22nd, up to the afternoon, she was resting. At 5 p.m., she got up and started moving again towards the Deulbharani khal. On 23rd too, she did not shift to any other location. It was only on the 24th evening (7 p.m.) that the incoming signals were being strengthened due to her quick movement. She was in dire need of flesh now. At 9 o'clock, it came within a range of only 200 m from the monitoring boat. The signal was almost fixed, i.e. she was resting in the forests. Silence prevailed during the next one and half hours. It was 10.40 p.m., when the signal showed rapid up and down movement. At about 11 p.m., she could locate and select the prey for hunting. But the targeted prey could sense presence of its camouflaged enemy nearby. The surrounding forest was reverberated at the repeated, loud and sharp alarm calls 'ack-ack' of the Chital deer (Axis axis) and within a minute the sound of splashing water was heard ahead within a distance of 10-70 m. The predator started chasing her prey in the flooded terrain. It was already high tide then, engulfing the lower forest floor at a rapid pace. It was a full moon night too. Still neither the predator nor the prey could be seen, though a tough struggle seemed to take place between the sturdy tigress and the lanky deer, as you often see in the popular telechannels. Whereas the chaser was galloping, the forerunner was trying to escape by leaps. But Priyadarshini was in vain and the deer won the race. A great relief for the panting even-toed ungulate! Although Priyadarshini was utterly frustrated, she did not give up hope and regained sufficient energy for the second chance. At 11.20 p.m., again the sound of splashing water was heard. The tigress relocated her prey and this time ran after the target in a zigzag way. The prey was really very tired and this time could not compete successfully. Suddenly it gave out a loud cry followed by a thud. At last, Priyadarshini could seize the prey (average weight 50 kg). It was exactly 11.24 p.m. She dragged the kill through water towards a high dry land for the dinner. It was past midnight, 00.09 a.m., i.e. 25th December. Another chital cried afar. But the tigress had then concentrated on her kill. While feeding, she was relaxed and gave out low growls in pleasure. A great feast of X-mas indeed!

In the forenoon of 26th, she moved with her half-eaten kill about 1 km away for relishing during the remaining few days of 2006. However, nothing could be seen as dense fog rolled into the entire forest and over the waterbodies. In the afternoon, the continuous monitoring was called off.

Based on data collected over 45 days, the home range of the tigress was estimated at 40 km² and 28 km² using the 100% and 95% Minimum Convex Polygon method, respectively in Panchamukhani and Pirkhali forest blocks, located side-by-side with similar type of vegetation and prey-base, where the tiger density was seasonally high. However, the radio-collar of the tigress was operative for four months (up to April next year) and then ceased operation, but the tigress with collar was seen physically many times afterwards in the same location, showing the sign of territoriality, i.e. within a definite area of operation in comparison to that of the strayed tigers.

Remote sensing

Due to inaccessibility of the hostile Sunderbans terrain, it was almost impossible even for the Forest Department staff to carry out land based survey of the mangrove forest and enumeration of the crop. To overcome these physical problems, remote-sensing technique was tried out for the region in 1999 to test the effectiveness of the technology in monitoring the mangrove ecosystem of Sunderbans.

Methodology

IRS ID, Liss-III data of 22nd December, 1999, in four bands (green, red, Near IR and SW IR) was requisitioned from NRSA, Hyderabad for the scenes (row-path nos) 107-57, 108-57 and 108-56, which covered the entire Sundarbans mangroves of Indian portion and major parts in Bangladesh. After applying the histogram stretch and appropriate brightness contrast, the standard FCC was classified through supervised classification using "Maximum Likelihood classifier" on the basis of known ground truth points (A.O.I.). The entire imagery was classified with the following: mangrove vegetation, water with high sediment, clear water, non-forest uses, sands and muddy channels/mud flats. Extensive Ground Truth Verification was carried out by the Working Plan & GIS Division with the help of STR, between 13 and 15 September, 2000 during full tide (*Kotal*) on 13th, a full Moon day.

With a view to clearly delineating the open and degraded mangroves created due to eroding effects of ebb tides. The RF boundaries of erstwhile 24-Parganas Forest Division and STR were digitized, geo-referenced using ARC/INFO software and overlaid on the classified imagery, both the raster and vector coverage being in polyconic projection with 88°E and 24°N as central meridian / projection origin.

Analysis of the classified imageries shows that the effects of erosion by ebb tides and function of muddy channels extending to the interior of the islands is very high in Goasaba 1, 2 and 3 blocks of TR. Goasaba-3 is the most affected compartment, which contains large patches of mud flats at the heads of these muddy channels. Though there is no harvesting or felling of trees in these parts of TR, considerable parts of the block is either devoid of trees, or is open type of forest with sporadic growth of bushes or dwarf trees like *Ceriops decandra*.

Critical tiger habitat (CTH)

In 2007, an area of 1,699.62 km^2 (by including 369.50 km^2 RF to the previous core area) in STR was notified as CTH, the inviolate space required for conservation of tigers, determined on the basis of the parameters like estimated tiger population over years, size of the territory and population viability analysis.

Smart patrolling

- Use of specially designed app for patrolling: Along with use of RT system, E-patrol/Smart patrolling was introduced in TR in 2015. In this new system, every camp has been given a cell phone having an android operating system with a compatible mobile application called 'Hejje' (Programmed in Java) installed in it for monitoring and patrolling purpose. With the help of this application the frontline staffs are recording their everyday activities like patrolling, monitoring the condition of fences, night patrolling, offence detections and wildlife sightings.

- Use of Unmanned Aerial Vehicle (UAV): UAV or Drones has been introduced as a part of the Smart Patrolling initiative in TR. This system has been proved to be a remarkably useful tool in patrolling those areas in TR which are otherwise inaccessible. This tool is also effective in case of locating a strayed animal in a locality especially tiger. UAV is also being used to monitor an animal's post release into the wild up to considerable distance inside the forest at close vicinity.

Nylon-net fence has been found to play an important role in preventing the straying out of tigers into village's areas from forest. A protocol for maintenance of the nylon-net fencing has been designed with an aim of carrying out thorough checking and proper maintenance. The Protocol includes involvement of local Stakeholders in FPC/EDC members also along with forest staff.

Transboundary tiger corridor

The National Tiger Conservation Authority (NTCA) has developed a three-level strategy to manage negative human-tiger interactions under the Centre's long-term tiger conservation programme. The strategy involves providing material, logistical and financial support to the tiger reserves to deal with big cats dispersing out of the protected areas. It also seeks to ensure safe dispersal of tigers by limiting habitat interventions (facilitating dispersal to other rich habitat areas). The NTCA's standard operating procedures have to be followed to deal with the human-animal and livestock conflict after dispersal.

NTCA in collaboration with the Wildlife Institute of India (WII) has published a document titled "Connecting Tiger Populations for Long-term Conservation" (2014), which has mapped out 32 major corridors across the country, management interventions for which are operationalised through a Tiger Conservation Plan, mandated under section 38V of the Wildlife (Protection) Act, 1972. But, even if, the Sundarbans is the prime tiger conservation area in the country and WII worked there for long during the 21st century, it was not included in this document.

Tiger corridor

A tiger corridor is a stretch of land linking tiger habitats, allowing movement of tigers, prey and other wildlife. Without corridors tiger habitat can become fragmented and tiger populations isolated leaving the tigers vulnerable to localized extinction. Corridors are used by other wildlife also. The corridors across states are important as big cats travel long distances. A tiger travelled 1,300 km over two states (Maharashtra and Telangana), six districts and four wildlife sanctuaries in about 150 days exploring a new area to set up its territory.

Justification

It is vital for the long-term survival and viability of tigers Corridors are under threat from development projects. The greatest threat here is the expansion of road and railway networks which cut through the corridor leading to highway killings of tigers, and human animal conflicts. All India tiger estimation 2018 results released in July that showed there were 2,967 big cats in the country, and large population ventures outside protected areas, this needs mapping of tiger corridors for their enhanced conservation. Need to sensitize villagers along these corridors about conservation of wildlife.

The entire region is connected by a mesh work of innumerable creeks, channels, rivulets and tigers are often found swimming across 50 m - 70 m wide water channels. In the west, the river Matla acts as an interdivisional corridor and on the eastern side, the river Raimangal acts as an international corridor as far as tiger movements are concerned. Though rare, there are occasional reports of tigers swimming across these big rivers. The Big Cats also swim freely across rivers and creeks from one island to another between India and Bangladesh along the Raimangal-Harinbhanga stretch.

According to Mallick (2011), "Migration of the tigers between these two segments was anticipated, but there was no proof in the past. But during 2010, evidence of such territorial shifting was known. The straying Khatuajhuri male, blind on the right eye with signs of territorial fights, which was captured on 20th May, radio-collared and released in the nearby forests on 22nd May, crossed the river Harinbhanga and strayed into the Bangladesh Sundarban, where it remained in the island of Talpati till his collar stopped functioning on 5th August. Hence, no further information regarding its movement was available."

It was stressed on delineating corridors for tiger movement across the Sundarbans by controlling ship traffic and human disturbance. The Sundarbans is totally cut off from any other nearby tiger population and gene pool. As a result, it is important to maintain this as large single transboundary population. A study report (2018) recommended that India and Bangladesh should carefully use the water-channels in the Sundarbans in a bid not to disturb the free movement of tigers between the two countries, which may ultimately affect their gene flow. Increased continuous use of these water channels inside the forest as a conduit for commercial boat traffic can transform the rivers to barriers to tiger movement. Despite efforts by forest departments of both the countries, joint patrolling and joint management activities remain a non-starter.

However, cargo movement through the mangrove forests of Sundarbans has been stopped on the Indian side since 2011. Now, the ships and barges take alternate water routes, avoiding the Sundarbans, to reach Bangladesh.

Promotion of the management of forests

As stated in the foregoing paragraphs, the forest department in both India and Bangladesh has aimed to manage the Sundarbans in an equitable, sustainable and effective manner through co-management by involving and empowering the local resource users around the Sundarbans, by introducing tools for patrolling and law enforcement (SMART = Spatial Monitoring and Recording Tool), by conceptualising Ecological Monitoring using Remote sensing and techniques on the ground and by formulating a long term Integrated Resource Management Plan for the Sundarbans.

Status of the key species

Despite their worldwide symbolic value and status as an icon of India and Bangladesh, a "charismatic megafauna," the Bengal tiger is considered endangered species since 1986, suggests a range decline much greater than 50% over the last three generations (7 x 3 = 21 years). The largest population of the Bengal tiger lives in the undivided Sundarbans, particularly the PAs. As human populations increase, so does conflict between the humans and the Bengal tiger. Development and poaching have contributed to a reduction of the tiger and its natural prey animals, whereas in the tiger land the humans have become targets for the hungry predators.

Factors affecting the wild tiger status are:

- Targeted killing on account of Human-Tiger Conflicts (HTC)
- Poaching owing to vulnerable geographical location, with proximity to several international borders
- Depletion of prey base of on account of subsistence poaching.
- Forced forest resource dependency of locals for want of livelihood options, viz., wood and nontimber forest produce
- Proximity to human-dominated land parcels

The ecological integrity of tiger habitat is also threatened by rise in sea-level on account of climate change, apart from reduced availability of fresh water owing to diversion.

Sundarbans protocol

The essence of success of conservation oriented forest management lies in protecting the forests from the onslaught of human misuse and overuse, proper management of habitat and habitant, and developing harmonious relationship between management and local people.

A Sundarbans Protocol was drafted on February 1, 1989, which was adopted on March 10, 1989 and widely circulated for awareness building. The protocol reads as follows:

Whether the relationships between man and nature has become complex and problematic;

Whereas the nature's endowments are being overdrawn to meet the demands of growing population for fuel, housing, cultivation, raw materials for industries and revenue;

Whereas the food chain and the natural cycles of water, energy, carbon, nitrogen and minerals, etc. are being disrupted;

Whereas the life support systems inbuilt in nature is getting strained;

Whereas the extinction of plant and animal species is disturbing the ecological balance;

Whereas the depletion of ozone layer and gradual global warming due to build-up of carbon dioxide and other greenhouse gases is causing alarm;

Whereas the ecology stands at the cross-roads of politics, science and economics;

Whereas income inequalities and illiteracy lead to indifference and hospitality to efforts for conservation of natural resources;

Whereas pollution of air, water and soil is threatening health and productivity;

Whereas noise, radiation, waste, toxic chemicals, etc. are worse confounding;

Whereas the nature has started taking revenges as manifested by repeated droughts and floods, storms and earthquakes;

Whereas the younger generation in general and future generation in particular as a consequence of present development initiatives for short-term gains are left to suffer from consequences of desertification, acid rain, climatic changes and biological losses;

Whereas the ill-effects of pollution are suffered largely by the poor;

Whereas environment defies any division – natural or man-made;

Whereas we can go in for a planned development in harmony with nature;

Whereas we cannot be helpless spectators of history to be written by thermonuclear ink or its geography to be pen-picture of seething inferno or freezing under an icy desert;

Whereas based on a common understanding, we can mobilize united and unified social, political and economic activities for conservation of nature, its life support systems and control population.

We, the people of India, resolve to constitute a "Code of conduct and plan of action"- call it Sundarbans Protocol- and address ourselves in voluntarily work in unity and unison as follows:

-   Cultivate a culture for cleanliness
-   Conserve the congenial environment
-   Cordon the treasures of nature
-   Combat pollution and combine the development activity with recycling of resources and appropriate low-cost technology
-   Clean our rivers and green our land
-   Challenge the systems which exploit Nature
-   Constitute an association of citizens of India for creation of a homeland for all living beings.

We do hereby adopt this Sundarbans Protocol for conservation of nature, for rejuvenation of environment and give it to me, to my family, to my neighbour, to my countrymen and women within the frontier and to the brothers and sisters beyond it.

Prospects

It is projected that the current trends of climatic variables, geomorphology, soil nutrients and anthropogenic factors may lead to massive degradation of the forest coverage within 2050. Under the circumstances, the most important ethical solution aiming at the biodiversity conservation is to promote human transition into living in harmony with nature. Sustainable plans and actions are, therefore, required integrating conservation and climate change adaptation strategies, including promotion of alternative livelihoods. Thus, interdisciplinary approaches are required to address the future climatic disasters and better protection of invaluable ecosystem.

High level organisational transboundary collaboration is sought to achieve effective conservation and restoration of the mangrove ecosystem. In this case, both the Indian and Bangladeshi government will require an integrated vision for approaching human well-being, sustainable alternatives to forest-based livelihoods and ecological sustainability, effective regeneration, meaningful participation of all stakeholders in biodiversity conservation and comprehensive monitoring, evaluation and reporting on indicators of sustainability.

There are still some signs of optimism during the 21st century like decadal stabilisation in mangrove extent, negligible rate of mangrove clearing, only a small proportion of the forest classified as degraded and recent increase of the tiger population (rose to 96 in 2020-2021 from 88 in 2018-2019) indicating some positive impact of the management of the ecosystem, although under resourced, and the conservation efforts are laudable.

However, there is a need to put in place an integrated management plan for maintenance of ecological character of the Sundarbans and addressing drivers of adverse changes. Continuation of the time-series and long term monitoring is crucial to evaluate the status of biodiversity and risk of ecosystem collapse over a threshold across multiple bio-indicators for rectification of the gaps and restoration by the stakeholders including the forest department.

Seven tasks have been identified for improving the ground situation.

1. Develop capacity in the Forest Department (FD) for effective wildlife and habitat conservation
2. Reduce community dependency on forest resources
3. Strengthen local institutions
4. Reduce wildlife-human conflict
5. Reduce tiger and prey species poaching and consumption
6. Deploy an effective and efficient man-power for wildlife conservation including their habitats
7. Establish an institutionalized system to curb transboundary trade and poaching of tigers and other wildlife.

Selected Bibliography

Adhikari, A, Mondal, A. 2019. A leucistic Collared Kingfisher *Todiramphus chloris* in the Sundarbans, West Bengal. *Indian BIRDS* 14(6): 191-192.

Adhri, AS, Sultana, A, Rahman, S. 2015. Habit and Habitat of Squirrels in Bangladesh. *International Journal of Recent Research in Life Sciences (IJRRLS)* 2(1): 41-44.

Alam, ABNS, Akhtar, F, Ahmed, S, Azmiri, KZ, Fahad, ZH, Haider, FH, Sharmin, N. 2020. *Dolphins of Bangladesh: A conservation effort in Sundarban*. Bangladesh Forest department, IUCN Bangladesh and UNDP Bangladesh.

Ansari, AA, Trivedi, S, Saggu, S, Rehman, H. 2014. Mudskipper: A biological indicator for environmental monitoring and assessment of coastal waters. *Journal of Entomology and Zoology Studies* 2(6): 22-33.

Aziz, MA. 2016. Threats to Olive Ridley Turtles on the Sundarbans Coast of Bangladesh. *Indian Ocean Turtle Newsletter* 24: 2-4.

Aziz, MA. 2018. Notes on Population Status and Feeding Behaviour of Asian Small-Clawed Otter (*Aonyx cinereus*) in the Sundarbans Mangrove Forest of Bangladesh. *IUCN Otter Spec. Group Bull.* 35(1): 3-10.

Aziz, MA, Smith, O, Barlow, A, Tollington, S, Islam, MA, Groombridge, JJ. 2018. Do rivers influence fine-scale population genetic structure of tigers in the Sundarbans? *Conservation Genetics* 19(5): 1137-1151.

Aziz, MA, Tollington, S, Barlow, A, Goodrich, J, Shamsuddoha, M, Islam, MA, Groombridge, JJ. 2017. Investigating patterns of tiger and prey poaching in the Bangladesh Sundarbans: Implications for improved management. *Global Ecology and Conservation* 9: 70-81.

Aziz, MA, Smith, O, Jackson, HA, Tollington, S, Darlow, S, Barlow, A, Islam, MA, Groombridge, J. 2022. Phylogeography of *Panthera tigris* in the mangrove forest of the Sundarbans. *Endangered Species Research* 48: 87-97.

Bahuguna, NC, Mallick, JK 2010. *Handbook of the Mammals of South Asia with special emphasis on India, Bhutan and Bangladesh*. Natraj Publishers, Dehra Dun, India, pp. 541+31 plates.

Baker, EB. 1887. *Sport in Bengal and how, when, and where to seek it*. London. Pp.266-268, 274, 282-283.

Bandyopadhyay, S, Bandyopadhyay, N, Kar, S, Dasgupta, S, Mukherjee, D, Das, A. 2022. Island area changes in the Sundarban region of the abandoned western Ganga–Brahmaputra–Meghna Delta, India and Bangladesh. *Geomorphology* 108482 https://doi.org/10.1016/j.geomorph. 2022.108482.

Barik, J, Chowdhury, S. 2014. True mangrove species of Sundarbans delta, West Bengal, Eastern India. *Check List* 10(2): 329–334.

Barlow, ACD. 2009. The Sundarbans tiger: adaptation, population status and conflict management. *Ph.D Thesis*, University of Minnesota.

Basu, D. 2019. Rice Field Spiders (Araneae: Arachnida) of South 24 Parganas, West Bengal. *PhD diss.*, Agricultural Biotechnology, Ramakrishna Mission Vivekananda Educational and Research Institute, 292 pp.

Bhowmik, M, Ghoshal, P, Salazar-Vallejo, SI, Mandal, S. 2021. *Sigambra sundarbanensis* sp. nov. (Annelida, Pilargidae) from the Indian sector of Sundarbans Estuarine System, with remarks on parapodial glands. *European Journal of Taxonomy* 744(1): 49-66.

Biswas, O, Das, AK, Chakraborti, U, Chatterjee, S, Khan, NH. 2013. On a collection of butterflies from Bali island, Sunderban, West Bengal. *Bionotes* 15(3): 83.

Biswas, O, Chakraborti, U, Roy, S, Mazumder, A, Mallick. K. 2016a. A new citation record of *Horaga onyx onyx* (Moore, 1857) from Sunderban Biosphere Reserve, West Bengal, India. *Journal of Entomology and Zoology Studies* 4(4): 12-14.

Biswas, O, Modak, BK, Mazumder, A, Mitra, B. 2016b. Moth (Lepidoptera: Heterocera) diversity of Sunderban Biosphere Reserve, India and their pest status to economically important plants. *Journal of Entomology and Zoology Studies* 4(2): 13-19.

Biswas, O, Chakraborti, U, Roy, S, Modak, BK, Shah, SK, Panja, B. 2016c. First record of *Amerila eugenia* Fabricius, 1794) [Lepidoptera: Erebidae: Arctiinae] from Eastern India. *Entomology and Applied Science Letters* 3(3): 6-9.

Biswas, O, Shah, SK, Modak, BK, Mitra, B. 2017a. Description of one new species of genus *Ramila* Moore, 1867 (Lepidoptera, Crambidae, Schoenobiinae) from Indian Sunderbans with a revised key to the Indian species. *Oriental Insects* 51(4): 409-416.

Biswas, O, Shah, SK, Modak, BK, Panja, B, Roy, S, Chakraborti, U, Mitra, B. 2017b. Additions to the Moth Fauna of Sundarban Biosphere Reserve, India. *Bionotes* 19(2): 58-59.

Biswas, O, Panja, B, Garain, PK, Roy, S, Shah, SK, Modak, BK, Mitra, B. 2018. *Hyblaea purea* (Crammer, 1777)[Lepidoptera : Hyblaeidae]

Infestation on *Avicennia alba* Blume in Sundarban Biosphere Reserve. *Proceedings of the Zoological Society* 71(4): 331-335.

Biswas, Olive, Suresh Kr. Shah, Balaram Panja, Apurva Das, Biplob Kr. Modak, Bulganin Mitra. 2019. Moths (Insecta: Lepidoptera) of Jambudwip Island, Sundarban Biosphere Reserve, India: a Preliminary Report. *Ambient Science* 6(2): 43-46.

Biswas, T, Bandyopadhayay, PK. 2016. First record of protozoan parasites, *Tetrahymena rostrata* and *Callimastix equi* from the edible oyster in Sundarbans region of West Bengal, India. *Journal of Parasitic Diseases* 40: 971-975.

Chowdhury, GR, Roy, K, Goyal, N, Warudkar, A, Raza, RH, Qureshi, Q. 2020. On the evidence of the Irrawaddy Dolphin *Orcaella brevirostris* (Owen, 1866) (Mammalia: Cetartodactyla: Delphinidae) in the Hooghly River, West Bengal, India. *Journal of Threatened Taxa* 12(8): 15905-15908.

Chakrabarti, R. 2001. Tiger and the Raj: Ordering the Maneater of the Sundarbans 1880–1947. In: Chakrabarti, R. (ed.). *Space and Power in History: Images, Ideologies, Myths and Moralities* Calcutta, Penman.

Chakrabarti, R. 2009. Local People and the Global Tiger: An Environmental History of the Sundarbans. *Global Environment* 3: 72-95.

Chakraborty, P, *et al.* 2018. Notes on the record of *Gymnothorax pseudotile* Mohapatra *et al.*, 2017 (Muraenidae: Muraeninae) from the Sundarbans, West Bengal, India. *Rec. zool. Surv. India* 118(3): 318-321.

Chakraborty, P, *et al.* 2020. First record of two species of fishes from West Bengal, India and additional new ichthyofaunal records for the Indian Sundarbans. *International Journal of Fisheries and Aquatic Studies* 8(2): 6-10.

Chakraborty, P, *et al.* 2021. Ichthyofaunal integrity, hydrological and environmental features trade-off in the Sunderbans, India. *Ecological Questions* 32(2): 1-30.

Chatterjee, TK, *et al.* 2013. Mangrove Associate Gobies (Teleostei: Gobioidei) of Indian Sundarbans. *Rec. zool. Surv. India* 113(Part-3): 59-77.

Chaudhuri, AB, Choudhury, A. 1994. *Mangroves of the Sundarbans. Volume 1: India.* pp.xii + 247 pp.

Choudhury, AK, Bhadury, P. 2014. Phytoplankton study from the Sundarbans ecoregion with an emphasis on cell biovolume estimates- a review. *Indian Journal of Geo-Marine Sciences* 43(10): 1905-1913.

Chowdhury, S. 2014. Butterflies of Sundarban Biosphere Reserve, West Bengal, eastern India: a preliminary survey of their taxonomic diversity, ecology and their conservation. *Journal of Threatened Taxa* 6(8): 6082-6093.

Cotton, E. 1907. *Calcutta, Old and New: A Historical and Descriptive Handbook of the City.* W. Newman, Calcutta.

Curtis, SJ. 1933. *Working Plan for the Forests of the Sundarbans Division.* Calcutta, India. Bengal Government Press.

Das, CS. 2013. Declining snake population-why and how: A case study in the Mangrove Swamps of Sundarban, India. *European Journal of Wildlife Research* 59: 227-235.

Das, SK, Sarkar, PK, Saha, R, Vyas, P, Danda, AA, Vattakavan, J. 2012. *Status of Tigers in 24-Parganas (South) Forest Division, West Bengal, India.* World Wide Fund for Nature-India, New Delhi.

Dey, TK. 2001. *Conserve biodiversity of Sundarbans Reserved Forest- The World Heritage Site.* Khulna, Bangladesh. 16 pp.

Dey, TK. 2007. *Deer population in the Bangladesh Sundarbans.* The Ad. Communication, Chittagong, Bangladesh. 112 pp.

Dey, TK, Kabir, MJ, Ahsan, MM, Islam, MM, *et al.* 2015. *First phase tiger status report of Bangladesh Sundarbans.* Wildlife Institute of India and Bangladesh Forest Department, Ministry of Environment and Forests, Government of Bangladesh, Dhaka.

Ganguly, D, Ray, R, Majumder, N, Chowdhury, C, Jana, T. 2014. Monsoonal Influence on Evapotranspiration of the Tropical Mangrove Forest in Northeast India. *American Journal of Climate Change* 3: 232-244.

Gani, MO. 2000. The olive ridley turtles (*Lepidochelys olivacea*) of the Sunderbans coast. *Tigerpaper* 27(3): 7-11.

Garga, DP. 1947. How far can a tiger swim? *Journal of the Bombay Natural History Society* 47: 545-546.

Ghosal, S. 2017. Population Dynamics and Feeding Potentiality of a New Species *Cunexa terminalae* sp. Nov. Collected from Sundarban Biosphere Reserve. *International Journal of Current Microbiology and Applied Sciences* 6(9): 289-292.

Ghosh, S, Saha, S. 2016. Seasonal diversity of butterflies with reference to habitat heterogeneity, larval host plants and nectar plants at Taki, North 24 Parganas, West Bengal, India. *World Scientific News* 50: 197-238.

Gokul, A. 2017. Taxonomy and ecology of brachyuran crabs in Sunderbans. *Rec. zool. Surv. India* 117(2): 131-139.

Gupta, S, *et al.* 2016. Indigenous ornamental freshwater ichthyofauna of the Sundarban Biosphere Reserve, India: Status and prospects. *Journal of Threatened Taxa* 8(9): 9144-9154.

Gupta, SK, Biswas, H, Sammadar, I. 2017. Life cycle of *Oligonychus iseilemae* (Acari:Tetranychidae), a new pest of *Avicennia germinans*, a mangrove plant of Sundarban under laboratory conditions. *Ann. Entomol.* 35(2): 85-89.

Haque, H, Choudhury, A. 2015. Ecology and behavior of *Telescopium telescopium* (Linnaeus, 1758), (Mollusca: Gastropoda: Potamididae) from Chemaguri mudflats, Sagar Island, Sundarbans, India. *International Journal of Engineering Science Invention* 4(4): 16-21.

Hasan, MAU, Neha, S. 2019 The Himalayan Crestless Porcupine *Hystrix brachyura* Linnaeus, 1758 (Mammalia: Rodentia: Hystricidae): first authentic record from Bangladesh. *Journal of Threatened Taxa* 11(12): 14624-14626.

Hassan, ME, Biswas, B. 2014. Record of some Hemipteran fauna from the Sunderban Biosphere Reserve, West Bengal. *Bionotes* 16(1): 29.

Hasan, MAU, Neha, SA. 2019. The Himalayan Crestless Porcupine Hystrix brachyura Linnaeus, 1758 (Mammalia: Rodentia: Hystricidae): First authetic record from Bangladesh. *Journal of Threatened Taxa* 11(12): 14624-14626.

Hazra, S, Samanta, K, Mukhopadhyay, A, Akhand, A. 2010. *Temporal change detection (2001- 2008) study of Sundarban.* Jadavpur, Kolkata, India: School of Oceanographic Studies, Jadavpur University. Retrieved December 7, 2015, from http://www.iczmpwb.org/ main/pdf/ebooks/WWF_ FinalReportPDF.pdf

Hendricks, H. 1975. The status of the tiger *Panthera tigris* (Linne 1758) in the Sundarban Mangrove forest (Bay ofBengal). *Saingetierkundliche Mitteilungen* 23 (3): 161-199.

Hunter, WW. 1875. *A Statistical Account of Bengal Volume I. Districts of the 24 Parganas & Sundarbans.* Trübner & Company, London.

Hussain, N, Khanam, R, Khan, E. 2017. Two and Half Century's Changes of World Largest Mangrove Forest: A Geo-informatics Based Study on Sundarbans Mangrove Forest, Bangladesh, India. *Modern Environmental Science and Engineering* 3(6): 419-423.

Islam, SN, Reinstädtler, S, Gnauck, A. 2019. Invasive Species in the Sundarbans Coastal Zone (Bangladesh) in Times of Climate Change: Chances and Threats. In: Makowski, C, Finkl, CW (eds.) *Impacts of Invasive Species on Coastal Environments*, 63-85. Coastal Research Library 29, https://doi.org/10.1007/978-3-319-91382-7_2.

Jhala, YV, Qureshi, Q, Gopal, R, Sinha, PR (eds.). 2011. *Status of the Tigers, Co-predators, and Prey in India.* National Tiger Conservation Authority,

Govt. of India, New Delhi, and Wildlife Institute of India, Dehradun. 302 p.

Jhala, YV, Dey, TK, Qureshi, Q, Kabir, J, Md, Bora, J, Roy, M. 2016. *Status of tigers in the Sundarban landscape Bangladesh and India. Bangladesh Forest Department.* National Tiger Conservation Authority, New Delhi, & Wildlife Institute of India, Dehradun. TRNO -2016/002.

Justin, JN, Routh, A, Ghosh, D. 2022. First photographic report on occurrence of Madras Tree Shrew from Mangrove-entangled Indian Sundarbans. *Small Mammal Mail* 37(5): 26-28.

Kar, A, Karmakar, K. 2021. Description of three new species of phytoseiid mites (Acari: Mesostigmata) from Sundarban, West Bengal, India. *International Journal of Acarology* 47(1): 51-60.

Karanth, KU, Nichols, JD. 2000. Ecological Status and Conservation of Tigers in India. *Final Technical Report* to the Division of International Conservation, US Fish and Wildlife Service, Washington DC and Wildlife Conservation Society, New York. Centre for Wildlife Studies, Bangalore, India.

Mahabal, A, Sharma, RM, Patil, RN, Jadhav, S. 2019. Colour aberration in Indian Mammals: A review from 1886-2017. *Journal of Threatened Taxa* 11(6): 3690–13719.

Mahmood, H, Ahmed, M, Islam, T, Zashim Uddin, M, Ahmed, ZU, Saha, C. 2021. Paradigm shift in the management of the Sundarbans mangrove forest of Bangladesh: Issues and challenges, *Trees, Forests and People* 5: 100094 https://doi.org/10.1016/j.tfp.2021.100094.

Maity, A, *et al.* 2016. Taxonomic notes on Tabanidae (Insecta: Diptera) of Sundarban Biosphere Reserve and associated mangrove ecosystem of coastal region of west bengal, India. *J. Adv. Zool.* 37(2): 96-113.

Mallick, JK 2004 Ungulates of West Bengal and its adjoining areas including Sikkim, Bhutan and Bangladesh (Jointly written). *Envis Bulletin* 7(1): 1-15, Wildlife Institute of India, Dehradun.

Mallick, JK 2007 Endemic mammals of West Bengal. *West Bengal* 49(10–11): 53–55.

Mallick, JK 2009 Endemic Marsh Mongoose *Herpestes palustris* (Carnivora: Herpestidae) of East Kolkata Wetlands, India: a status report. *Journal of Threatened Taxa* 1(4): 215-220.

Mallick, JK 2010a New national aquatic animal, Ganges dolphin 2010. *West Bengal* 52: 27-31.

Mallick, JK 2010b GPS tracking of the tigers in Sundarban Tiger Reserve. *Annual Report 09-'10* (WWF, West Bengal State Office, Kolkata). 1: 21-22.

Mallick, JK. 2010c. Endangered fishing cats inhuman dominated landscape. *Environ* 10(3): 40-43.

Mallick, JK. 2011a. New records and conservation status review of the endemic Bengal Mongoose *Herpestes palustris* Ghose, 1965 in southern West Bengal, India. *Small Carnivore Conservation* 45: 31–48.

Mallick, JK. 2011b Ecology and Status of the Ganges Dolphin (*Platanista gangetica gangetica*): India's National Aquatic Animal in Southern West Bengal. Animal Diversity, Natural History and Conservation 1: 277–305.

Mallick, JK. 2011c Status of the mammal fauna in Sundarban Tiger Reserve, West Bengal– India. *Taprobanica* 3(2): 52–68.

Mallick, JK. 2011d The National Aquatic Animal Ganges Dolphin. *Environ* 11: 54-60.

Mallick, JK. 2013a Ecology, status and aberrant behaviour of Bengal Tiger in the Indian Sundarbans. In: Gupta VK, Verma AK (eds). *Animal Diversity Natural History Conservation. Vol.2.* pp. 381-454. New Delhi, India: Daya Publishing House.

Mallick, JK. 2013b Ecology and Conservation of Endemic Bengal Marsh Mongoose in East Kolkata Wetlands, a Ramsar Site in West Bengal. In: Singaravelan N. (ed) *Rare Animals of India*. Bentham Science Publishers, Sharja, UAE. Pp. 204-241.

Mallick, JK. 2014a Wildlife conservation in West Bengal- An overview, 244-256. In: *Towering Presence: 150 years of Forestry 1864-2014*. Directorate of Forest, West Bengal.

Mallick, JK. 2014b Wondrous wildlife: 1. Tigers of the Sundarban mangroves. *Tambourine* January Issue.

Mallick, JK. 2014c 2. Tigers of the Sundarban mangrove: II. Adventures of the first satellite-tracked tigress Priyadarshini. *Tambourine* April Issue.

Mallick, JK. 2015 Transformation of Sundarban tiger to opportunistic omnivore. *Tigerpaper* 42(2): 17-21.

Mallick, JK. 2019a An updated checklist of the mammals of West Bangal. *J New Biol Rep* 8(2): 37-123.

Mallick, JK. 2019b Primates in the Sundarbans of India and Bangladesh, 110-123. In: Nowak K, Barnett AA, Matsuda I (Eds.) *Primates in Flooded Habitats: Ecology and Conservation*. Cambridge University Press, UK.

Mallick, JK. 2020 Shadow in the swamp: The aberrant tiger of Sundarbans. *roundglass sustain* 1-6. https://sustain.round.glass/conservation/swamp-tiger-sundarbans/

Mallick, JK. 2022 *Ethics of Biodiversity Conservation: An Ecological Study*. Ethics International Press Ltd, UK.

Mandal, AK. 1988. Addition of the Fawn-Coloured Mouse, *Mus cervicolor* Hodgson, 1845 [Rodentia: Muridae] to the Fauna of Sunderbans in West Bengal, India. *Rec. zool. Surv. India* 85(2): 363-365.

Mandal, B, Mukherjee, A, Banerjee, S. 2013. A review on the ichthyofaunal diversity in mangrove based estuary of Sundarbans. *Reviews in Fish Biology and Fisheries* 23(3): 365-374.

Mitra, A. 2019. Brackish Water Fish diversity in Indian Sundarbans in the backdrop of climate change induced salinity alteration. (Chapter 9: 124-150) in Sarkar,U. K., Das, B. K., Mishal, P.and Karnatak, G.(eds). *Perspectives on Climate Change and Inland Fisheries in India*. ICAR-Central Inland Fisheries Research Institute, Barrackpore, 428 pp.

Mitra, B. 2017. Insect faunal diversity and their ecosystem services in Sundarban Biosphere Reserve. In: eds. Chandra, Alfred, Mitra and Roy Chowdhury, *Fauna of Sundarban Biosphere Reserve*. Zoological Survey of India, Kolkata, pp. 137-173.

Mitra, S, Chowdhury, MR. 2018. Possible range decline of Ganges River Dolphin *Platanista gangetca* (Mammalia: Cetartodactyla: Platanistdae) in Indian Sundarban. *Journal of Threatened Taxa* 10(13): 12738-12748.

Mandal, S, Deb, S. 2018. *Ancistrosyllis matlaensis* n. sp. (Polychaeta: Pilargidae) from the Sundarban Estuarine System, India. *Zootaxa*. 4531(3): 419-429.

Manjrekar, MP, Prabu, CL. 2014. Status of Otters in the Sundarbans Tiger Reserve, West Bengal, India. *IUCN/SCC Otter Specialist Group Bulletin* 31(2): 61-64.

Mitra, B. 2017. Insect faunal diversity and their ecosystem services in Sundarban Biosphere Reserve. In: eds. Chandra, Mitra and Roy Chowdhury *Fauna of Sundarban Biosphere Reserve*. Zoological Survey of India, Kolkata, pp. 137-173.

Mitra, A, Pal, S. 2002. *The Oscillating Mangrove Ecosystem and the Indian Sundabans*. (Banerjee, Sand Tampal, F. Ed.). World Wide Fund for Nature-India, WBSO.

Mitra, B, *et al.* 2016a. Insect Faunal Diversity of the Sunderban Biosphere Reserve, West Bengal, India. *International Journal of Current Research and Academic Review* 4(9): 87-98.

Mitra, B, Roy, S, Biswas, O, Chakraborti, U, Jehamalar, EE. 2016b. New records of aquatic bugs (Insecta: Hemiptera) from Sunderban biosphere reserve, West Bengal, India. *Journal of Entomology and Zoology Studies* 4(4), 8-11.

Mitra, B, *et al.* 2018. Insect faunal diversity of Chintamani Kar Bird Sanctuary and other protected areas of West Bengal. *International Journal of Entomology Research* 3(2): 180-189.

Mitra, B, *et al.* 2019a. First exploration on the insect faunal diversity of Haliday Wildlife Sanctuary, Indian Sundarban. *Research Journal of Life Sciences, Bioinformatics, Pharmaceutical and Chemical Sciences* 5(6): 11-19.

Mitra, B, *et al.* 2019b. New records of the terrestrial bugs (Insecta: Hemiptera) from Sundarban Biosphere Reserve and their diversity in Indian mangroves. *I3 Biodiversity* 4(403): 1-15.

Mitra, B, Panja, B. 2017. A brief note on moth diversity in Sundarban Biosphere Reserve with special reference to their change in host preferences from non-mangroves to mangroves. *Banabithi* 55-65.

Mitra, B, *et al.* 2018. Insect faunal diversity of Chintamani Kar Bird Sanctuary and other protected areas of West Bengal. *International Journal of Entomology Research* 3(2): 180-189.

Mitra, B, *et al.* 2019a. First exploration on the insect faunal diversity of Haliday Wildlife Sanctuary, Indian Sundarban. *Research Journal of Life Sciences, Bioinformatics, Pharmaceutical and Chemical Sciences* 5(6): 11-19.

Mitra, B. *et al.* 2019b. New records of the terrestrial bugs (Insecta: Hemiptera) from Sundarban Biosphere Reserve and their diversity in Indian mangroves. *I3 Biodiversity* 4(403): 1-15.

Mondal, MSQ. 2017. Population and land cover dynamics of Sundarbans impact zone in Bangladesh. *Landscape & Environment* 11(1): 1-13.

Mondol, S, Karanth, KU, Ramakrishnan, U. 2009. Why the Indian Subcontinent Holds the Key to Global Tiger Recovery. *PLoS Genet* 5(8): 10.1371/annotation/9f8748f6-300f-450e-bbed-63e66e1b6661.

Mukherjee, AK. 1975. The Sundarbans of India and Its biota. *Journal of the Bombay Natural History Society* 72: 1-20.

Naha, D, Jhala, YV, Qureshi, Q, Roy, M, Sankar, K, Gopal, R. 2016. Ranging, activity and habitat use by tigers in the mangrove forests of the Sundarban. *PLoS ONE* 11(4): e0152119.

Nandi, NC, Das, AK, Sarkar, NC. 1993. Protozoa fauna of Sundarban Mangrove Ecosystem. *Records of Zoological Survey of India* 93(1-2): 83-101.

Nandy, T, Mandal, S, Chatterjee, M. 2018. Intra-monsoonal variation of zooplankton population in the Sundarbans Estuarine System, India. *Environmental Monitoring and Assessment* 190(10): 603. https://doi.org/10.1007/s10661-018-6969-8.

Nandy, Tanmoy and Mandal, Sumit. 2020. Unravelling the spatio-temporal variation of zooplankton community from the river Matla in the Sundarbans Estuarine System, India. *Oceanologia* 62: 326-346.

Neumann-Denzau, G, Denzau, H. 2010. Examining certain aspects of human-tiger conflict in the Sundarbans forest, Bangladesh. *Tigerpaper* 37(3): 1-11.

Nishat, B, Rahman, AJMZ, Mahmud, S. 2019. *Landscape narrative of the Sundarban: Towards collaborative management by Bangladesh and India.* The World Bank Report (No. 133378): 1-207.

Pandit, PK, Guha, A. 2015. Peculiarities in food chain of Sundarban Tiger Reserve: Recent case studies. *Indian Journal of Biological Sciences* 21: 17-22.

Pennant, T. 1798. *The view of Hindoostan, Vol. II: Eastern Hindoostan.* London, Henry Hughs, pp. 153-154.

Poddar-Sarkar, M, Ray, S, Chowdhury, P, Samanta, G, Brahmachary, RL. 2013. On the body odour of wild-caught mangrove marsh Bengal tiger of Sundarban. In: East, ML, Dehnhard, M (eds). In: *Chemical Signals in Vertebrates 12.* Springer; New York.

Poddar-Sarkar, M, Brahmachary, RL. 2014. Pheromones of Tiger and Other Big Cats. In: Mucignat-Caretta C (ed). *Neurobiology of Chemical Communication,* Chapter 15. Boca Raton (FL): CRC Press/Taylor & Francis.

Poncins, E.de. 1935. A hunting trip in the Sundarbans in 1892. *Journal of the Bombay Natural History Society* 37: 844-858, pls. 1-4.

Prabu, CLJ, Naha, D, Roy, M, Manjrekar, MP. 2013. A recent sighting and photographic record of Goliath Heron (*Ardea goliath*) in Sundarban Tiger Reserve, West Bengal, India. *Zoo's Print* 28(10): 14-15.

Prater, SH. 1926. The occurrence of the Giant Heron *Ardea goliath* in the Khulna District, Bengal. *Journal of the Bombay Natural History Society* 31(2): 523.

Prater, SH. 1937. Black Tiger. *Journal of the Bombay Natural History Society* 39(2): 381-382.

Purkayastha, J, *et al.* 2019. A new species of *Polypedates* Tschudi, 1838 (Amphibia: Anura: Rhacophoridae) from West Bengal State, Eastern India. *Zootaxa* 4691 (5): 525-540.

Purushothaman, Jasmine, *et al.* 2020. Diversity of Ctenophores in the Sundarban Mangroves, Northern Indian Ocean. *Rec. zool. Surv. India* 120(2): 133-140.

Rahman, MM, Vacik, H. 2016. Recruitment of invasive plant species in the Sundarbans following tropical Cyclone Aila. *American Geophysical Union, 2016, A54B-A52709.*

Rainey, HJ. 1878. Note on the absence of a horn in the female of the Sundarban rhinoceros and Javanese rhinoceros (*Rh javanicus*, Cuv). *Proceedings of the Asiatic Society of Bengal* 1878 June: 139-141.

Rakshit, D, *et al.* 2016. Diversity and Distribution of Microzooplankton Tintinnid (Ciliata: Protozoa) in the Core Region of Indian Sundarban Wetland. *CLEAN - Soil Air Water* 44(10): 1278-1286.

Rookmaaker, K. 1997. Records of the Sundarbans Rhinoceros (*Rhinoceros sondaicus inermis*) in India and Bangladesh. *Pachyderm* 24: 37-45.

Roy, S, Chakraborty, SK, Parui, P, Chakraborti, U, Biswas, O, Mitra, B. 2016. Re-description of Cadrema pallida var. bilineata (de Meijere, 1904) (Diptera: Chloropidae) and its role as pollinator and carrion feeder from Indian Sunderbans. *Ambient Science* 3(2): 93-94.

Roy, S, *et al.* 2016. Sundarbans mangrove deltaic system- An overview of its biodiversity with special reference to fish diversity. *Journal of Applied and Natural Science* 8(2): 1090-1099.

Roy, S, Parui, P, Mitra, B. 2017. Plagiostenopterina sagarensis sp. nov. (Diptera: Platystomatidae: Platystomatinae) from Sunderban Biosphere Reserve, India with a key to Indian species. *Zootaxa* 4294 (4): 487–493.

Roy, S, Chakraborty, SK, Parui, P, Mitra, B. 2018. Taxonomy of Soldier Flies (Diptera: Stratiomyidae) of Sunderban Biosphere Reserve, India. *Proceedings of Zoological Society, Calcutta* 71(2): 121-126.

Roy, S, *et al.* 2019. A new species from the genus Astochia Becker,1913 (Diptera: Asilidae) from the Sundarban, India with an updated checklist of Indian species. *Oriental Insects* https://doi.org/10.1080/00305316.2019.1593254.

Roy Chowdhury, DK. 1984. Birds in and around the Calcutta Metropolitan area. *Naturalist* 1: 7-16.

Saha, D. 1984. Olive Ridley Marine Turtle from the Sundarbans, West Bengal. *Environment & Ecology* 2(3): 236-237.

Saif, S. 2016. Investigating tiger poaching in the Bangladesh Sundarbans. *Doctor of Philosophy (PhD) thesis*, University of Kent, Durrell Institute of Conservation and Ecology.

Sankhala, K. 1978. *Tiger.* William Collins and Co Ltd, Glasgow, 220 pp.

Sau, Susovan, *et al.* 2017. Species composition and habitats of macro-benthic crustaceans in the intertidal zones of Sundarban, West Bengal, India. *J. Exp. Zool. India* 20(2): 1103-1107.

Seidensticker, J, Lahiri, RK, Das, KC, Wright, A. 1976. Problem Tiger in the Sundarbans. *Oryx* 13(3): 267-273.

Seidensticker, J, Hai, MA. 1983. *The Sundarbans Wildlife Management Plan: Conservation in the Bangladesh Coastal Zone.* Gland, Switzerland: International Union for the Conservation of Nature and Natural Resources.

Sen, K, Mandal, R. 2019. Fish diversity and conservation aspects in an aquatic ecosystem in Indian Sundarban. *International Journal of Zoology Studies* 4(4): 16-26.

Sharma, D. 2013. The message of the Sundarban (p. 19). In: Khan, R. (Ed.) *SUNDARBAN: Rediscovering Sundarban The Mangrove Beauty of Bangladesh.* Nymphea Publication, Dhaka, Bangladesh.

Simson, EB. 1886. *Letters (No.47) on sport in Eastern Bengal.* London. Pp. 190-191.

Singh, SK. 2017. Conservation genetics of the Bengal tiger (*Panthera tigris tigris*) in India. *Ph.D. Thesis,* University of Oulu Graduate School.

Sohel, MSI, Hore, SK, Salam, MA, Hoque, MA, Ahmed, N, Rahman, MM, Khan, HM, Rahman, S. 2021. Analysis of erosion-accretion dynamics of major rivers of world's largest mangrove forest using geospatial techniques. *Regional Studies in Marine Science* 46(101901): 2352-4855.

Tamang, KM. 1993. *Integrated Resource Development of the Sundarbans Reserved Forest: Wildlife Management Plan for the Sundarbans Reserved Forest, Khulna, Bangladesh.* FAO/UNDP Project BGD/84/056. Rome.

LIST OF ABBREVIATIONS

AD: Anno Domini
BoB: Bay of Bengal
BP: Before Present
CRZ: Coastal Regulation Zone
EKW: East Kolkata Wetlands
HTL: high tide line
KMDA: Kolkata Metropolitan Development Authority
LTL: low tide line
MoEFCC: Ministry of Environment, Forests and Climate Change
NP: National Park
SBR: Sundarban Biosphere Reserve
SRF: Sundarban Reserve Forest
STR: Sundarban Tiger Reserve
WLS: Wildlife Sanctuary

* 9 7 8 1 8 0 4 4 1 1 0 1 8 *